New Science Theory
By Vincent Wilmot

This is basically the substantial New-Science-Theory.com website as on 6 April 2022 – for changes since then visit the website where its Sitemap notes any further updates.

Especially good for those interested in physics, it concentrates especially on four great physicists - William Gilbert, Rene Descartes, Isaac Newton and Albert Einstein.

It also has fine sections covering History of Science, Philosophy of Science, Galileo, Kepler, Tesla, Information Physics, String Theory, Standard Model Physics, Probability Science, Gravity, Light, Solar System Problems and General Image Theory Science.

Published by Lulu.com

© Copyright 2022 Vincent Wilmot.
ISBN : 978-1-4467-9538—5

PLEASE NOTE, this website has run for years and has had no user complaint on security or privacy.

April 06, 2022.. - Hear briefly about this website >>

New Science Theory - *exemplified chiefly by physics theory*

William Gilbert . Rene Descartes . Isaac Newton . Albert Einstein .... Science History . General Image Theory .... Sitemap . About us
- Site Search at bottom v

Science is basically the combination of good logical reasoning with good practical knowledge of actual natural phenomena. All humans do some logical reasoning and have some practical knowledge of some actual natural phenomena, but most have to busy themselves with feeding themselves and their families as best they can. Few have been able to devote much of their time to reasoning and/or gaining better knowledge of nature, and only some of these have made small or big contributions to science. But poverty reduction has helped boost science, as could a proposed Universal Basic Income, though the quality of science matters more than just its quantity. This website did start by challenging the failings of modern physics in a non-confrontational limited politic way but, that seemingly not working, we have moved to our now more fully honest confrontational non-politic direct challenge that seems really needed. If this may tend to prompt stronger opposition from the prejudiced, it may also better convince the relatively unprejudiced.

In considering science theory, this site concentrates on physics theories from the now entirely untaught ideas of William Gilbert, Rene Descartes and Isaac Newton to Albert Einstein and beyond - and we also have good related sections on Galileo Galilei, Johannes Kepler, Nikola Tesla, Gravity, Light, Probability Science, Standard Model Physics, String Theory and physics now, Information Physics and Science Philosophy.

Get this website as a Zoomable, Searchable and Printable pdf Ebook with helpful Bookmarks - about £2 at New Science Theory PDF Ebook
our Sitemap shows any sections updated since its 6 April 2022 (Or for £8.99 get the nice A4 paperback version at New Science Theory book)

PHYSICS NEWS. The biggest physics news for many years was maybe the vastly expensive 2008 CERN Large Hadron Collider (LHC) for 'atom-smashing' experiments on 4 July 2012 reporting discovering a new 125GeV particle claimed to be the Standard Model predicted Higgs boson or God particle supposed to explain gravity, though that had been predicted to be around 500GeV. On the data published to date it is maybe at best just 'some new particle' - but now the collider's power has been doubled so that it might produce a bit more information but not really new physics experiments still. But interestingly 2017 saw Sarah Charley saying 'Colliders do not collide anything, as subatomic particles are largely empty space' (at http://www.symmetrymagazine.org/article/whats-really-happening-during-an-lhc-collision). However the first reported detection of gravitational disturbance waves by the US LIGO was called the biggest physics news of 2016 and claimed to suppoprt Einstein's physics. It is good though gravitational disturbance waves are not really an essential of, or peculiar to, Einstein's physics as many are claiming . Gravitational disturbances really also follow from the physics of Newton, though he did not specifically discuss them, and are really of minor significance to theoretical physics. The 2017 binary-neutron-star-merger 'gravitational wave' detection has also been reported as being accompanied by wide multi-band electromagnetic EMR radiation, seemingly being preceded by 6 minutes by a gamma ray burst detection that could maybe have happened 2 seconds after it, and about 11 hours later by a light burst and other EMRs, but Miguel Zumalacarregui claimed it proved Gravity travels at the speed of light ? (see IOP). Astronomy has also produced some more new information in recent years, though maybe nothing really major.

> Banned by the Catholic Inquisition, but published in 1600 Protestant England and then pirate-published twice in Protestant Holland,
> see a new improved English translation of William Gilbert's Latin heretical science De Magnete still of some real interest.
> The real physics of William Gilbert that Newton put as one of two physics options and privately favoured, in print On The Magnet.

November 2012 saw some modern 'mainstream physicists' now pushing to abolish the teaching of classical experimental physics in schools as 'obsolete'. They want the experiments by Newton on light, by Galileo on gravity and by Gilbert on magnetics/electrics to be deleted from human history. It seems that only modern thought-experiment mathematical physics, or conjectural-physics, should be taught. This is being pushed at the US President through YouTube in a video "Open Letter to the President : Physics Education", which seems to be from the 'Perimeter Institute for Theoretical Physics' of Canada. The physicists concerned were obviously taught classical physics at school in the awful way it is always taught now, with no study of the actual works of Newton, Galileo or Gilbert and doing no actual experiments. This attempt at killing real experimental physics and its associated theories can be seen on YouTube at http://www.youtube.com/watch?v=BGL22PTIOAM

But all children from age 3 or 4 should first be generally encouraged to experiment and explore with 'let us try this' and 'let us try that' without any particular science in mind and allowing failures and later move on to study William Gilbert and later study Einstein. Mathematics should likewise be taught in steps from an early age, and later its use in science also.

And 2015 saw the UK openly move from 'science test' passes requiring a 'theory test' pass AND a 'practical test' pass, to 'science test' passes requiring a 'theory test' pass ONLY for A Level tests. (See https://www.iop.org/news/14/apr/page_63036.html) And now 'science theory' tests below PhD level are effectively 'narrow coverage science history without dates'. While now 'science practical' or 'experimental science' tests below PhD level are effectively 'drawing, surgery or plumbing depending on the science'. Neither now really test abilities to theorise OR abilities to experiment, even with the best of test marking, so it is doubtful if any who now make it to a PhD have any real science ability. And more notice should maybe be taken of IQ tests now.

These science trends fit with education, TV and the internet all being generally dumbed-down now. Good science websites are being dropped in search engine results 'because the average person is not interested in intelligent stuff'. So for science searches all are increasingly served with either completely dumb websites or half-dumb sites like Wikipedia, and science theories now 'win' more by dumb votes than by facts. This website has now basicly moved from an Isaac Newton early-edition more-constrained publishing style to a later-edition or a William Gilbert less-constrained publishing though none were really unconstrained and none worked well for them.

The basis of science theory.

Those who have specialised only in logical reasoning have often been called philosophers, and some of the best of these first emerged in Ancient Greece. The most rigorous logical reasoning, as with Euclid, has often been in the field of mathematics. Those who have specialised only in gaining better knowledge of nature have often been artisans or nature lovers, and their studies often have been concerned with their work or their leisure. Here metallurgy and astronomy were two fairly significant fields of study, with many others. The chief scientific advance in gaining better knowledge of nature came with the realisation that it chiefly needed the precise measurement of natural phenomena so that the rigours of number could replace vagueness and be better amenable to logical reasoning so that the two chief elements of science better combined.

Early ideas on the natural world generally took some vague magical or religious form of theorising, as that natural bodies had life forces or that god caused everything. In line with this, the widely accepted though entirely unproven explanation of gravity by the philosopher Aristotle was that all bodies had 'a natural tendency' to move to their 'natural place'. Such unproven opinion was to be challenged by the emerging experimental science method, chiefly in getting rigorous factual descriptions of more natural phenomena and then in developing all kinds of theories to try to explain the known facts. The many science theories came in two basic types - Black Box theories of laws of universe behaviour like gravity to explain what happens, but not trying to explain why things happen, and full-explanation theories that did seek to explain why things happen.

Hence the chief tools of science are observation and experiment, and the main secondary tools of science are theory and mathematics. And the chief results of science are truths and technologies. Human knowledge of natural phenomena has undoubtedly always been increasing to some extent since our species began, though often in accidental or ad hoc ways and some discoveries have been lost and re-discovered again later. Yet on average human history has involved progress in factual knowledge of nature and in technology deriving from that knowledge as in producing first farming and then industry. But theories of nature showed little or no progress in our early history, and indeed have struggled to show progress in modern times also.

The Black Death first hitting Europe badly from around 1350 and the mini-Ice-Age from around 1400, destabilising life and government and religion, probably encouraged the questioning and innovation that led to the Enlightenment and the rise of science and of modern industry in Europe. But it was maybe not until the 1500's that real planned science emerged first in Europe, with the chief requirement that both good logical reasoning and good practical knowledge of actual natural phenomena must be combined to try to produce valid descriptions of natural phenomena and valid science theories. Though there were earlier neo-science developments such as Alchemy in different parts of the world, the real emergence of science was driven first by Europe wanting to explore and exploit the wider world, and then by Europe's developing industrial revolution. World exploring required use of the astronomer's stars and of the magnetic compass. After his death in 1543 Nicolaus Copernicus published an improved description of heavenly bodies where the Earth correctly orbited the Sun, and a basic compass was in some use from the 1200's.

William Gilbert in 1600 (shortly before his death) published his many science experiments and his physics chiefly concerning magnetism and improved compass use but deriving a rarely understood full-explanation effluvia signal theory of physics relating to the Earth and bodies generally. Another major early scientist then, Galileo Galilei (1564-1642) experimented chiefly in mechanics and astronomy with a little on a push physics theory and had a lot of trouble from the catholic church and governments for that and for backing Copernicus, but William Gilbert (1544-1603) working mainly on magnetism in protestant England openly dismissed Aristotle and all philosophising or theorising that was not directly substantiated by scientific experiment, and practised what he preached with his one early publication concentrating on his many experiments and a little on a signal-response attraction physics theory. Galileo supported Gilbert's experimental work but dismissed his theory and Francis Bacon pushed a false-Gilbert no-signals attraction physics, while Johannes Kepler (1571-1630) working in mathematics, optics and astronomy developed a 'forcefield push' version of Gilbert's physics and also backed Copernicus. In response to emerging science attacking different aspects of Aristotle, like Gilbert's 1600 De Magnete and Galileo's 1623 Assayer, the catholic church and its Jesuits reluctantly began dropping Aristotle for ancient greek Atomism theory (of the atheists Leucippus and Democritus, and then chiefly supported in Europe before Galileo adopted it by some alchemists) from around the 1620's to 1640's slowly, while slightly modifying its terminology. Their new position then encouraged many scientists to try to comply with that theory.

Then the philosopher Rene Descartes (1596-1650) produced his mechanical push physics theory that impressed many as fitting with much of the emerging science - and it was later falsely claimed also fitted with that of the mathematician and physicist Isaac Newton (1643-1727) though his work chiefly favoured Gilbertian attraction theory but settled for a black-box physics theory like a few other physicists then. While advances continued in other sciences, physics theory had to wait about 200 years before Albert Einstein produced his new partial-explanation forcefield spacetime theory. One basic advance in physics then had been the discovery that the originally supposed elementary particles 'atoms' seemed basically mini-solar-systems with smaller particles and mini-action-at-a-distance. Strong evidence that solids are far from solid supported the conclusion that at least some 'pushes' may not be contact pushes and so maybe at least partly supports either a field type physics or a signal type physics where signals establish contact between separated bodies but do no pushing ?

After Newton, physics theory seems to have somewhat sidelined experimental study in favour of mathematical study, so that increasingly universities located theoretical physics in their mathematics departments rather than in physics departments. And certainly new physics theory since Einstein, such as 'string' and 'loop' theory, seems to largely have been on the mathematics and structure of fields and/or of 'elementary' particles as possibly explaining everything somehow though it perhaps is muddy water - and 'fields' may yet be shown to not exist and/or the 'elementary particles' may yet be shown to be mini-mini-solar-systems themselves. In physics the big may be as reasonable a model of the small as vice versa, or not, and a signal physics may yet prove of some use also.

Many have been involved in the development of science, and many more in supporting or opposing it, covering all countries. But the key science theory ideas around physics can perhaps best be seen by going backwards from Einstein. Einstein considered that the theory that he chiefly had to face up to was Newton's, and Newton considered that the theories that he chiefly had to face up to were Descartes' and Gilbert's though Newton was guarded in commenting on Gilbert's attraction physics or remote-control physics. It seems the key physics theories were indeed those of Gilbert, Descartes, Newton and Einstein which this site examines further on other pages in an interrelated way rather than entirely separately. On this site you can start with William Gilbert and somewhat simpler early physics theories and journey on to rather more complex modern physics theories.

While Newton considered various possible explanations of gravity and other 'forces', he ended up publicly supporting none and insisting that physics should support none. He concluded that black-box mathematical behaviour laws were enough for science, and that any explanation must involve untestible unseens and be 'outside science'. This basic conclusion of Newton can certainly be challenged, but Einstein and others ignoring it and wrongly pretending that Newton's theory was a simple billiard ball push theory was one of the worst mistakes in physics theory history. It meant that no physicist has worked from or built on Newton's actual physics position - only on a simplified false 'Newton position' ?

Although Gilbert, Descartes and Newton took science as not allowing contradictions, Einstein and others later adopted 'duality physics' for light and for particles requiring them both to be 'wave' and be 'not-wave' and so allowing contradiction in their science. Not just allowing contrary interpretations and contrary mathematics, but allowing actual contradiction in experiments and in actual nature. This became possible by rejecting earlier strict definitions of 'wave' and 'particle' and basically using no strict definitions, and its acceptance by governments has halted big sciences advances though lots of small technology advances do continue for now at least.

The interest of Gilbert and Newton in at-a-distance force theory or signal physics theory was perhaps before its time and has really been developed by nobody since. Many physicists from Galileo to Einstein ridiculed action-at-distance or remote-control physics as 'impossible', but the invention of the TV Remote and the computer supported Gilbert and Newton against that silly 'disproof'. (Gilbert-Newton physics had forces acting just like the TV Remote and the computer act, but many opponents lying or in ignorance took Gilbert-Newton physics as having forces acting like a TV Remote that emits no signals!) Gilbert and Newton were less interested in the physical nature of any signal emissions, be they particle emissions or energy emissions or wave emissions, than in how bodies experimentally responded to natural signals. Some modern physicists are now talking of a 'quantum-information' physics, a 'quantum computation' physics or a 'digital' physics involving maybe a 'cellular automaton universe' - including among others Pablo Arrighi and Jonathan Grattage affiliated with the University of Grenoble and ENS de Lyon, France (see http://membres-lig.imag.fr/arrighi/). And the possible relevance in physics still of Gilbert-Newton 'attraction signal-response physics' is maybe also even suggested by a recent quote of Google on them letting application developers for their Android phones use C or C++ code "as in signal processing, intensive physics simulations, and some kinds of data processing". For a 'quantum gravity' theory some kind of quantum signal-response theory physics is now maybe looking of more promise ? And a main still-unresolved problem for physics is ARE atomic particles actually some kind of push balls, or do they emit signals and respond for gravity and magnetism ?

It is maybe of some small interest that Einstein was the only one of these four major scientists to marry (and indeed twice) and to have children, suggesting that having a family to feed or other major activities can hinder the development of substantial new science !? Most having no descendants is unfortunate but more positive is the fact they all seem to have retained their mental capacities well in old age - maybe an old-age IQ fall from 100 to 95 gives poor mental functioning but an IQ fall from 165 to 160 still leaves excellent mental functioning when older ? All four were European men, which was also necessary in Europe in those days. It has been claimed seemingly rightly that Einstein's first wife was strongly interested in and competent at mathematics and physics. But in that case it seems that Einstein failed to help his first wife much with that, maybe because he could not afford a nanny or other help for her ? Of course probably more likely because they followed in those times male dominance and female submission in marriage being a strong social norm. (Einstein's wife) Though maybe unclear in Einstein's case some male scientists did have a wife who helped with their science, and in some of those cases the wife help has been perhaps underrated but in some cases perhaps overrated. And when these four produced their main physics ideas none of them was employed in physics, all being 'hobby' or 'amateur' physicists. Gilbert was a royal physician and hobby-physicist, Kepler was an astrologer and hobby-astronomer, Newton was a mathematician and hobby-physicist, and Einstein was maybe lucky that working as a patent official he was accepted as a physicist which would almost certainly not be accepted by todays physics 'professionals'. And these four key scientists are half from England, with Descartes being the sole Catholic though there were other early Catholic scientists. In early Europe it would have been much harder for a woman to fund a science career and maybe also harder for a woman to resist the marriage and child-bearing that could also make a successful science career harder.

The ideas presented on this site are based on extensive studies of William Gilbert and of much of Descartes, Newton and Einstein and others relating to their theories. Currently the internet offers little of these four to read online, and much of their work has still not been translated, so this site will be trying to help with that over time. Science histories often have serious weaknesses, and for basic physics history this website's interpretations are the best and should be studied first, but you may also like a look at a mostly not too unreasonable summary of science history at http://faculty.kirkwood.edu/ryost/chapter1.htm

Good physics experiment and good physics theory.

Physics experiments and physics theories have at times come from very different types of sources, some good and some not. Early good physicists, like Galileo or William Gilbert, often had no physics training and some were self-taught hobby physicists or anti-establishment physicists.

Today some insist that every good physicist must have a physics degree, and that everybody with a physics degree is a good physicist (but we certainly do not have 900,000 Isaac Newtons today, and on physics Newton considered himself self-taught). It may seem more accurate to say that today a good physicist should probably have a physics degree, and that some with a physics degree are probably good physicists.

1. But this issue maybe needs clarifying somewhat to account for the fact that physics involves basically two different aspects - experiment and theory - and useful physics experiment seems to have somewhat less need of formal training than physics theory. Hence most technology advance has been independent of theory, so a computer engineer working for Google may produce some good physics experiments.

2. A further issue concerns the nature of formal physics theory training, in earlier times including substantial philosophy and history of science - but today seeming entirely confined to post-Einstein physics theory. This may suggest that most of today's formally trained physicists may have too narrow a focus to their physics theory ideas, so a philosopher or historian might be better on physics theory.

We should of course still expect most good physics today to come from those with a physics degree, but should not be entirely surprised if some good physics ideas comes from a philosopher or engineer. A modern William Gilbert is possible.

But some are getting very worried that there now seems to be massive backing for eg many-universes 'fiction physics' theories and no backing for 'real physics' theories ? And the real concern with this should be less of right or wrong theories, than of significant physics experiment option areas being closed off now !

Great scientists and great skills

All great scientists do need to have some great skill or skills, but all great scientists do not need to have every possible great skill. But highly skilled people perhaps tend to be one of three skill types ;

1. Mathematicians and rule followers
Some great scientists like Isaac Newton have had great mathematical skill, and have been great at mathematical rule-following reasoning. Of course some of them, maybe also including Isaac Newton, have also had some great artist-artisans rule-breaking experimenting skills.

2. Artist-artisans and rule breakers
Some great scientists like Galileo Galilei have had great artist-artisan skill, and have been great at rule-breaking experimenting. Of course some of them, maybe also including Galileo Galilei, have also had great mathematical rule-following reasoning skill.

3. All-rounders or multi-skilled
Some great scientists may have had great mathematical skill and great artist-artisan skill and equally used both, but some of these may have employed one strength more than the other. These may have been great at rule-following reasoning and great at rule-breaking experimenting, but some of these employed one more than the other. This might depend on their own view of science or of priorities at the time, and some great thinkers and scientists have had different views on that.

Most of the big leaps in science has been the work of great individuals working alone, while many of the smaller advances have come from team collaboration - smaller maybe partly due to teams often being composed of too narrow a range of skill types ? Technology advances have mostly come from experiment and not from theory, as with Galileo inventing the refracting telescope and Newton inventing the superior reflecting telescope being both based on light experiments rather than light theory. Experiment is often more useful than theory, but better theory can lead to better experiment. And honest science has always been the more useful, as in not putting up a false simplified-Newton to knock down. Newton certainly never claimed that a light ray would not bend towards the sun, nor that a gyroscope some miles above Earth would hold a perfectly stable spin. And Newtonian physics does not imply either of these claims. Many modern physicists can seem to show a perhaps low regard for truth at times, and much overplay conjecture theories ? But in science the problem should not be do you experiment, but what are the best actual experiments for you to do of the billions of possible experiments ? Often like now pressures predominate towards doing less useful kinds of experiments, and against doing more innovative 'non-mainstream' kinds of experiment. And well established science prejudices can be almost impossible to shift.

While artist-artisan based skills often show culture differences - as in Egyptian, Roman and other art/science/technology - mathematics has generally developed as one mathematics involving the following of one set of rules. And while science does seem to require that there can be only one actual truth of anything, it can reasonably be claimed that science does not also require that there can be only one valid description of one truth. So modern physics dependence on mathematics only may be inadequate. Art often describes the same thing in different ways successfully, and a science with one mathematics may still validly allow of different image-theory explanations. But a one-truth science does not seem to really allow of contradictory explanations such as Duality Theory in current physics ?

While we do consider science theory generally, this site is the very best at examining the fundamentals of physics. If you want to really learn physics then this website really helps people with mastering physics online, and can also point you to some of the best other online physics sources.

PS. Some might say that the last 50 years has maybe seen no significant new physics theory published and no really new physics experiments, and maybe generally business and government hijack any new science to their own ends anyway, leaving little real value to any new science ? Some incremental technology change has certainly continued, though maybe giving more new problems in pollution and medicine than answers. When in 1959 I was offered a physics place at Imperial College London they showed me some of their cloud-chamber 'atom collision' experiments. I saw those nice cloud-chamber spirals as unlikely to be due to any collisions or pushes as commonly claimed, and as more likely being responses to some signals as moths respond to light from a light source. But for the last 50 years I have been sitting on a new general science theory and new physics experiments, that I cannot afford to run, developed after the first BSc degree I took in Biology and Chemistry. Then for a second BSc degree when I took year 1 Philosophy, I part ran it past the Professor of Philosophy who had been a Physicist, in a 1985 essay for him on the history of physics. He gave that top marks and promptly made several attempts to get me to switch to majoring in philosophy under him (which I would have done but at that time I could not see it as a practical career option for feeding my new wife and baby and owed some loyalty to my then City University mentor Andrew Mott). But being satisfied that the basics of my new general science theory may possibly be worth at least a temporary publishing rather than just all dying with me, I have now put the basics of it on this website - in the hope that you may find it interesting (and this website is all interrelated so studying all of it should help you understand it). Additionally, this site simply tries to clarify some of the basics of science theory history to date as I see it - though many do interpret science history differently and often very wrongly. Some of the problems involved in the history of science are discussed in our Science History, or you can check our Site Map.

(Two websites to slightly help inform you on what physicists and astronomers are up to lately are Physics World at http://physicsworld.com and Universe Today at http://www.universetoday.com)

AND get the must-have book for anyone really interested in science, William Gilbert's 1600 "On The Magnet ..." ;

+ TRY our great Newtonian gravity Android App - 'Sun Pull' - to help you study or re-design the solar system better, in our Solar System section, which also discusses what is probably chiefly needed for real actual contact with 'alien' people from other worlds. Hopefully more useful science Apps may follow, though the Google Play app store, Microsoft store and Apple store will probably all ban them in their bureaucratic rule changes ?!

You can do a good search of this website below ;

Search on this site www.new-science-theory.com, with Google.

Or do a search of the web better with DuckDuckGo -  Type web search then Enter

NOTE. If you use quotes in your search you may get a more accurate search, as "..."

otherwise, if you have any view or suggestion on the content of this site, please contact :- New Science Theory
or write Vincent Wilmot 166 Freeman Street Grimsby Lincolnshire DN327AT UK.

OR if you would like to make a donation to help with site development, and just possibly with some crucial basic physics experiments long planned but never afforded, then see the bottom of our other main sections. The main science funders today are governments and businesses, but they have their own aims and rarely fund the best science. They are largely funding 'safe science' experiments, which are basically old experiments using a bit more power or a bit more accuracy, and avoid funding any 'unsafe science' original experiments like ours.

You are welcome to link to this website homepage, eg www.new-science-theory.com/

 - On Twitter, see @vwilmot.

© new-science-theory.com, 2022 - taking care with your privacy, see Privacy Policy

Hosted by :-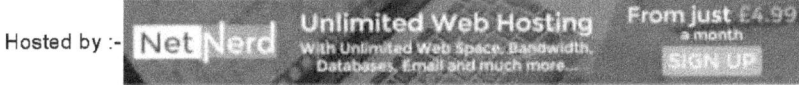

History of Science - *problems with the history of physics*

Homepage . **William Gilbert** . **Rene Descartes** . **Isaac Newton** . **Albert Einstein** .... **Science Philosophy** .... **General Image Theory** .... **Sitemap**

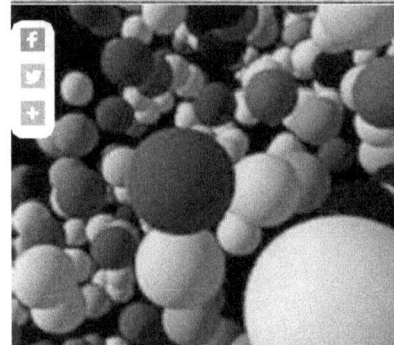

Science teaching, in schools and in universities, now often includes bits of science history - and there are some courses specifically on History of Science. Much good science history work has been done and is being done, the major exception being much history of early physics up to and including Isaac Newton which is generally very bad and includes major errors. The education of most physicists in earlier physics theories explained here is probably at its most limited ever today, with teaching on dumbed-down textbook Newton etc not actual Newton etc. While in eg Biology to date fraud has been occasional, intentional and soon exposed, in physics theory fraud has largely been predominant, unintentional and unexposed from prejudice built up by the 1600's as Newton noted and still ongoing. But physics history and textbooks are largely written by 'winners' who claimed to have but did not have the best physics, totally failing on William Gilbert and largely failing on Newton also and sticking us today with spacetime-continuum multiple-universe fiction-physics.

And while most old fiction works like Shakespeare's are freely available to all on the internet, old science works generally have restricted availability and can involve substantial costs. As old science claimed often wrongly to be 'disproved' can still have good bits, this website will try to help on that also and on todays science which should be helping everybody being controlled chiefly by and for a powerful few. (see World Poverty)

History and science history.

If basic science is basically experiment, then that started very early in human history, and science progress developed onwards from the earliest experiments that produced mankinds earliest tools and aids. But some historians chose to see 'science' as starting with alchemy etc by around 1200 or earlier while others see 'real science' as starting later in the 1500s with William Gilbert and Galileo who certainly helped establish modern developed science. Some earlier like English philosophers including William of Ockham saw causal knowledge as best obtained from experience or experiment, and he saw the Sun heating Earth, object perception and magnetism as best explained in line with his Occam's Razor by some action-at-distance physics rather than involving medium-action, God-power or motivated-objects as some others claimed then. (See Medieval Science) In earlier times churches of various religions were often among the social groups that most encouraged learning, if often in their case with some strict limits mostly with regards to complying with the wordings of their particular 'holy books'. However the Black Death first hitting Europe badly from around 1350 and the mini-Ice-Age from around 1400, destabilising life and government and religion, probably encouraged the major questioning and innovation that led to the Enlightenment and the emerging of modern science and of modern industry in Europe. But certainly Middle Ages philosophy and alchemy were not generally opposed by European governments or the prevailing Catholic Church then, probably largely because, unlike those the church called 'heretics' or 'pagans', they did not significantly challenge the words of the Bible that the church relied on. But when real science finally emerged around 1600 and it did begin to really challenge some words of the Bible, and did then draw strong opposition from the Catholic Church of that time who branded some leading scientists 'heretics'. The early Catholic Church official position on their Bible was that every word of it was absolute factual truth, and anyone disputing that was a heritic and not a true Catholic, but there have always been some who considered themselves Catholic but did not agree with that official position. The modern Catholic Church official position on their Bible is basically that most words of it are absolute factual truth, but a few words of it are 'poetic' and 'true at heart' but are not actual fact, and anyone disputing that is a heritic and not a true Catholic. But there are still some who consider themselves Catholic but do not agree with that official position. Also other churches, and indeed other religions, have often showed a similar range of positions which has often meant them conflicting with some science at some times. And of course science has often involved differing and changing positions on what it has claimed as being facts on different matters.

England's 1588 defeat of the Spanish Armada with a much smaller navy confirmed the fact that by then Protestant England's science and technology, with William Gilbert and without Francis Bacon, was ahead of Catholic Europe with their Galileo. England basically already had a more developed steel industry. And in early modern England an Agricultural Revolution and then an Industrial Revolution involved many new inventions, such as the 1730 Iron Plough of Joseph Foljambe and the 1733 Flying Shuttle of John Kay, and soon pushed England ahead of the world. And early English science from William Gilbert to Isaac Newton undoubtedly helped inspire that, along with English rulers like Queen Elizabeth 1 encouraging military improvements as in guns and ships and exploration. Yet on early science most science historians have concentrated entirely on the science of Galileo and mainland Europe and foolishly ignored the significant early English science of the time. The actual emergence of science was based on contributions from many countries but this website tries to present what is probably a somewhat more real science history than others. Human history can perhaps be best seen generally in terms of religion evolving from nature-magic with supposed spirits to multi-god religions to one-god religions, and science basicly evolving from nature-magic and technology change to astrology and alchemy spirits-science to astronomy and no-spirits natural-law-cause-science, with the 'religions' and 'sciences' often competing for support from the rich and the rulers. But it was earlier simpler magical thinking basicly growing more complex into both religion and science, and encouraging astronomy, mathematics and even philosophy. And astrology did involve accurate observation of the skies, while alchemy involved accurate chemical experiments - a limited advance on earlier 'simple magic'. But it was probably chiefly the development of technologies like fire, farming, wheel, metals etcetera that basicly drove the development of both new religious and scientific ideas. The early European science societies like the Royal College of Physicians of England had generally been preceded by earlier craft or technology bodies. Though philosophy was one of several things that somewhat helped with the emerging of experimental science, just confusing philosophy with science is both wrong and unhelpful. Those who wrongly insist on calling unsubstantiated notions 'science' instead of philosophy, just help undermine both philosophy and science. Hence often ideas like the idea of inertia did first begin as unsubstantiated philosophy notions, but later were backed by observational or experimental scientific evidence and so only then really became science. But the somewhat difficult 1600 Latin 'De Magnete' of William Gilbert is maybe the nearest early science equivalent of 'The Golden Bough' magic study of Sir James Frazer, or maybe that is this website ?

In reasonable defence of alchemy.

From around 1597-1625 in Germany, France, Holland and Sweden controvertial 'Rosicrucian' neo-christian and neo-alchemist texts advocated revolution in philosophy, religion and science, backed in England by Robert Fludd. This was opposed to traditional scholars and their book study, as was William Gilbert, and favoured direct investigation of nature 'supporting God'. Major proponents of change in Europe around then were perhaps Luther, Calvin and Descartes and in England maybe the dubious Francis Bacon. (see 'Universal Reform' Eds Hotson & Urbanek Vol 3, 'Reformation, revolution, renovation' by Lyke de Vries, Brill 2021.) Unlike most others of the time William Gilbert did distinguish chemists from alchemists, but many early scientists had studied alchemy and the like to at least some small extent enough to be partly influenced by some of its ideas. Alchemists were commonly said to have two main aims, firstly to change substances into other substances and especially change base metals like lead into noble metals like gold - and secondly to make or find an elixir of life to cure ageing and all disease. These two alchemy aims could perhaps be basically said to be aims to do chemistry and to do biology. Some alchemists did do some successful substance-changing chemistry mostly using fire and solvents, but they achieved nothing on changing elements since that can be done only using nuclear fission and/or nuclear fusion which are very complicated and expensive processes though the former has been used to make some radioactive elements. Alchemists attempts to do biology were even less successful, since handling much disease needs much detailed knowledge of the body and of germs, and to get anywhere with ageing almost certainly needs much better knowledge of the body and of genetics and evolution. Todays biology 'body-clock theory' seems quite inadequate, in requiring the evolution of one biological system that both takes a baby to a fit young adult and then also to an unfit old adult. That does not look like something evolution could do, and there is no other similar example at all. And some body systems like the immune system clearly function to keep a fit young adult fit and young, though genes seem to suffer from cumulative mutations, like computer files can, but seemingly have no body system to fix them. But maybe the body does actually have an evolved 'ScanGene' repair system similar to computer 'ScanDisk' repair systems. If so then ageing would just need the evolving of a switch that turns 'ScanGene' off, and that might possibly be somehow reversible ? So maybe both of the alchemists main aims might really be achievable, if not very easily, and with a human race afflicted by ageing and illness surely trying to fix these is not any bad aim ? And that some new substance from their first aim might help with this second aim maybe does not seem impossible either so that appropriate areas of experimental science today still have one or both of these alchemy aims while leaving out the gold-making claim that was used chiefly to lure rich sponsors. And if 'alchemists' from early witch-doctors and metal-workers helped develop experimental science, some early attempts at establishing science theory can be found in ancient-Greek atheist Atomist writings which were almost entirely destroyed by the early Christian church, but its ideas were basically saved, though in a less atomist and a dualist form allowing of Gods, by the Roman poet Lucretius (c99-55 BCE) in his De Rerum Natura (On the Nature of Things) published about 60BCE to a limited extent. This was discovered by Catholic scholar Poggio Bracciolini in a German monastery in January 1417 and after the circulation of maybe 50 handwritten copies it was first printed in 1473 and helped in encouraging the emergence of science. It was widely wrongly interpreted as presenting Greek atomism, but did present a rather different early attempt at presenting science-inclined ideas and is referenced 7 times in William Gilbert's 1600 De Magnete. But actually the chief support for ancient-Greek Atomism in Europe before Galileo adopted it was probably some alchemists including Daniel Sennert.

In general history it is the kings and queens who make the news and seem to be only people who matter, though they more often act against progress than in support of progress. And it is often the many smaller people who most advance progress by adopting newer technologies and ideas. The masses and the voters will at times support general progress, though not always regularly or smoothly. And new ideas are favoured chiefly in times of extreme necessity like war, partly explaining the emergence of Enlightenment ideas and early science in Europe when war began between the new Protestant Christian church and the Roman Catholic Christian church. While in Europe the Middles Ages involved the domination of the Catholic Church and that being challenged and weakened allowed the Enlightenment and Science to emerge. The domination challenge basically involved religion-against-religion war into which parts of science were dragged sometimes significantly. But generally religions have somewhat reluctantly since become more tolerant of science as they have come to realise that they cannot disprove science, though sometimes only after doing significant harm to some areas of science. Hence in Europe the Enlightenment included the study of the ideas of other cultures like ancient Greece. So early science emerged with Copernicus, Gilbert, Galileo and Kepler in parts of Europe that then had substantial political and religious instability or whose stability was under strong threat. Catholic-Protestant warring was extreme from about the 15th century to the 18th century but did eventually moderate. And the early scientist prepared to doubt everything then had to support himself with paid work of some kind that was often very unscientific, or else find a rich private patron of maybe science.

In science history it is the science 'giants' who mostly produce the big new technologies and new theories, though with most of them more readily accepting new technologies than accepting new theory change. Often the many smaller science people more readily adopt newer technologies while also resisting theory change. Resistance to theory changes was taken by Newton as being chiefly due to a currently accepted theory creating mental theory 'prejudices' that prevented fair consideration of alternative theories. So science technology has progressed reasonably smoothly, while science theory has often advanced very patchily and included steps backwards. The science masses and the career-scientist peer voters will often oppose theory progress.

Historians of science generally do good work on clarifying the development of science ideas and inventions, though their work is not always good. Scientists themselves are mostly slaves of their time so that sixteenth century science in Europe is mostly based on predominant sixteenth century ideas coming mostly from the ancient greeks. But a few scientists have managed to successfully think 'timelessly' or 'ahead of their time' and historians of science often 'explain' these few scientists quite wrongly in terms of ideas of their time. The time rule used by historians of science, which correctly helps explain the work of the majority of scientists, is for a few key scientists a false prejudice grossly misrepresenting their science. William Gilbert was one early scientist widely misrepresented, as was the other notable Protestant English scientist Isaac Newton who supported a similar physics theory, though neither seems to have much supported any of the main established churches of their time. And science historians have also at times not correctly understood the science that they try to explain. Science history is certainly done no good by the many modern 'science historians' who like to put down substantial scientists like Gilbert, Newton and Tesla by citing numbers of obscure alternative scientists who had done some related but much inferior science ! Of course science translators have also mostly done good work, but they have shared the same problems as science historians.

In early modern Europe the socially challenging Black Death was followed by the socially unsettling emerging war between the catholic christian church and new protestant christian churches which saw a sixteenth century mainland Europe becoming more concerned with philosophy and mathematics and an England becoming perhaps more concerned with technology and experiment. Big social challenges or projects like Pyramids or Great Walls have prompted social progress in other societies. But while the basic idea of experimental science was developed early in England as by Robert Grosseteste (1175–1253) from early Greek ideas though only as ideas and not really practiced till rather later with William Gilbert in England and Galileo in Italy. Gilbert, Galileo, Kepler, Descartes and Newton all noted and emphasised the constancy of the basic natural forces as with gravity and magnetism following some constant natural laws. Galileo, Kepler and Descartes followed ancient greek Atomism and its simple mechanical push physics. But Gilbert, maybe combining the best of greek Atomism with the best of Aristotle as shown by his own experiments, produced his attraction physics necessarily includes signal emission, signal transmission, signal reception and signal response, that might also allow of some affects by the environs and so allow of some multifactoral aspects to give variation in such forces. Though wrongly discredited by catholic church Jesuits and others Gilbert's action-at-distance remote-control physics was basically backed by Newton, though not fully openly, but was little studied by most early physicists, yet when modern experiments seemed to show some variations to the basic natural forces, physicists rushed instead to produce new theories to explain such. Most of the new physics theories have basicly been versions of Cartesian-theory-like push physics and nobody has really tried to develop the original and perhaps more amenable attraction physics theory. But a strong case can be made that it was really only in William Gilbert's physics that Isaac Newton's real physics could be seen.

Some see science as progressing in a manner similar to social progress as seen by Karl Marx, and in line with Thomas Khun - by periods of steady experimental progress ending in theoretical paradigm revolutions also progressive. From this Khun saw eg Einsteinian paradigm theory as incompatible with Newtonian paradigm theory so that they could not really support or disprove eachother - though experiment might disprove either or both. But for an experiment to really disprove eg attraction physics theory, the theory must be shown to be invalid in an attraction physics theory interpretation of that experiment. The fact that some theory X interpretation of an experiment fits theory X, cannot disprove theory Y. All possible theory Y interpretations of the experiment need to be disproved. Firm disproofs like this are rarely attempted in physics, and have yet to be attempted for the Gilbert-Newton 'attraction physics' theory that is widely wrongly claimed to have been disproved. Older physics theories have mostly just become 'disapproved' with new theories being given more publicity and hype. So, unlike much science, physics theory has not been very progressive. And there are still supporters of religious churches backing brands of physics theory now and holding their own often biased views of its history.

The motions of physical objects.

The first serious consideration of things like the motions of physical objects was probably by many early philosophers including ancient-Greeks Aristotle and Leucippus around 400 BC. Limited observational evidence on the motions of the Earth and other planets was put by earlier astronomers like Copernicus and Kepler around 1543-1619 AD. Then better evidence was added by the first actual experimental studies by scientists William Gilbert and Galileo around 1580-1640 AD with some theory ideas. Isaac Newton and Rene Descartes (or Des Cartes) then chiefly developed theory ideas further around 1640-1680 AD, though Newton's were largely misunderstood and mispresented as Cartesian theory with Newtonian maths. Einstein and others added various further conflicting theory ideas from especially around 1910-1930 AD, some reasonably widely supported though no one theory getting very wide support among all physicists. And around 1962-2022 AD Vincent Wilmot tried to develop a single theory or single set of compatible theories which is maybe still not yet finished.

Science and science theory.

Many early attempts were made to establish a theory-led modern science as in ancient Greece by Plato, Socrates, Aristotle and others, but these attempts all failed. Real science only emerged when William Gilbert and Galileo Galilei showed that science needed to be technology/experiment based - as most clearly shown in Gilbert's published polemic arguments and disproofs against mere-theorising. The compass and the telescope beat the Platos and Aristotles, not any science theory - though in science the mere-theorisers never fully accepted defeat and today commonly present science wrongly as being a theory-led progress. Compared with a reasonably smooth technology-experiment progress, the history of science theory is actually very messy and perhaps involves little real progress. Newer science theories have often not been better science theories despite claimed 'proofs' of such. Ancient Greek theories basically divided between Aristotlean 'active matter with God motivation' and Atomist 'dead matter with law determinism push-physics' theories. The early Christian church from the fourth century backed Aristotlean science against Atomist science which it then saw as Godless. But the catholic church Jesuit Order was founded to counter non-Catholic ideas, and with its backing modern science from around the seventeenth century largely adopted the Atomist push-physics with some minor modifications. Interestingly Gilbert produced a largely not understood new 'active matter with law determinism' signal physics that was basically backed only by a few in England like Robert Hooke and Isaac Newton.

Early forms of experiment alchemy emerged perhaps independently in the Chinese, Indian and ancient-Greek/Roman civilizations usually combined with some mysticism, more as an alternative development from magic like religion. It died out there but only to re-emerge in the Arab world where around 1200AD Europe found it and developed it further. Experiment alchemy in Europe then developed further as a chiefly-experiment method to try to determine a range of truths alternative to religion's inadequate preached truths and philosophy's inadequate argument truths. Some 'alchemy' was really good chemistry and good experiment but just with weak interpretation - as http://www.conciatore.org/2017/01/transmutation-of-iron.html). Unfortunately some alchemists claimed that its truths would enrich people materially, as in giving easy gold production and a cure for aging, and some alchemists still gave support to religious-type mysticism as did 'witches'. Europes alchemists of this period were rebels and were to varying degrees oppressed. Only after the Christian church in Europe fragmented, with battles between Catholic and Protestant branches, was occasional logical argument raised against religious-backed 'truths' as by Copernicus, and then experiment finally emerged in Europe as the championed means to determining truth for truths sake with the science of Galilei Galileo and William Gilbert displacing alchemy. And while Gilbert distinguished chemists from alchemists before 1600, science history has generally failed to differentiate these till after 1700.

Different parts of the world had seen the development of Alchemy as an early experimental science or experimental magic which became a neo-religious secret cult that believed that experiment might produce wonders and rightly or wrongly was strongly opposed by churches. While it did encourage experiment, the dubious trappings it often had may have also encouraged opposition to early experimental science. But from early ideas that some experiments might give some truths, science soon required that every claimed truth about the physical universe must be supported by appropriate nature observations or experiments. This was basically opposed by all religion though maybe somewhat more by the Catholic Church more strongly backing belief im miracles and holy relics and a view that truth best comes from the words of particular holy books or holy men. Of course individual scientists while committed to science did often also perhaps perversely have some religious beliefs.

From evidence early experimental science generally held that, though the universe or parts of it may change over time, there are some laws of nature governing any change that are constant. For evidence experimental science observes 'the present', some of which is logically deduced to involve effects on the present that some laws of nature produced from past states. And experimental science chiefly has tried to define laws of nature, at least some of which are hoped to be basic and constant, from evidence obtained chiefly by close observation of nature. Of course there have also been some involved in religion, philosophy and/or theoretical or even experimental science who have taken different views of nature and laws of nature.

Problems with science history.

As a separate taught subject 'History of Science' tends to be chiefly concerned with people and especially with ;

1. who first produced a new science idea.
2. who helped with or helped inspire that new science idea.
3. who opposed that new science idea.

This can certainly be very interesting, but an excessive concern with people can mean that the actual science ideas are not examined closely enough and so can include major errors. And a science idea itself can also have significant actual problems from ;

1. a scientist publishes a new science idea, but then develops and amends it.
2. other scientists develop and amend that science idea.
3. others opposing that new science idea misrepresent it (unintentionally or intentionally).
4. others merging that new science idea with their own amend it or misrepresent it.
5. any new terms, or new use of existing terms, involved in a new science idea can be misunderstood.

When science was all books written and read in Latin, that had some big advantages and disadvantages for scientists in allowing publication to be international but often very limited and censored so that the education of early scientists mostly involved Plato, Aristotle and Euclid. Some of the better early scientists, like Gilbert and Newton, did widen their education with substantial home-study. But after Newton local natural language science journals took over and scientists were soon addressing only recent journal articles in their own language and were generally poorly educated on wider science theory. Hence, Einstein was chiefly knowledgeable only on other German language physics of his time and grossly misunderstood Newton and earlier physics theory ideas - and almost all modern physicists are similarly ignorant. And relativity, quantum mechanics and modern physics theory were basically developed before remotes and computers became common and they and other 'modern physics' have failed to incorporate the main modern technology ideas that were anticipated in Gilbert's physics and its part-development by Newton. This was like Hero of Alexandria, a Greek or Egyptian in Roman Egypt, who published the basic ideas both of steam-power and of wind-power about 100 AD and so long anticipating their much later full development. So windmills only became common from around 850 AD in the Arab middle-east, to around 1200 AD in Western Europe, and steam-engines only became common from around 1700 in England.

Also now scientists publish their theories in ad hoc articles, encouraged by government funders and science journals wanting newsworthy briefs. But science theory write-ups need to be comparable to show where they are compatible or incompatible to identify and evaluate their proof issues - in physics compatible write-ups 'Principia'-style are really needed but are rarely produced. Of course Newton's Principia was essentially a write-up of three theories in one - Newton blackbox physics, Gilbert attraction physics and Descartes push physics - though they would maybe be more useful as three separate write-ups. Without a comparable write-up a new physics theory may seem to explain some claimed cosmology issues but hide the fact that it cannot explain two marbles colliding.

This website presents a lot of history of physics, trying to concentrate on theory ideas as the final published thoughts that you may be able to read especially of four specific famous physicists - Albert Einstein, Isaac Newton, Rene Descartes and William Gilbert. There are of course many others, but these four give a good range of basic theories of the universe worth considering here in an interrelated manner. And here it is the scientific ideas that are examined, ignoring whether part of Descartes optics may have come from Snell or parts of Gilbert, Newton and Einstein ideas may have come from others. Not the textbook physics history of rubbish and lies pretending smooth scientific advance, but the real history of physics of both religious non-scientists limiting its advance and of ignorant unscientific scientists limiting its advance.

(While early science faced strong opposition from some powerful religious factions, some early scientists were themselves strongly religious and those that were not often tried to present their science as not challenging religion. Even today some religious factions argue strongly against evolution. But the mechanism of evolution is genetics and its basics were the work of hobby geneticist Gregor Mendel (1822-1884) who was an Augustinian Christian monk who did his genetics experiments in his monastery. Mendel was a skilled hobby scientist like William Gilbert, at least until he gained promotion to being in charge of his monastery and his increased religious work made him too busy to continue his science. He did not see science and religion as having any basic conflict, and his church may have seen him as 'just growing flowers', but some do still see science and religion as having basic truth conflicts.)

It can seem natural or necessary to judge the science of a scientist from their publications and mostly that works. But sometimes a scientist can significantly mispresent their science ideas, as especially often happened in early science from the 1500s to the 1700s. That was a period when other scientists, church people, political people or others often attacked a scientist and his science ideas like they attacked 'heretics' perhaps very improperly and even misusing or threatening the law against a scientist. Hence in England initially William Gilbert presented his science as 'just about magnets' when he actually had a much bigger physics and Isaac Newton presented his physics as 'not judging hypotheses' while disproving some physics hypotheses, and in Italy then Galileo was presenting some of his science as fictional Greek dialogues.

Of the chief four sets of physics ideas examined around this website, those of Albert Einstein and Rene Descartes seem somewhat less problematic in that their ideas were generally taught reasonably accurately though often not very clearly. Of course Descartes simpler physics is not taught now, and modern General Relativity is taught with key aspects not compatible with Einstein's theory. But the theories of both Isaac Newton and William Gilbert have both been long taught as differing very substantially from the ideas that they published, often robbing Newton's theory of its Black Box base and support for Gilbert attraction physics and robbing Gilbert's theory of its Robot Matter signal-response base. Since there is no good reason for Newton's and Gilbert's theories being so grossly misrepresented, extra efforts are made to ensure that they also are presented as correctly as possible on this site.

Though Galileo in Italy and Gilbert in England had given experimental physics a strong start by 1600, it was maybe 1700 before experimental physics and experiment-based physics theory were more widely practiced. Till then 'science' was chiefly astronomy, mathematics and philosophy which more suited Cartesian 'certain knowledge' mere-theorising physics theory. The long failure of experimental 'Alchemy' or 'Chemistry' seems to have worked against experimental science and so strongly hindered a wider early acceptance of the experimental physics advanced by Galileo and Gilbert, and the basing of physics theory on experiment. Of course the study of physical materials, or chemistry, does have some real relevance to physics so many early physicists having interest in it was not necessarily unscientific as many at times wrongly claim.

Robert Boyle (1627-1691) was typical of a good experimental physicist and chemist of the time and visited Galileo who was under house-arrest in Catholic Italy, and Boyle was born in Catholic Ireland but preferred to work in Protestant England. Like many scientists then Boyle supported Cartesian mechanical physics, though backing a vacuum and also a particulate view of both gasses and 'effluviums', as shown in his 1673 'Essays of the Strange Subtilty, Great Efficacy, Determinate Nature of Effluviums', his 1674 'The Hidden Realities of the Air' (with Air as being an aggregate of 'particulate effluviums' from differing bodies), and his 1676 'Experiments and Notes about the Mechanical Origin or Production of Particular Qualities, including some notes on Electricity and Magnetism'. Boyle was seemingly unacquainted with Gilbert's physics and his interest in Chemistry was shown in his 1661 'The Sceptical Chemist', which maybe confirmed him as being strong on science experiment but weak on theory. He was a non-denominational Christian and did not marry.

Gilbert had formulated from experiments his 'attraction physics' involving matter responding to signals travelling through space from other matter and with no special place for god or humans, in the 1580's when religions and governments with their scholars and philosophers backed Aristotle's non-experimental 'logical divine science'. But experimental science only really began to be accepted from around 1650, when the semi-experimentalist philosopher Descartes won wide backing (including often by religions and governments) for his 'logical semi-divine science' with a mechanical push universe including a matter ether that filled space - and with god and humans having a separate special place outside science.

In the Gilbert-Newton era, Protestant England easily beat Catholic Europe in technology development - leading the Industrial Revolution. But Catholic Europe and its greater number of physicists refused to admit that they had been bettered in physics theory also by their much fewer Protestant peers. On theory, unfortunately the greater numbers wrongly won out - largely by using cheap name-calling and misrepresenting rather than by any scientific disproofs. Early supporters of catholic Galileo and Descartes claimed that Gilbert-style signal response physics was 'unscientific' because it required bodies to be 'animate' - but animals are animate and at least now most certainly agree still obey the laws of physics, so those claiming that the animate was unscientific were being idiotic and Gilbert's physics was dismissed without being disproved. Not the only case of physicists being idiotic. Later attempts by Newton to disprove with experiments the strongly entrenched Descartes' logic-physics, especially on its ether, were so fiercely opposed by what were then peer scientists that he had to moderate his opposition to Descartes physics and moderate his support for Gilbert-style action-at-distance attraction physics. That was enough to allow Newton's physics to be falsely presented as being an improved-Descartes push-physics including Descartes' ether that Newton considered he had actually disproved.

In the modern era, emerging economies like Russia, India and China have basically taken up modern Western science-journal physics without considering early physics theory much. Latin is not so big in these countries and they have inclined to simply adopt prevailing Western physics theory prejudices. But still if physics today is somewhat less constrained about experiment, do physics theory errors and limitations actually limit physics experiments today ? The answer to this seems to be a definite yes, since the physics experiments planned by me in the 1960's still in 2022 seem to have been done by nobody. The best of religions and of governments can certainly act against aspects of science that they come to think may adversely affect their religion or politics, hence aspects of nuclear science have long been opposed by the CIA, MI5 and other such government bodies. And despite technology today being all signal technology, physicists are still ignoring Gilbert-Newton style signal physics. But emerging-economies science theory prejudices may be somewhat less deeply engrained, so hopefully 'come on China'?!

To date there have been 4 basic types of causal theory explaining the behaviour of physical bodies, including gravity behaviour, that have had some substantial support. These have had variants, and there have been some other less well supported physics ideas also, but the 4 main theory types are characterised in the diagrams below;

1. God/Magic physics

2. Gilbert physics .... 3. Descartes physics

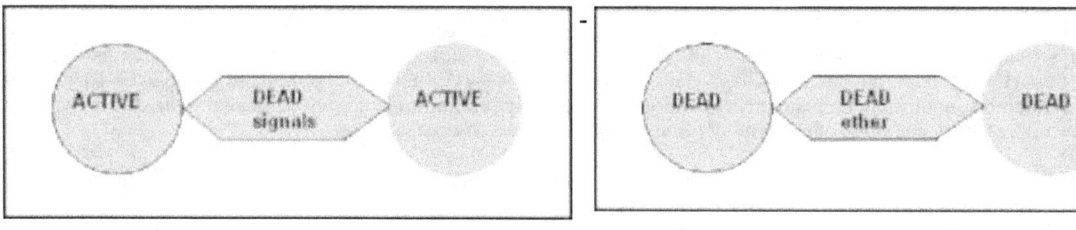

4. Einstein physics

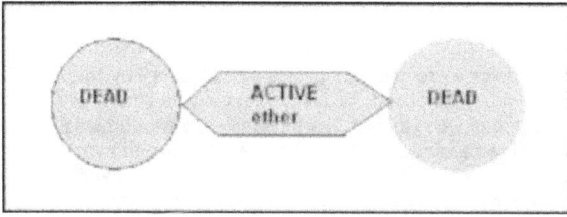

Of course there have long been some preferring 'non-causal universe' physics often based on views of the universe having been created by a God having chosen to create eg. a musical universe or a mathematical universe. Hence early Kepler creationist physics included a Geometry Mathematics Physics and a Music Mathematics Physics. And post-Einstein physics today maybe also seems based on the view that mathematics is primary in the universe, with its Wave Mathematics Physics and Quantum Mathematics Physics. Of course after Kepler studied William Gilbert, he concluded that the universe is NOT really Mathematical but is 'Experienceal', though Kepler's physics then went with a Descartes 'touch-push' experience rather than a Gilbert 'signal response' experience. And such creationist type physics requires not just a God creating the universe using existing laws of nature, but a God creating the universe and also creating the laws of nature as Descartes supposed in 'The World'.

Newton, the probably most astute of physicists ever if not always entirely correct, concluded that NO causal explanation fell within science proper and that EITHER of the causal theory types 2 or 3 above might be correct and give compatible mathematics. And comparison of the diagrams above suggests also that a post-Newton type 4 theory Einstein physics could be a mathematical mirror image of Gilbert type 2 physics.

Some of Newton's specific ideas were readily accepted, like his almost-new conclusion that terrestrial gravity and planet motion involved the same thing, but were not immediately taken as disproof of anybody. Newton's experimental disproofs of chunks of Descartes' logical physics were only fully accepted some generations on, and his main blackbox theory was never widely accepted by scientists. When there was a developed body of scientists their 'peer review' generally worked well for smaller bits of science, but often worked badly for big science ideas that required a complete rethink of established science. Now to many, physics is itself a religion and they fiercely fight against ideas that they see as 'against the mainstream' or ATM. Just like many religions witch-hunt Inquisitions against ideas that they see as 'against the gospel' or Heretic. So much physics debate has now perhaps really shrunk to small details only, though with them often being falsely presented as being 'fundamental'. And the strong prioritising of real experiment on physical systems by Galileo and Gilbert, became much weakened by those supporting instead model 'experiments' on physical models of physical systems, simulation 'experiments' on computer simulations of physical systems and thought 'experiments' on thoughts of physical systems. Since thoughts are really the opposite of experiments, the term 'thought experiment' is clearly a nonsense abuse of science language and should be more honestly termed 'imagined experiment'.

As an alternative chiefly to Descartes 'The Optics' 1637 push-physics optics theory, Christiaan Huygens (1629-1695) produced an early light wave theory, and later James Clerk Maxwell (1831-1879) developed light and 'force field' wave theory. And using such ideas, Michael Faraday (1791-1867) unsuccessfully tried to link gravity and electricity, saying 'The long and constant persuasion that all the forces of nature are mutually dependent, having one common origin, or rather being different manifestations of one fundamental power, has often made me think on the possibility of establishing, by experiment, a connection between gravity and electricity ... no terms could exaggerate the value of the relation they would establish.' Einstein also unsuccessfully tried the same later, though of course early physics had produced two theories each of which gave a common basic mechanism to all forces - in Gilbert/Newton attraction physics and in Descartes push physics. Both of these are now ignored by modern physics, though many physicists did make unsuccessful attempts to develop Descartes push physics while only Newton made any real attempt to develop attraction physics.

When Einstein arrived on the science scene later, he started with some smaller science ideas like the photoelectric effect and built-up to his bigger ideas like relativity. But special relativity had been basically developed before Einstein by others including Fitzgerald, Lorentz and Poincare though maybe with not as good writeups. And when Einstein claimed to disprove 'Newton's ether theory' - he was actually disproving Descartes, or rather the Lorentz modification of Descartes' ether, and was really ignoring and not disproving Newton's real physics or indeed Gilbert's. Descartes physics was the one with a rigid space and matter requirement and had no energy other than matter motion - unlike Gilbert and Newton physics which as such was really closer to Einstein's physics. Often a majority of scientists have rejected a better theory to support a weaker theory. And if England was one of the earliest centres of emergence of experimental science from before 1600, it was also early in the bureaucratising of science from around the 1700's that was to help limit its development for hundreds of years.

Since the start of science, some significant supporters of religion have attacked science though that has maybe been somewhat reducing recently. But in the past religion has mostly been more popular than science, though that is now reversing in more progressive countries. Still supporters of science have to date resisted attacking religion. But science has been closely studying many areas and aspects of the universe, and ;
1. has found no real actual evidence for the existance of any God.
2. has found no real actual evidence for the existance of any Heaven.
3. has found no real actual evidence for the existance of any Hell.
4. has found no real actual evidence for the existance of any Miracles.
The logical conclusion from this is that the real actual existance of any God, Heaven, Hell or Miracles is very unlikely and religions are most likely based on fictions, lies and cheating. Of course science can also have substantial doubtful areas as in theories by Einstein and others especially it seems concerning physics and astronomy where particular interpretations of some experiments or observations may be taken when contrary interpretations may be as feasible. Scientists will often strongly defend some very doubtful pieces of science that they should be doubting, though much of science may be well founded with good evidence.

Interestingly, 2005 saw an attempt in the USA to introduce a new law called The Restore Scientific Integrity to Federal Research and Policymaking Act", requiring that science be controlled by government science agencies rather than by central government ! There is now increasing pressure in favour of 'Open Science' or 'Perfect Science' and against normal science. So 2020 US government ruled that only 'Perfect Science' could be used by US government. As there actually is no 'Perfect Science', US government now can use no science and so some support for 'Perfect Science' is actually opposition to actual science. Increasingly governments seem to support only 'government-approved science' which with governments commonly favouring selective secrecy and lying is likely to chiefly mean fake-science becoming predominant. And increasingly in more countries the media is basically supporting such fake-science as in Fox TV's much-pushed 'Ancient Aliens' show. But the internet by 2009 did look close to opening up the long-closed shop of science publishing, as 41 Nobel laureates call for 'open access'. Especially helped by one open-science website, Cornell University's arxiv.org, and maybe a little by this website and others. However 2010 raises some concern with Cornell considering some possible charging policies for future users of ArXiv.org to cover its rising running costs, hopefully limited to charging bigger institutional users and/or publishers or maybe it carrying paid advertising ? 2007 did see the UK's Channel 4 TV disprove the theory being supported by some environmental scientists that the Global Warming weather changes that Earth is getting now are NOT mainly due to man-made CO_2 from burning oil, gas and coal. It certainly seems to be due at least partly to some other cause - natural or man-made ?

William Gilbert and others had strongly argued for science theory to be based on direct deduction from experience and experiment on natural phenomena only. But Kepler and to a lesser extent Newton supported a wider validity of general logical deduction as from mathematics in science theory, and Descartes even allowed religious deduction a basic role. (Newton did privately try but failed to develop his physics to fit with his religious ideas, and to develop its effluvia/spirits side and to develop chemistry.) And Einstein's adoption of 'thought experiments' has perhaps encouraged many physicists now to confine themselves to only logical theorising, now perhaps mostly based on manipulating equations or mathematical language ? Much modern physics theory now rests basically on 'mathematics experiment'. Of course as actual experiments have revealed more complex natural phenomena needing more complex maths, it is maybe understandable that real physics explanation has become more problematic. And experiments (like the Mitchelson-Morley experiment in our Albert Einstein section) are designed to try to demonstrate something specific, and strongly tend to being interpreted only in that regard even when they might more realistically be demonstrating something quite different in fact.

The Unteachability of Science

It seems well proven that many people can be correctly taught small bits of science. But it is not proven that many people can be correctly taught major science theories, and substantial doubts regarding that have been expressed by all four of the key physicists considered on this website.

Gilbert, Galileo, Descartes and Newton were all slow to publish their works and the latter three certainly claimed in some cases at least to be publishing only after major pressure from others to do so. In England, Gilbert waited until he felt that he had gained some sufficient support from Queen Elizabeth, and in France, Descartes' science waited until he felt that it had been made sufficiently acceptable to the prevailing catholic church.

Gilbert, Descartes and Newton certainly all saw one major problem to the teaching of any major new science as being previously learnt wrong thinking - or, as Newton explained in his Principia's introduction to Book 3, physicists having 'prejudices to which they had been many years accustomed'.

But they all seemed to also conclude that most people would never be able to correctly understand any major science. Gilbert specifically wrote that his work was not for the 'common person' or 'common scholar', while Newton basically said that his science rested only on the work of 'science giants'.

Einstein explicitly said numbers of times that he did not believe that anybody fully understood his physics.

While small bits of science are certainly teachable, the history of physics theory certainly supports the conclusion that major science theories are actually almost unteachable. And that casts major doubt on the modern view that science generally can progress by 'peer review'. Clearly peer review should work fine for small bits of science, but might not work for major science theories.

And the history of physics theory does indeed seem to confirm just that. Gilbert's theory was correctly understood by almost nobody, as was the case with Newton's black-box physics and with Einstein's physics. Of course there are always lots of people who will falsely claim that they do correctly understand those theories. Science has always had lots of fools and liars posing as experts successfully, chiefly by understanding some smaller bits of science.

This maybe backs Gilbert's trusting chiefly in nature experience and experiment, more than in merely deductive or mathematical reasoning. But the biggest case of experience or experiment being itself misleading is of course the fact that on Earth it clearly appears that the Sun orbits the Earth every 24 hours - though we now know that it is actually the Earth revolving every 24 hours. Some of the supposedly 'key' or 'crucial' experiments of physics are probably open to different interpretations than those normally being assumed for them. And though useful human invention began BEFORE science developed, science ideas have helped motivate useful invention - even science ideas that were later fully disproved !

Is modern physics dumbed-down or what ?

In more recent years, developed countries governments have taken a strong lead in greatly dumbing-down and politicising education - including science education - pushing to a-degree-for-everybody policies that have cut the average IQ of modern 'scientists'. And in science, governments are now also pushing views of everything being relative and of assorted theorised ideas being as valid as fact based ideas - or non-science being as valid as real science. Physics theory, like most science theory, is being driven backwards to mere government-sponsored philosophy as governments have concluded incorrectly that science theory is unimportant and has little effect on technology. See Science Teaching Today, Cold War Science, and UK Science Funding.

2009 saw two 'physicists' claim proof that 'the LHC was disabled by a bird from the future' ;
"Sometime on Nov.3, the supercooled magnets in sector 81 of the Large Hadron Collider (LHC), outside Geneva, began to dangerously overheat. Scientists rushed to diagnose the problem, since the particle accelerator has to maintain a temperature colder than deep space in order to work. The culprit ? 'A bit of baguette' says Mike Lamont of the control centre of CERN, the European Organisation for Nuclear Research, which built and maintains the LHC. Apparently, a passing bird may have dropped the chunk of bread on an electrical substation above the accelerator, causing a power cut. The baguette was removed, power to the cryogenic system was restored and within a few days the magnets returned to their supercool temperatures. While most scientists would write off the event as a freak accident, two esteemed physicists have formulated a theory that suggests an alternative explanation: perhaps a time-travelling bird was sent from the future to sabotage the experiment. Bech Nielsen of the Niels Bohr Institute in Copenhagen and Masao Ninomiya of the Yukawa Institute for Theoretical Physics in Kyoto, Japan, have published several papers over the past year arguing that the CERN experiment may be the latest in a series of physics research projects whose purposes are so unacceptable to the universe that they are doomed to fail, subverted by the future."
- Quoted from November 2009 Stealthfusion.com

The number of people entering science professions in more recent years is much greater than a hundred years ago, but in some respects the range of people entering science professions has been greatly narrowed. Hence though much good Physics has been done using relatively simple mathematics, and now a physicist will commonly have a computer or an assistant that can do more complex mathematics for them, but physics exams lately have generally been designed to fail all whose main interest is not mathematics. And much good Biology has been done using no art drawing, and now a biologist will commonly have a camera or an assistant that can do art drawings for them, but biology exams lately have generally been designed to fail all who have little interest in art drawing. Exams needed to enter science professions often severely limit the range of entrants and help limit the scope of the sciences concerned. This compounds science funder restrictions and science teaching restrictions.

The windmill, compass and telescope saw Gilbert, Galileo and other emerging science driven by ideas driven by emerging technology. But some science ideas with seemingly strong proofs are not believed by a majority of the public, though other science ideas that seem to have weaker proofs may be widely believed. This can be due to poor science teaching or due to the science being actually wrong, or in some case just really due to the science having some conflict with popular cultural thinking of the time. The scientific revolution really needed restrictive church-government to be destabilised as happened in Europe's 'Reformation'. Even today the more ideas and science are controlled by governments the less is science trusted, because people now know that governments commonly favour lying. Physics pushing ever more speculative and maybe untestable theories about time travel and multiple universes does not help either.

PS. For a very interesting and good if imperfect recent work on some issues of science history and theory from a philosophical viewpoint, see Laura Aline Ward's Objectivity in Feminist Philosophy of Science PDF 0.25mb to load !

The physics time chart below for the chief physicists considered here, has bars for when they lived and filled when their science chiefly published ;

William . Johannes . Galileo .. Rene .... Isaac .... Nikola ... Albert .
Gilbert .... Kepler .....Galilei . Descartes . Newton .. Tesla .. Einstein

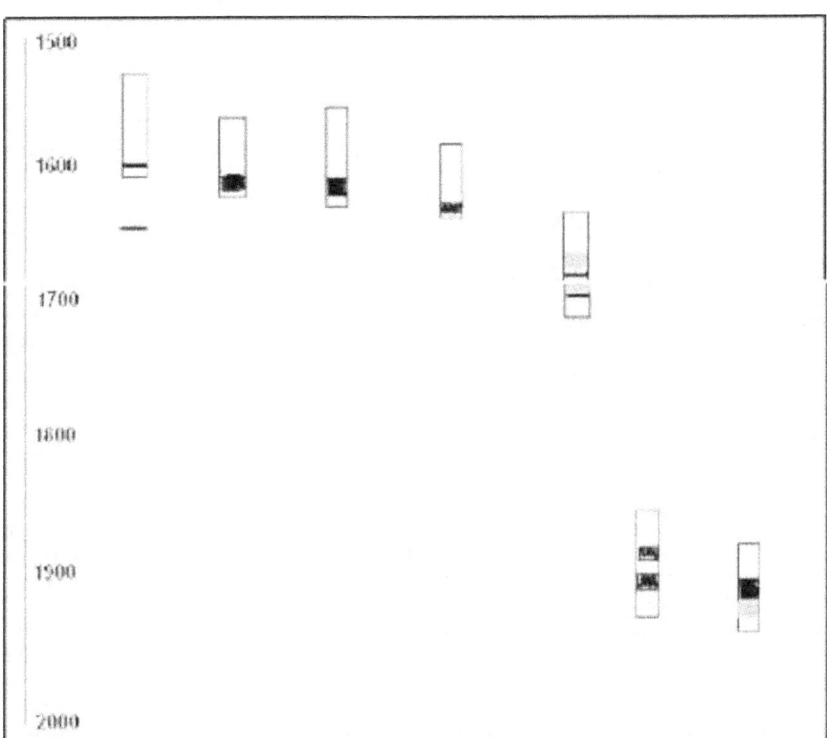

Of these six physicists, only Gilbert and Newton seem to have studied most physics theories available and Newton seems unique in being able to both understand and use very different types of theory. Hence for gravity Newton used Gilbert-like attraction theory but he also used particle and wave theories elsewhere - while using a blackbox theory and not committing to any one explanation theory. Gilbert publishing very late in his life, had little time for defending or further developing his theory.

It is common for modern physics theories to use terms like 'Mass' or 'Field' or 'Continuum' or 'Particle' or 'Wave' with no full or specific definition of the terms as applying to the theory, but with partial definitions or implied definitions that can contradict terms common or classical science meanings and can include logical contradictions. Definitions of 'Mass' for some theories have varied around 'amount or volume of matter', 'amount of inertia', 'amount of gravity production', 'amount of energy equivalence' and other meanings. So often the use of terms with no specified definition in modern physics means them having little or no real scientific meaning, try Google 'definitions'. Much modern physics can be taken as blackbox science where it is the mathematics of processes that is being defined rather than physical reality, and that may or may not be taken as generally being satisfactory. But mathematics can be taken as having no limits so that it can support anything, while actual nature has real limits. In principle experiment on nature should set limits to the mathematics acceptable in a science. But some particular science theory and its mathematics might fit well with the well established and understood experimental results of many common natural phenomena, while concentrating on the experimental results of some one abstruse natural phenomenon might not fit that theory well and may seem to fit some more abstruse theory and its mathematics better. It is not clear that this always disproves the first theory, though it may cast some doubt on it.

It is also common on modern physics websites to see comments asserted as being scientific like 'Revisionism is a serious offence'. (Google it !) This basically means 'Trying to disprove a current science theory is a serious offence' - and is of course what Galileo was put under house arrest for and other good early scientists were executed for. Current science's 'anti-revisionism' is really anti-science.

The death of science

All science basically rests on physics and physics theory now is certainly dying, having been reduced to physicists debating a bunch of poorly defined partial physics theories none of which seem to offer any realistic chance of a provable complete physics theory to explain the full physical universe. Physics theory is now looking unprovable and, unless physicists wake up, may well soon become widely accepted as having died.

Experimental science may not need specific theories but its motivation does rest strongly on it seeming possible that science can explain the universe better than religious or other explanations. But soon religion could with seemingly good reason be proclaiming victory over a science slowly grinding to a halt. The world looks to be now advancing to a new Dark Ages, unless physicists can put aside their current physics prejudices and be open to really rethinking physics theory fundamentally. And that is what this site is chiefly about.

Early scientists were often very afraid to publish their real ideas, as were often the 'alchemists' many of whom who did not publish and only wrote in code for their own use but with some of that writing published after their death without permission. They were basically idea-anarchists and of course some were a little bolder than others, though often still moderating or self-censoring what they published so that science historians and translators often cannot clearly see their real ideas. But now contrary pressures are building on modern scientists to be afraid of NOT publishing, for fear of losing their funding and/or jobs, though any form of pressure is unhelpful to real science and for the social good is best applied to technology development and marketing only. Of course many confuse technology with science, as many confuse theory with science.

In history what the facts really are is one thing, but what people wrongly believe the facts are can have much stronger real effects. History has often been driven as much by lies as by facts, and untruth has often driven religions and wars and even science. What is taught as being factual 'history' mostly gets written falsely by history's 'winners', and this certainly holds for the history of physics theory. While the Islamic religion made politics subservient to religion and successfully killed-off an earlier attempted emergence of Arabic science, Christianity and the Catholic church failed to do the same in Europe as governments there followed ancient Greece, Rome, Egypt and China in making religion subservient to politics. Though science can be based on facts, in reality for any place and time, the 'mainstream science' will generally be whatever the mainstream goes with - for whatever generally undetermined actual reasons that may in fact differ little from those supporting religions or pseudosciences. In Chemistry and Biology the best theories seem to have generally won, but in physics maybe not. Certainly that was the conclusion of Isaac Newton when he decided to walk away from physics. It might be nice to think that physics theory has improved since then, but has it really ? Science today finds itself stuck with too much fake physics and fake medicine often based on mere theorising and mere statistics, and often going with fake governments or fake religions of course posing as the truth. No matter how many facts are shown as seemingly contrary to Einstein's physics or Standard Model physics they are still taken as remaining right, but one fact seemingly contrary to Newton's physics is taken as completely disproving all Newton ? Though science was not just developed by some few 'science giants', there has been a few who made notably more significant contributitions to its development. Hence if asked to name a 'King of Physics' many in physics might name Copernicus, Galileo, Newton or Einstein - but evidence given across this website might seem to really support William Gilbert as much as any of them.

On the fake 'disproof' of action-at-distance physics theory

William Gilbert's new action-at-distance physics struggled in the very cut-throat Europe of the time. Some in chiefly Catholic Europe resented any ideas such as action-at-distance coming from non-Catholics, so Galileo and some others vaguely praised Gilbert's experiments while totally ignoring Gilbert's physics theory. Some in Protestant Europe like Francis Bacon and Robert Hooke maybe wanted to steal some of the ideas of action-at-distance promoters for their own self-promotion, so they also vaguely praised Gilbert's experiments while falsely claiming bits of Gilbert's physics theory ideas as their own.

What began as Protestant action-at-distance physics theory developed through William Gilbert, Isaac Newton and Nikola Tesla, but from its start was strongly opposed by early catholic church Popes and their Jesuits who somewhat reluctantly instead backing Galileo-Descartes push-physics theory. While anti-protestant physics theory seemingly won, it was a hollow victory with no fruits of victory - for the Industrial Revolution and later remote-control technologies then developed in protestant North-West Europe and protestant North America, and not in catholic South-East Europe or catholic South America. Nikola Tesla's work on remote action-at-distance physics led to remote radio and tv technologies, remote phone and radioastronomy technologies and to other remote technologies. But the work of Einstein on his relativity physics, even 100 years later, has still produced little or no new technology, and certainly not the time-travel it implied. So today almost every person carries a remote-communication device and almost every home has at least one remote-control device - and nobody has time-travel. Relativity, Quantum Mechanics and String Theory have produce little new technology and a science that produces little or nothing is surely little or nothing.

But there are still one or two physicists who today basically support action-at-distance physics theory as the best and correct physics theory basis, while lesser modern physics theories fight uselessly among themselves. Some of the bad mistakes of early physics theory history still basically persist and are still significantly restricting 21st century physics today, and chiefly stem from failures to properly understand and evaluate William Gilbert's physics and the major Kepler and Newton advances that really developed from that. Hopefully the new improved English trantslation of William Gilbert's 1600 Latin 'De Magnete' may help explain it correctly. It is available as an A4 book or an ebook at 'On The Magnet', or is currently free to read online here at On The Magnet. The path to real science truths is undoubtedly still chiefly the observation/ experiment/ deduction path so strongly advocated by William Gilbert, who everyone claiming to be a scientist or a science historian should really study.

The fight for truth and for true science

It has long been the case that governments and religions have strongly opposed some truths that they found uncomfortable - as by those talking about nuclear dangers, war dangers, pollution dangers, government dangers or religion dangers etcetera. At the extreme they have had a few people supporting such uncomfortable truths killed, but they have more often found a few years prison enough to silence opponents especially when combined with some further action to constrain their lives. Governments and religions can also use their secret services and apologists as to invent or exaggerate a criminal charge and to press a judge to a guilty verdict and maximum sentence, or to press a person's boss or professor to limit their career progress. Italy's Galileo and France's Descartes were Catholics who despite some initial opposition from the church helped ancient-Greek Atomist push-physics suppress the really better action-at-distance-physics of Protestant England's William Gilbert and Isaac Newton who themselves also faced opposition from some fellow English Protestants like Francis Bacon and Robert Hooke. So supporters of uncomfortable truth have most often been successfully silenced as their lives were adversely affected and also the lives of their partners and children if they had such. These very underhand actions are at times additional to supporting false claims that the uncomfortable truths are untruths and falsely debasing supporter reputations. And of course with thought often being seen as less uncomfortable than actual science experiment, it is notable how modern physics basically now follows the chief requirement of the Catholic Inquisition trial of Galileo that he "must present his science as being only thought". Today even funding for physics experiment is now chiefly concerned with dealing with theoretical thought issues rather than being more uncomfortable basic experiment. False popular physics theory has long blocked physics progress, as one piece of false popular theory in medicine has blocked one area of medical progress, see Medicine Frozen. The same has long applied in physics and in physics funding, with private funding largely organised by universities and charities whose science already gets much government funding.

Google Books - a new growing resource

New Science Theory has to commend Google Books on becoming a good new growing resource for older and rare books - and increasingly so for early science books that are not readily available otherwise. To search them yourself go to Google, More, Books and then to Advanced Search and click FullView with an author or book name. But unfortunately governments have allowed Google Books to become substantially frozen in legal shackles, only very partly circumvented by some Google-supporting parties like The HathiTrust Research Centre.

New Science Theory will be keeping a keen eye on Google Books for good new additions that we can offer freely to you, this often depends on good universities or others helping Google - unfortunately far too few to date. You might do some real good for this world, by helping Google Books, if you have a good older science book that they do not now have in FullView or if yours is a better copy than Google Books have. Of course Google Search seems to favour websites with second-rate content that are popular like Twitter, Youtube and Wikipedia so Google Books may tend to do likewise for books. Hopefully this trend will be opposed.

For now, thanks to Google Books, you can download below from this site three great physics books in PDF ;
(if you need one, a good FREE PDF reader is available (from www.Adobe.com)
1. download Isaac Newton's Principia (1848 English 24.1mb -imperfect),
2. download Isaac Newton's Opticks (1730 English 16.2mb),
3. download William Gilbert's De Magnete (1600 Latin 27.6mb).

PS. This site strongly believes that much more published science should be freely available to all on the internet - now there is regrettably too little available even on the many subscription sites. The 2012 UK government commissioned Finch Report gives some backing to support for 'Open Access' science publishing, to current government approved or funded science (see http://www.researchinfonet.org/publish/finch/). But this is being promoted as just part of the increasing control over science by funders entitled to do so, though the best science like the best art maybe really requires more freedom for scientists and for artists. But a scientist refusing any conditional funding is now dismissed as 'amateur' and has a big struggle to get his science anywhere, though there have always been a few people happy to struggle for their science. And science chiefly impacts society through technology which is where social controls should chiefly be applicable. Science publishing should be somehow rewarded but not be enforced. Maybe the big search engines could be made to make small per-view payment to all websites they carry who could then pay royalty payments on anything they carry ? The internet should certainly try to have more science books and papers, and also more science computer models - nice working computer models of Gilbert's terrella, of Kepler's rudolphine tables, of Newton's tide forces, big bang models and more ? Ideally computer models that allow user inputting and good numerical result reporting. Some could use a spreadsheet like Excel that can do iterative calculation on equations to some accuracy. Our Android gravity app 'Sun Pull' in the Google Play app store is a basic example.

Unfortunately today many get their 'science' from Wikipedia and Discovery Channel which do have some good bits of truth, but with big chunks of rubbish mixed in regurgitating poor modern science textbooks - though not quite as bad as the anti-science History Channel with its repeatedly claimed false 'proofs of aliens' and 'proofs of conspiracies'. The first substantial and fraudulent poor science textbook was undoubtedly Mark Ridley's 1613 magnetism textbook 'A short treatise of magneticall bodies and motions' which stole the published experimental work of William Gilbert's 1600 'De Magnete' and published it as his own without the relevant science reasoning and theory that Gilbert had presented it with and so grossly misrepresented Gilbert's science. But poor science textbooks today are being consolidated into our modern Wikipedia-style computer systems tending to freeze science history to wrongly make permanent any false past ideas of science? So this helps prevent todays scientists from considering the physics of Gilbert and Newton objectively, forcing scientists to take them as permanently disproven and so making any past science errors unfixable ? Universities have now basically stopped teaching Gilbert's and Newton's physics and are really leaving that to Wikipedia which does a very poor biased error-ridden job of it so that any science error that has become widespread is almost uncorrectable in Wikipedia because of how Wikipedia works. And modern physics theories are mostly based on some one interpretation of some one supposedly 'key' or 'crucial' experiment which is probably open to different interpretations than those normally being assumed for them. But thousands of observations and experiments still support Gilbert-Newton physics theory, and relying on some one experiment is not good science proof but is really just tricky argument. But most modern physics theory is basically just tricky argument and is bad science. So even professional government-approved science in the West is now often looking increasingly dubious and seems to be increasing being challenged by a somewhat more enquiring Third World science, see Innovation in Asia. Today's science peer-review systems and science government-grant systems reward those producing lots of short speculative papers with fashionable headline titles, and severely punish those wanting to do real science work. So now physics is stuck producing adhoc bits of speculative 'theory' and repeating the same old 'atom-smashing' collider experiments at slightly increasing velocities. Generally modern times are seeing the internet producing a paradigm shift in our relationship to knowledge, from an 'information knowledge age' to more a 'reputation knowledge age'. So modern physics time-travel and multi-world ideas look much more based on peer-review inflated reputations, increasingly influenced by government funding and regulation, than on physics facts - and modern physics 'peer-review' is now almost entirely just weak academics wrongly backing eachother. For others now seeing modern physics as having abandoned reality see 'Farewell to Reality: How Modern Physics Has Betrayed the Search for Scientific Truth' by Jim Baggott 2013 and 'Lost in Math, How Beauty Leads Physics Astray' by Sabine Hossenfelder 2018 June. Science now having done some millions of experiments, there is almost certainly some experiment that disproves or appears to disprove every science theory but which are being ignored by scientists who are studying ever more narrowly ? And in physics now experiment funding is confined to repeating old pre-1940s atom-smashing collision experiments at slightly increasing energies only so that real physics has halted. Some real biology and chemistry is still continuing now, but they will undoubtedly soon follow physics and bring all real science to a halt. And they will likely soon also follow physics theory where today 'whatever is proved is true' has been replaced with 'whatever is popular now is true'. Unless maybe somehow some lottery funders or some other new funders can start to fund real science experimenting ?

Technological and economic progress has commonly rested on basic physics and other science experiment. The greatest advance for medicine was basic physics experiment producing the microscope, and medicine most probably still needs such basic physics experiment though nobody is funding that now with physics dominated by thought-experiment and particle-colliding experiment and today medical research funding is not helping that. Climate science probably also needs such basic physics experiment as is not being done. Of course political parties rarely seriously discuss science policies and unfortunately neither do most scientists, so that I having planned some basics physics experiments in the 1960s when I was expecting to do a physics degree at Imperial College London find now in 2022 those basic physics experiments have still not been done by anybody. Science experiment now is very restricted and can produce only limited small advances.

Different countries at different times have seen varying levels of social censorship or regulation of art, religion and/or science. Generally religion and science competed for peoples support, often mediated by or using the arts. It did look as though over the last century or two the most progressive countries were increasingly backing science more and religions less. But more recently in USA and elsewhere this is apparently being reversed with increasing criticism and controls chiefly being applied to science. This is not helped by some areas of science today being poorer science as with theoretical physics, statistical medicine and maybe climatology. But the main reality is that mounting population pressures, generally encouraged by religions, mean that good science is needed now more than ever before. But there is now maybe little chance of stopping Global Warming from soon ending the human race, because Global Warming is being boosted by population growth pushed by governments boosting GDP and by some big churches pushing population and by improving medicine and humanitarian aid cutting deaths. And helped by governments costly funding of repeated atom-smashing experiments that offer little, instead of funding new basic science experiments on stuff like gravity that might give a big breakthrough that could save us from global warming and need cost nothing ? Politicians may say that they will deal with it, but they may have left it too late so that now it has probably gone too far for politicians to be able to stop it with current technology. So maybe governments now need to moderate religions more, and maybe now need to encourage charities to support new basic science more ? And maybe governments could for example make science websites in their country free to run as well as to read, like this UK science website maybe ?

Translators of science often concentrate on doing the most accurate language translation, and historians of science on doing the most accurate history dates etcetera. This is certainly itself good as their chief job but can commonly be entirely ruined by a failure to fully understand the science that they are presenting, which can easily happen because often textbooks get areas of science wrong especially in physics and the scientists themselves can often be poor at presenting their science ideas clearly. In science the biggest such problems are in physics where science historians and science translators most commonly wrongly take 'widely-approved' and 'widely-disapproved' as implying 'proved' and 'disproved' when no actual proofs or disproofs exists. So we should of course doubt 'science historians' and 'science translators' if they don't well understand the science they discuss, as is commonly the case for those discussing action-at-distance physics relating to William Gilbert, Kepler, Newton and Tesla ? For this especially read free online at Gilbert's 'On The Magnet', or buy the paperback at 'On The Magnet'. Of course this problem also applies to many discussing more modern physics, who also should be doubted.

Religions are now pushing governments worldwide to enchain scientists with increasing 'Open Science' or 'Open Access' and other science work restrictions. Currently it rests on a claimed justification of science being publicly funded justifying the public imposing restrictions on scientists, but undoubtedly soon no justification will be deemed needed other than a claimed 'danger' of science. But really it remains that religions and their anti-science are a greater danger to the world as they have long been and on whom greater public restrictions should be applied. And religions still have many supporters who can make false claims, hoping their great number can supress the truth as it often has. If some science can destroy and is embraced by the world. And if some science could help the world but is ignored. Should a good scientist quit ? Of course science history may not have gone entirely that way, but may have gone substantially that way. Certainly it seems that society now needs to take a better approach to science ? History does have some tricky problems, as shown by for example one 'conspiracy theory' regarding Nikola Tesla.

Isaac Newton was much concerned with the problem of unseens in physics and they remain an issue. Unlike in Newton's time we can now detect and control xray signals and radio signals, though maybe not all of the electromagnetic spectrum including its extremes. And if gravity, magnetism, electric charge and like forces are effected by signals as Gilbert and Newton seemingly favoured, or by push fields such as Einstein theory's spacetimegravity continuum, these have not to date been directly detected or controlled either. But the history of physics indicates that in the future hopefully soon more of them will probably also be detected and controlled, and at some point maybe the Gilbert-Newton action-at-distance signal-response physics might finally be proved right or not. And maybe we will see that no object or atom is solid and that push does not exist, but to date physics seems still stuck with some of its big basic early mistakes ?

Two sites to help inform you on what physicists and astronomers are up to now are www.universetoday.com and http://physicsworld.com
Or for free online non-Google Latin translation (of course still very imperfect), see www.translation-guide.com/free_online_translations.htm

IF you like this site then you could maybe make a donation ;

It will help with site development, and just possibly with some key basic physics experiments long planned but never afforded.
[PS. and you may perhaps help make history for science ?]
(It is over a hundred years after Einstein's science and still we have no time machine today, though some would prefer a gadget that produces gravity which also we still do not have ? Maybe today most governments education and science setups and funding have become too bureaucratic to allow really big science advance, so we now get only small incremental science despite many loud hype claims to the contrary ? More than any other area of science, experimental physics has in recent times become entirely institutionalised so that now an individual physicist is only acknowledged or funded if they are part of some instititution. Of course institutions have some real benefit (akin to mass-production) as in being good at 'getting stuff done', if it is basically standard conventional mainstream stuff rather than real original science stuff. The many supporters of institutional science now increasingly push views that individual scientists should be acknowledged less and the gangs of institutional scientists more. But substantial science really rests on individual unconstrained originality which in physics is now dead, as those working in institutions are working constrained, and real physics can only return if at least some science funders can be changed to backing some real non-institutional individual science though that seems very unlikely now. And the fictional time-travel and multi-universe type ideas of modern physics theory have long totally discouraged certain lines of physics experiment despite there being strong reasons to believe them to be very promising if not essential lines of experiment. Some such lines of experiment considered here identified as early as the 1960s seem still to have had no work done on them and there is maybe not much more time here for this. Science funding both government and private unfortunately now all goes to basically safe standard mainstream science, and no money at all goes to any really innovative risky science that might pay a thousand times greater. Maybe people writing their wills should now consider leaving something to some basics science experiment programs ?)

Today many of the well-paid career-scientists including career-physicists seem unable or unwilling to produce much of use for computer or internet users, but if YOU have or know of some good science that this site could host or could link to, then do tell us.

OR, if you have any view or suggestion on the content of this site, please contact us at :- New Science Theory
Vincent Wilmot 166 Freeman Street Grimsby Lincolnshire DN32 7AT.

You are welcome to link to any page on this site, eg www.new-science-theory.com/history-of-science.php

© new-science-theory.com, 2022 - taking care with your privacy, see New Science Theory HOME.

Philosophy of Science - *problems in philosophy of physics*

Homepage . William Gilbert . Rene Descartes . Isaac Newton . Albert Einstein .... Science History ...... General Image Theory

Science, or 'natural philosophy', emerged in the 1500's as a new way of establishing truths relating to our universe - and as a challenge to the philosophy which till then had considered that as its domain. Religions also often considered truths relating to our universe as their domain, but while scientists often presented their ideas as 'theories', most of them certainly considered that they were dealing with provable fact - unlike the 'mere theorising' of philosophy and religion though such might also have truths. Certainly the truths of science should be proven by facts, and should not be imposed by any 'peer opinion' or governing powers as is often really done. So some significant early science ideas like those of William Gilbert were improperly disapproved though never disproved.

But little is taught on the basic conflict of science, philosophy and religion, and how they deal with the basic questions of truth and error discussed below. Can any science be definitely proved true, and if that is possible then exactly how can any science or science theory be definitely proved true or be definitely proved untrue ?

Problems with science philosophy.

Science developed as a new means of proving truths, against the many errors presents as truths by religions and by the older philosophers like Aristotle that were often backed by churches and governments. William Gilbert clearly showed that nature and the universe do not act in line with beliefs, logic or mathematical conjectures - but act as they actually act and which can be proved only by accurate observation and experiment. Those like Galileo and Newton who chiefly studied simple gravity could perhaps more easily support a 'logical or mathematical universe' but Gilbert's study included also more complex magnetism and electricity and he correctly insisted in his 1600 'De Magnete' that experiments were the surest science proofs, which remains true although experiments are not without significant problems that are commonly not well considered. And while science can often reasonably dispute truth claims of philosophy and religion as 'just philosophising', philosophy and religion often less reasonably dispute some truth claims of science as 'just a theory'. So to look into this requires first examining the four basic ways in which people have considered that a truth might be demonstrated ;

1. GOD. Some hold that there must 'definitely' be a god, and that therefore what are necessary consequences of that must also be definite truths A, B, C which can be used in combination with some logic and observation to demonstrate a wider range of definite truths. In this philosophy the god truths are the fundamental truths to which universe truths are secondary, perception and thought being uncertain and god coming before and creating the universe. Rene Descartes took this as the general philosophy of his physics, and some others have taken this general position which has often been backed by religions. Of course by now science has closely studied much of the universe and might claim to have found no evidence of the existance of any God, nor of any Heaven or of any Hell. But religion might counter with the spiritual argument that 'God, heaven and hell are spiritual things and science cannot detect the spiritual' if 'spiritual' does not really mean 'imaginary'.

2. LOGIC. Some hold that starting from some few 'definite' logical truths, logical deduction can be used to demonstrate a wider range of definite truths. This can involve seeing both god and perception as being uncertain, and argued logic and mathematics as being more reliable. Measured observation of nature often shows that mathematically definable laws seem to apply in nature, so mathematical logic seen as reliable could be reliably used in science. The early Kepler and Albert Einstein perhaps basically took this as the philosophy of their physics, involving logic in combination with a little observation, and others have taken this general position. Some extend Newton's blackbox physics position to 'the only thing science theory needs is the best mathematics'. Support for logic and mathematics in science has mostly involved a requirement of logical consistency in theories, though Einstein and some others conclude that logic does not require logical consistency and support for example light both being a wave and being not a wave.

3. OBSERVATION. Most scientists have held that what is 'definite' is basically what you can see or touch, and that only verifiable observation or experiment can really demonstrate truths. William Gilbert and Galileo Galilei took this position strongly and experimental observation in combination with minimal deductive logic became central to early science in demonstrating a wider range of proved truths. Basically this position takes confirmed perception as most certain, and a theory is to be proved or disproved only by appropriate experimental observations fitting with it or conflicting with it, with minimal deduction. An experiment may be significantly sensitive to just one or two factors like temperature and pressure, but may also have some sensitivity to some less obvious factors making exact replication difficult. And only observables such as finite distances that are measurable can really be used in proofs but not unmeasurables as zero or infinite distances. But as measured physical observations showed that mathematics seemed to have a strong place in nature independent of observation, mathematical logic was increasingly taken as allowing of more than just minimal deduction in science. But the common assumption that minimal deduction would allow only one possible interpretation of any observation to all observers was and is very doubtful - different people can think differently and can make differing deductions from the same observation or experiment. So observation or experiment cannot be absolute proof of any deduction though being evidence partially supporting any number of deductions that are consistent with it, and exact replication may give proof for the observation itself only. Hence the chief problem with observations and experiments, even if exactly replicable, is that different people can interpret the same thing differently. And to be exactly replicable observations or experiments must be fully and precisely specified, which many may not be, and even then there have been plenty of good observations or experiments where some claimed universal law of nature appears disproved but that has been shown to be wrong.

4. POPULARITY. Some hold that what is believed by a majority is true. So social norm beliefs, traditional beliefs, government policies or laws, religious church beliefs or rules, and ideas generally with more popularity are more often taken as being true. This common truth mistake was strongly opposed by some early scientists like Gilbert, Galileo and Newton but still persists in science under the cloak of 'peer review' and 'mainstream science'. But a science truth is not really proved by the number of its supporters or their popular reputations. And science 'peer review' has really long become 'clique review' with eg string-theory science reviewed only by string-theory scientists who favour it.

Each of these four things on their own can either be shown to contain some uncertainties or can demonstrate only a limited range of truths. This is why many scientists have supported using combinations of two or three of them, while often perhaps taking one as being more fundamental. The various positions taken on this by physicists have depended chiefly on their evaluation of three issues ;

A. On the use of minimal logic and simplicity in science. Early science generally supported observation with 'close logic' or 'minimal logic' involving only deductions that seemed to derive directly from experiments. This was seen as according with the fact that repeated observation of our varied and complex universe seemingly showed that its fundamental behaviours were relatively simple, and with Occam who concluded that logic works best when it involves minimal assumptions. But this perhaps best suited small physics theories, as one theory for mechanics and another theory for planet motion etcetera with each small theory needing fewer assumptions or deductions. Early scientists tended to concentrate experiments on particular areas, such as mechanics or magnetism etcetera, and this helped them to conclude that nature basically followed simple laws. This supported the conclusion that a more simple science theory with fewer assumptions is more likely true than a less simple theory with more assumptions.

Any scientific experiment was commonly taken as requiring the simplest interpretation, but assumptions have differed widely on what is simplest. Hence simple 0/1 binary computer programs can certainly give computer computation complexity. Gilbert saw the simplest assumption for his experiments on Magnets and Iron that Iron must respond to signals from Magnets, but others thought it simpler if Magnets somehow push or pull Iron. But for gravity at least Newton experimentally proved that push mathematics do not hold, though most physicists even today strangely have still not bought this clear proof (neither a spinning disc not a spinning bucket of water match the mathematics of planet orbits). Clearly simple to some can be complex to others.

Also needing eg 3 small science theories each needing 3 assumptions was seen as less simple than needing eg 1 big science theory needing 5 assumptions. And both William Gilbert and Rene Descartes tried producing one Theory Of Everything (or TOE) to explain everything physical. And then Isaac Newton showed that some one theory could explain both terrestrial gravity and planet motion - which till then had involved two different theories. If one theory with 4 assumptions could explain all that two theories each with 3 assumptions explained, then one 'more complex' theory seemed relatively simpler than two absolutely 'simpler' theories. So to some a more complex theory seemed acceptable as long as it explained more, and absolute science simplicity would have to be sacrificed to a greater science coverage. So even Rene Descartes in trying to produce one full-coverage physics theory from a simple mechanics base only, had to add complexities to try to cover everything including gravity, magnetism and electricity.

B. On the fundamentals of science. Early physics first split into two camps as to what was really fundamental in our physical universe. Galileo, Descartes and others concluded that the universe was fundamentally mechanical - where its key properties were only matter structure, matter solidity, matter motion and matter contact and impenetrable pushings unaffected by any energy or activity that might exist independent of matter. But some like Gilbert and Newton could see the physical universe as fundamentally energetic or active, with its key behaviours being matter attractions and other motion responses to gravitational, magnetic, electrical and maybe other signals or non-push penetrable energies. Kepler, Einstein and others held maybe a third neo-mechanical 'fields' position and that somehow such 'half-active' entities were fundamental if not exclusive in our universe. But in physics when considering the ability or inability of things to penetrate 'solids' such as glass it is often assumed that things that are apparently solid like glass are actually solid and are not composed largely of empty space as we now know they in fact are. So may some things penetrating a solid be only penetrating empty space, and may some things seemingly pushing a solid be in fact pushing empty space ?

Of course some physicists have concluded that nothing was exclusively fundamental, and accepted some two or three of these as being different aspects of our universe that are compatible. Physicists holding different positions on what is fundamental in the universe, have supported very different types of theories - as a push-physics TOE, an attraction physics TOE or general relativity theory plus electromagnetic field theory. And the issue of what is fundamental in the universe can be entwined with the issue of what is fundamental in observing the universe. So there is Isaac Newton and a few others holding that as science experiment can only observe appearances and apparent behaviours, and not the actual causes of those, then science theory should omit all unseen causes and must leave discussion of such fundamentals to philosophy. And more specifically Newton showed that if physics was to include unseen causes then there may be no scientific experiment way to choose between two such different physics so that one theory seeming right could not itself disprove an alternative theory which might also seem equally right as the only scientific proof is replicable experiment. But many scientists have rightly or wrongly held the view that science theory can somehow validly extend science beyond experiment to some greater or lesser extent.

C. On mathematics and science. While ancient philosophers considered logic to be certain, it seems many scientists now consider mathematics to be certain. Most physicists now consider mathematical laws as fundamental to science, though mathematics itself can clearly have some problems for science as William Gilbert feared. Hence Newton did not put his three laws of motion as mathematical equations for good reasons. Action and reaction being equal and opposite is generally handled in mathematics with positive and negative where perhaps nature has no negative. So gravity pulls on a body by two bodies either side of it can be termed positive and negative and may yield zero, while no actual negative gravity is involved and the reality is quite different if bodies are closer than if they are farther though the mathematics for both may yield the same zero. The mathematics of different physics theories also often involves constants or other elements whose actual physical meaning is quite unclear or ambiguous so that they effectively represent physical unseens or unknowns. Opposite electric charges are undoubtedly both positive forces that can 'cancel out', but something having one of each is not the same as something having neither. And mathematics also can really only deal with futures or non-existents, like the idea of 'potential energy' ('energy which will exist if'), as if they actually exist when they do not. (this particular example achieved no mention in the classic laws of motion and laws of thermodynamics, and maybe only really fitted Gilbert active-matter theory, but is now taken by some as an actual existent rather than a potential existent.) Of course special forms of mathematics like vector mathematics and others can seemingly 'solve' some of these issues, but generally mathematics and nature actually go together much less easily than many imagine though some do extend Newton's blackbox physics position to 'the only thing science theory needs is the best mathematics'. And undoubtedly the chief need of experimental science is precision and precise definition which the involvement of mathematics undoubtedly aids - vagueness and ambiguity are certainly chief enemies of real science to be avoided at all cost, but they can arise both in logic and in mathematics also. The square root of four is two, and it is also minus two. And perfectly accurate measurement of things in nature under all possible circumstances and even when approaching infinitely big or infinitely small is not possible and never really will be possible. So no piece of science mathematics can ever be proven to be perfectly accurate and proof of a maths inaccuracy may not be a good disproof of a science theory. And some physicists including Einstein have stated that they do not beleive that anybody really understands their science theory correctly, but can a science theory that few, if any, correctly understand be a valid science theory ?

So what do these basic issues now indicate for our basic question, can any science theory be definitely proved true - and if so then exactly how can a science theory be definitely proved true or be proved untrue ? Consider three types of theory ;

1. If we take a small science theory saying only that on Earth all bodies tend to fall towards the Earth with an acceleration whose value decreases as the square of its distance from the centre of the Earth, then this says nothing about assumed causes, and only involves some generalisation of some verifiable observations. Most scientists would take that small science theory as fully provable, and perhaps as fully proved if many people had made many observations over many years. Would that still hold if observations were by only one person, and if observations were of only a few bodies, and if observations were only at a small distance from the Earth's surface and if observations were over only a small time period ? Most scientists would

probably say that the theory could reasonably be taken as proved after only a few observations for as long as no observation conflicts with it, and taken as disproved as soon as one verified observation does conflict with it. This position of course involves the small theory never being definitely proved, but many would say that it is reasonable to take it as being definitely proved if taken as applying 'generally now' and not 'always' or 'forever' as scientists would hope to prove.

2. If we take a somewhat larger science theory saying only that all bodies in the universe tend to move towards other bodies with an acceleration whose value decreases as the square of its distance from the centre of the other body and increases as the mass of the other body, then this again says nothing about assumed causes, and only involves a greater generalisation of some verifiable observations. Most scientists would take that somewhat larger science theory as fully provable, and perhaps as fully proved if many people had made many observations over many years concerning many bodies and distances. But again would that still hold if observations were by only one person, and if observations were of only a few bodies or limited distances and if observations were over only a small time period ? And some of the needed observation being of distant bodies, can their movements and masses truly be observed accurately ? Some would probably say that this theory is less easily taken as fully proved because observations cannot ever easily cover the whole universe, yet some would probably say that this theory is MORE easily taken as fully proved because it is logically simpler - it looks more 'inherent to matter' and to being 'apriori' and 'forever'. In fact the gravitation theory of Newton claimed to exclude explanation was taken as proved but then was later taken as disproved as some observations relating to distant bodies were claimed to conflict with it.

3. If we take a much bigger science theory and it also includes claimed explanations, then the proof question changes. But the changes are like the move from small theory to somewhat larger theory above, with observation proofs for a bigger explanation theory as for a bigger no-explanation theory becoming harder but with 'logical simplicity' proofs for an explanation theory as for a bigger theory seemingly becoming 'logically convincing'. A big explanation theory of everything may need only one set of assumptions and proof, where some no-explanation theory plus some explanation theory needs two sets of assumptions and proof.

EXPERIMENT. Now on disproving a science theory, we have noted the idea that observation conflicting with a theory disproves it. Of course an observation can be interpreted differently by different observers, and of course some observations may be less accurate and/or reliable than others. We can throw a ball at a wall and many people may conclude that they contacted, but contact needs the distance between objects to be actually zero - and we cannot now observe and may never be able to observe or measure infinitely small distances. The existence of contact between bodies has not yet been definitely proved and may never be able to be definitely proved, though some instances of claimed contact might be disproved. And if we observe a light in the sky - are we truly directly observing some moving star accurately or has the light perhaps undergone some aberrations of which we are unaware ? Such uncertainty may seem likely because of conflicting theories of light and perhaps limited knowledge of light and of space. Hence Einstein's theory seems to require that light be gravitationally attracted to massive bodies which is a process that generally accelerates bodies, yet Einstein's theory also required that light cannot be accelerated beyond 'the speed of light' - and currently there certainly also remain other tricky light issues like assumptions regarding light 'red shifts'. Yet there are current physics and astronomy theories entirely dependent on such perhaps uncertain distant light observations. And the same observation may be interpreted differently by different observers, as the Sun daily rising in the East and setting in the West being wrongly seen as the Sun orbiting Earth daily while we now know these observations are correctly explained by the fact that Earth is just a big sphere that revolves daily. Observations also strongly suggest that Earth is basically flat with some hills and valleys rather than the sphere that it really is. Things that observations or experiments strongly suggest are true, can be entirely wrong. The same observations or experiments will commonly be interpreted differently by people assuming different theories, so 'ball hitting wall' will be interpreted by supporters of Cartesian push-physics as contact push-force action but will be interpreted by supporters of Gilbert-Newton attraction-physics as proximity repulsion-force action (it being known that forces like magnetism and gravity increase with proximity or weaken with distance) and the same experiment may be interpreted in yet another way by supporters of some other physics theory. Hence the modern physics Pauli Exclusion Principle repulsions between fermion particles proposed as resisting matter collapsing at least till gravity gets excessive can be interpreted as the basis of matter 'push'. The impossibility of actually measuring infinitely-close-to-zero distances seems to make distinguishing actual contact from actual proximity-repulsion impossible. But this need not actually be the case as actual contact collision requires no effect until actual contact, while proximity-repulsion does require some proximity repulsion effect prior to the claimed collision time at some finite if close distance. If more-powerful collider experiments involve no actual collisions, but merely bring 'collision objects' to greater proximity, then at some point experiment results should give proximity effects that differ mathematically from actual collision effects. Likewise with attraction-slingshot reflection also considered by Newton as giving similar reflection angles but in an effect somewhat later than the 'collision' time. But such differences have probably not been looked for. And prhaps interpretation of no experiment can be definitive, but always in fact must rest on some theory assumption. Then no experiment can really prove or disprove any theory, but can only offer some **partial evidence** of consistency with or inconsistency with differenttheories.

It may be that some interpretation of 1000 experiments seem all to support theory A while some one interpretation of one other 'crucial' experiment seems to support theory B rather than theory A. From this some will take the one experiment as proving theory B and disproving theory A. Disregarding the mass of experimental evidence may be justified on the basis that for theory A to be true it must hold in every instance so that it failing in just one instance disproves it. Of course theory A is probably intended to hold only under some set of circumstances as 'in the absense of interfering forces' which may or may not have been fully specified, and whether the 'crucial experiment' is or is not fully within such circumstances may not always be clear so that this kind of claimed science proof can certainly be doubtful. It can be claimed perhaps that this is the basis of the possibly doubtful claims for Einstein theory against Newton theory for example. The nonsense of claimed 'crucial experiments' also rests on considering only two possible alternative specifically-defined theories with the experiment supporting one only, and refusing to admit that there may be one or more other applicable theories that actually might be better. But it is not possible to prove that there are no other applicable theories, so there are actually no crucial experiments and all experiments are equally valid though some may seem more interesting. So lazy scientists considering as few theories as possible and claiming 'a crucial experiment' do not help experimental proofs but actually confuse and weaken them.

INVENTION. With progress in science has generally come progress in useful inventions like TV and the internet, so that some might think of new inventions as proving some science theory true. However useful invention started long before science and 'blind' experiment has certainly given many new inventions, though sometimes a science theory has prompted a new experiment and new invention. Science encourages invention less by the truth of science theory, than simply by science encouraging experiment. So even false science can help invention, and invention cannot really be taken as reliable proof of the truth of any science theory. (hence one common false belief now is that nuclear power came from, and so helps prove, Einstein's general relativity theory - when it actually came from experiments on radioactivity as by Marie Curie that were making good progress before and without Einstein's theory.)

RELATIVITY. If things like light and gravity are taken as being signals carrying information about source objects or events, then they might carry correct absolute information or they might be liable to aberrations and then carry incorrect appearance information - and information may be modified by its observation and so be relative to the observer. Hence in considering a distant objects motion and mass, it may be necessary to consider its absolute, apparent and relative motion and mass. If as Newton noted it is not possible to absolutely distinguish a body being at rest from a body in uniform motion, that need not mean that there is no real difference or that velocity is of no significance. But things like light and gravity may be not only information signals, but also have some absolute effects on real objects so that perhaps the apparent or relative can also have absolute effects. And the effects on some bodies of signals that are in some respect relative, may be to some non-relative aspect of them. So normal debate underlying science theory, about taking things as being only absolutes as against taking things as being only apparent or relative, may be trying to distinguish the two too much - our universe undoubtedly includes both. And that fact can maybe affect how science theories should be proved or disproved. Unlike some physics theories, an action-at-distance signal-response physics requires that all present physical actions are necessary responses to previous signals so that time must exist and must basically be one-directional and also that change does not just happen probabilistically but must happen to some determined signal-response laws. And to the extent that some responses might be to some multiple set of signals, the link between some responses and individual signals can be statistical or probabilistic without any less actual determinism though allowing some apparent indeterminism. And in a Gilbert-Newton action-at-distance attraction signal-response physics, actions always involve at least one observer or detector responding to at least one prior signal so that an action can be relative to at least one observer and one signal. So for any observer or detector an action is relative to the relative directionality of a received signal to that observer or detector, and its preferred frame of reference will differ both for different received signals and for different observers or detectors. And quite unlike Cartesian physics with its basially supposed one universal preferred frame of reference, an action-at-distance physics can allow greater complexity involving multiple simultaneous relativities while remaining basically the simplest physics.

SCALE. The behaviour laws of large masses of things can appear to differ greatly from the behaviour laws of the individual things. An ocean does not seem to behave like one water molecule behaves, or at least their behaviours can be described quite differently. And two things very close to each other may behave quite differently to when they are far apart. Are such scale differences real differences or are some or all only apparent differences ? This issue may fundamentally concern how both 'small-scale' quantum physics and 'large-scale' relativity physics relate to 'medium-scale' classical physics theories. See one recent interesting Scientific American physics theory article relating to this by Renate Loll at www.signallake.com/innovation/SelfOrganizingQuantumJul08.pdf (though he maybe believes in some mathematical universe, like the young Kepler before he 'wised up').

DEFINITIONS. The extent to which a science theory has clear and complete definitions for the things that it deals with, determines the extent to which the theory is provable or is disprovable. A very vague theory is hard to prove or to disprove, and perhaps should not be considered a science theory at all. And is a mathematical definition of something physical a real definition or maybe not ? Hence for acoustics there has long been used a clear physical definition of a sound wave, but for optics the definition of light waves has varied and now is perhaps only mathematical and so physically undefined ? Some may define 'energy' as relating only to change in motions, while others define 'energy' relating to uniform motion and/or rest states also. Some may define 'force' accelerations as relating to change over time but many as change over distance or space (though a constant force accelerating a given mass in some direction against no resistance over some standard distance, will equal the constant force accelerating the mass in that direction against no resistance for some standard time). Some modern physics theories seem to have weak definitions of even their basics like mass, energy and space - and some seem to almost entirely avoid definitions.

Of course in reality most science theories will consist of some small set of basics essential to the logic and self-consistency of that theory, and some larger set of inessential correctly or incorrectly derived assumed consequential deductions. A theory may also include some explanation of its language terminology and usage which may or may not include all elements essential to it. The larger set of derived consequences in a theory is more likely to include some deduction errors that can be proved wrong by experiment or observation. But proving some small inessential bits of a theory wrong does not actually disprove the theory, only disproving one of its essentials or disproving its logic can fully disprove a theory. And sometimes it may not be clear exactly what the real essentials of a particular theory are, so that an apparent disproof of the theory may not be a real disproof. It will often be easier to just push a new theory rather than to try to really disprove an old theory, and often new theories have mainly gained support that way - in fact leaving an older theory disapproved but not actually disproved.

In at least their early stages most self-consistent science theory write-ups will generally be incomplete - the theory write-up will cover only some limited range of phenomena and give only some limited mathematics. So showing that it does not, in that early and incomplete stage, give an acceptable explanation of verified Experiment X is proof only of its incompleteness and not proof of it being a wrong theory. It may be possible to develop that incomplete theory in a way that is fully consistent with its basics so that it does give an acceptable explanation of the Experiment X. A theory should be taken as proved incorrect only if its basics are proved to actually contradict all reasonable interpretations of verified Experiment X.

But experiment and observation conflicts are not the only things that have been taken as proving or disproving a science theory. There are cases where a theory has been taken as disproved by a new theory, with no observation conflicts. This displacing of one science theory by another can be on good science grounds as when a new theory has fewer assumptions, or can be actually on non-science grounds as when a new theory wins support for being better in line with the religion, politics or prevailing attitudes of the time. Even scientists are human. And in the end public 'proof' is whatever some humans take as proof, and may not always be real definite proof. Definite proof or disproof may not always be possible in science, as elsewhere such as in religion ?

There have been many more imagined disproofs of science theories than actual disproofs of science theories. Claimed science theory disproofs very often themselves involve errors, often relating to the fact that theories commonly fail to clearly and fully specify their fundamentals and often also include inessential deductions that may be incorrect. Disproofs of science theories can be taken as generally falling into two basic categories ;

1. Experimental Disproofs. Experimental disproof of a science theory is generally taken as requiring some well verified experiment fact conflicting with some essential aspect of the theory, and not just some interpretation of an experiment conflicting with some inessential bit of the theory. Eg a theory requiring that the universe cannot expand is not disproved by some interpretation of light wavelength variations as indicating universe expansion, if the no-expansion theory can allow of such light wavelength variations without expansion. So experimental disproofs can only rest on actual experiment results, and not on any claimed explanation or interpretation of experiment results. The fact that some theory X interpretation of an experiment fits theory X, cannot disprove theory Y. All possible theory Y interpretations of the experiment need to be disproved, by showing that no theory Y interpretation of the experiment fits theory Y. Firm disproofs like this are rarely attempted in physics, and have yet to be really attempted for the Gilbert-Newton 'attraction physics' theory that is widely wrongly claimed to have been disproved.

If any basic required aspect of a theory is disproved then that theory is disproved, but disproof of an inessential aspect of a theory does not disprove the theory but only that inessential part of the theory. Science theories often include one or more deductions that are incorrect but that are also really inessential to the theory. Hence William Gilbert deduced incorrectly that the Earth's magnetic signals should not vary over time, but that was not any basic requirement of his theory that magnetism involves emitted signals and response to them. So claims to disproof of a theory when only some inessential part has been disproved is ridiculous 'throwing out the baby with the bath water' and unfortunately common in physics claimed disproofs. Even the disproof of alchemy involves that error with its whole experiment idea often wrongly ridiculed.

2. Compatibility Disproofs. Showing that 2 theories are incompatible, as by showing that their mathematics are incompatible, is generally taken as proving that one or both theories are invalid - but still allows that either theory may be valid. If one required aspect of a theory contradicts another required aspect of the same theory, that is generally taken as proving the theory is invalid, but if either aspect is not required in the theory then that

contradiction proves nothing about the theory's general validity but only that it requires a modification.

For 2 theories having different coverage but both covering some common area, as 1 theory of mechanics and 1 theory of mechanics and gravity,
A. if 1 of the theories is taken as being fully proved then the second theory can be taken as fully proved only if shown to be fully compatible with the first theory.
B. if 1 of the theories is taken as being disproved then the second theory can be taken as disproved only if shown to be fully compatible with the first theory.
C. if the 2 theories are proved to be incompatible, then 1 or both must be invalid.
D. if the 2 theories are both fully proved, then it must be possible for them to be somehow shown to be compatible.

Of course, it may be easier to show two theories to be compatible or incompatible than to fully prove or disprove theories. And while either 1 of 2 theories that are incompatible with each other might be valid, as Newton concluded, contradictions between 2 theories is generally taken as showing that at least 1 of them is not valid. But there are some who now see contradiction within 1 theory, within its mathematics, in results of experiments, and in actual nature, as being acceptable science. Current wide acceptance of particle-wave duality and of Einstein's general theory seems to require that position, though for most of the history of science it was considered unacceptable. Alternative science theories in the past have been required to produce proofs against eachother, but now those who see multiple science theories as acceptable also see them as not needing to produce any such disproofs. (They should of course instead produce acceptable evidence of consistency but generally do not.)

MIND AND/OR MATTER. Another basic issue much disputed by both philosophers and scientists is the issue of Mind and/or Matter in the material universe.

Early pre-science philosophers generally allowed that both 'mind' and 'matter' exist, but with some requiring that mind be associated with some matter or with all matter and others requiring that they exist separately only. Matter to some was the 'dead' aspect of the universe and mind the 'active' aspect of the universe, and to some the universe was basically only one or the other and not both.

As one of the first physicists William Gilbert concluded that his attraction theory experiments proved that all 'inanimate' matter possesses some simple 'mind' properties in being able to detect and respond automatically to magnetic, electric and gravity signals emitted by other matter. In this view allowing of simple mind in simple matter, and of complex mind in complex matter, allowed a complete 'mind from active-matter' physics and science had to be based on active behaviour laws of nature.

But, in line with ancient greek Atomism and Galileo, the early scientist/philosopher Rene Descartes claimed that there could be no mind beyond God and Humans, and that matter could not respond to anything and could only be pushed by contacts with other matter - a 'no mind' dead-matter physics. Science had to be based on dead-matter structure and dead-matter motion, generally unconnected to his separate spiritual and mental universe. He allowed a unique human mind to think and to relate to the human body, but he required animal brains and bodies to operate only as mechanical push-clockwork robots without any thinking.

But philosopher George Berkeley concluded that observing our universe showed that mind was certain and matter uncertain, allowing a 'no matter' Gilbertian science. Isaac Newton's blackbox theory basically concluded that any of these positions might be true but science could not prove which - a 'don't worry' science. (though Newton was widely suspected of privately favouring Gilbert attraction theory while publicly supporting his own blackbox physics as being the best physics possible only as long as there were no fully proved physics theories without unseens) Modern physics theory seemingly ignores the major issue of mind, vaguely claiming it is outside physics, though some modern information science does try to consider it as a physics issue. A minimum requirement for the claim that mind is outside of physics would seem to be a real physics disproof of attraction theory which nobody has really managed to produce yet and no modern physicist has even tried to disprove.

The Descartes, Berkeley and Newton positions on this general dispute was summed up in the philosopher or physicist ultimate phrase - "No matter ? Never mind !". In common English the phrase 'no matter' has a double meaning as 'don't worry', and 'never mind' also has a double meaning as 'don't worry'. (the phrase may derive from the joke 'What is matter? - never mind. What is mind? - no matter.' which was published in Punch and may have originated with Oscar Wilde.)

Of course although Gilbert's 'no dead matter' physics was somewhat in line with the later 'no matter' philosophy of George Berkeley and opposed by the 'no mind' mechanical physics of Rene Descartes, Gilbert physics does maybe better allow of the compatible existence and interaction of both in the universe. If any body can be a signal, relative to some observer body that can respond to it, and any body can be an observer relative to some signal body to which it can respond, then all physical observers, unlike intelligent observers, can always respond to signals in fixed predictable reliable manners. And that may be the real basis of experimental science data, not Descartes human 'certain knowledge' which seems far more uncertain ? And sensing data does not require any 'knowing' or thinking or intelligence, though such is certainly required to produce any science theory from given data. The chief requirement of a good science theory remains that it involve the least knowledge assumptions being added to the established data, and some science theories seem to involve much assumption. While science seems to have a strong case in disputing many truth claims of philosophy and religion as 'only philosophising', philosophy and religion do also seem to have a strong case in disputing truth claims of some science as 'only theory'.

THOUGHT AND SCIENCE. There are in fact 3 quite different but easily confused thought-related issues of basic concern to science theory.

Firstly on producing science theories, many philosophers and some scientists like William Gilbert have been concerned with science having errors due to the thought element involved in producing a science theory. But some philosophers perhaps often tend to overvalue the 'thought' part of science, as against the experience-experiment-data part, shown by eg George Berkeley and more recently Wilfred Sellars at http://ditext.com/sellars/epm.html For developing a scientific theory Gilbert repeatedly supported strongly an anti-philosophising/reasoning and strongly pro-experiment/experience position, requiring that a good theory must be as directly from the data as possible and so involve the least deduction assumptions. But experiment or experience regarding the natural world is NOT entirely dependent on the human senses direct detection of natural signals as some have assumed. Science has developed, and still is further developing, many different detectors of natural signals - many indirect alternative senses. These adding further confirmation of our own human senses add further to the proof value of experiment, and further reduce the proof value of mere 'logical' thought. So the experimental science method as advocated so strongly by Gilbert in 1600 has perhaps always had, and still now really has, a more solid base than the 'thought experiment' science method as advocated by Einstein and others. And really in science Experiment is more certain than mere Observation which itself is more certain than mere Thought.

Secondly on the content of science theories, some philosophers and many scientists like Rene Descartes have been concerned with science having errors due to human-like phenomena as especially thought-like phenomena being wrongly ascribed to the non-human part of the universe. Of course it is perhaps not certain that two exclusive universes exist, human vs non-human or spiritual vs material, and modern computer and remote-control technology does clearly demonstrate that thought-like thoughtless processes exist and could be widespread in the physical universe. So while rejecting theories that incorrectly ascribe thought-like phenomena to some physical processes as 'anthropomorphic' may be sound, labelling a science like Gilbert's signal theory physics 'anthropomorphic' is almost certainly a bigger science mistake.

Thirdly on the descriptions involved in science theories, a few linguistics theoreticians like Noam Chomsky have been concerned with science having errors due to their basically being descriptions of thoughts of a universe and description allowing of ambiguity or other linguistic error. This issue is considered more fully in our General Image Theory section.

Fourthly on the widely assumed conflict between 'determinate causation' and 'thinking choice' and the commonly ignored possible 'indeterminate causation' or 'causal thinking'. Erwin Schrödinger (1887-1961), in a BBC TV 1949 'Do Electrons Think ?' programme, considered causation, thinking and apparent choice. see https://fedora.phaidra.univie.ac.at/fedora/get/o:168238/bdef:Asset/view But on this he poorly considered only Descartes and ancient Greek 'science' confusing 'thinking' with 'non-causal' and 'free choice' and reaching no real conclusions. A more scientific Gilbertian approach to causation, thinking and apparent choice is possible. In nature, natural signals have some level of digital or statistical variation or 'noise' around means. Hence natural responses to such have some level of digital or statistical variation which is more significant at smaller or more localised levels. Of course many might conclude that simple automatic determinate responses to signals does not involve thinking, but if responses have a more complex or computational relationship to signals then many might conclude that is thinking ?

Science perhaps needs to be concerned with all four of these quite different thought-related issues and not just with some one of them.

CERTAINTY AND SCIENCE. Science has long had a double-edged sword problem on the question of certainty and certain knowledge. On the one hand science must oppose claimed certain knowledge about the universe, with the requirement that knowledge can be gained only after much scientific experience and experiment on all possible aspects of the universe. Galileo and Gilbert were two of the prominent early scientists pushing this need-more-experiments anti-certain-knowledge view of science. But science commonly also supports the idea that there can be only one set of truths, which some few science experiments can prove and so give certain knowledge of the universe. Like some early philosophers including Aristotle, some theoretician scientists such as maybe Descartes and Einstein have seemed to be offering certain knowledge. Certain knowledge tends to being popular, even with scientists, but also tends to being wrong knowledge. Newton and more recently Heisenberg argued that there are significant limits to scientific observation knowledge, and that basic 'unseeables' necessarily allow of alternative views of the universe and allow of no complete certain knowledge. Widespread support for any form of claimed certain-knowledge has actually always opposed new real science. Yet still today many in science defend the indefensible 'only one right theory' dogma with 'Theory X is proved' that can only hold science back. Others today unreasonably want multiple theories accepted with no logical consistency requirements. Blackbox science can be taken as aiming to explain how the universe behaves, without aiming to explain why the universe behaves as it does. So blackbox science can then be basically taken either as making-no-hypothesis science or theoryless science in line with Newton, or as science using hypothesis or theory as 'only idealization aimed at simplifying the mathematics' in line with Einstein. Yet either of these blackbox alternative-theory ideas perhaps still need some developing as along the lines of General Image Theory science ? That science based on observation and experiment is more factual than other ways of thinking does not mean that science is always fully correct.

CAUSATION, EXISTENCE AND CHANGE. Change to Newton requires some external force cause, and non-change requires no external force cause. Things exist eternally unless some external force cause produces change. It also seems likely that force causes are themselves produced only by changes, so that a change is produced only by a prior change. So change happening now probably also requires that changes have always happened. This seems to hold in both Gilbert-Newton attraction-physics regarding signal responses and in Galileo-Descartes push-physics regarding motion collisions - and probably also for any valid physics. This would imply an eternal and always-changing universe with no beginning or ending, unless something beyond natural physical laws can intervene. So some modern Big Bang Theory physics may not be as sound science as some think, or may need some further basic developing ? Also there is the issue of mutual causation and reversible causation. Both William Gilbert and Isaac Newton posited 'mutual causation' for multi-body systems in magnetism and gravitation respectively especially. If body A causes some response in body B then body B also causes some response in body A and, while the science theory may not need either generally preceding the other, a specific cause does precede its specific effect. So physics causation is as with chicken and egg causation, and an egg cannot produce the chicken that produced it but only some new chicken. So mutual causation is not necessarily reversible causation requiring that a system A change to B and then be changed back to A and the probably impossible confirmation that the final A is totally identical to the initial A. The fact that the state of a system may be generally reversible, need not require specific causation to be reversible. Actual causation need not always match apparent causation in relation to time, as per the example in our main Einstein section. And what would it even mean to claim observation of an effect occurring before its cause ? If pushing actually exists then it appears reversible, but if signal-response actually exists then it appears unidirectional. While it seems that all forms of push-physics as Cartesian and Einsteinian must be reversible, any kind of action-at-distance signal-response physics as William Gilbert and Newton must be one-directional. There is some evidence of reversible push-physics being wrong, and there is some evidence of unidirectional action-at-distance signal-response physics being wrong - but does the evidence really decide between them as many have claimed ? Some earlier philosophers had seen, and some scientists now see, distance effects like magnetism, gravity and object-perception as involving God-powers or motivated-objects or medium-action as including transmission or emission of pushes, energies or tension-pulls. However to William Gilbert evidence proved some emission of signals by some objects to which some other objects responded with motion, including evidence that vacuums fail to produce drag which mediums should produce ?

Basic issues for any piece of science

The strongest science theory is the most fully proved science theory, and a science theory is proved only to the extent that its observables are confirmed by multiple observations and by multiple observers. An observable event for scientific proof is a unique event that creates multiple direct effects that allow of multiple observations of them, or is one event in a class of multiple similar events which allows of multiple observations of the multiple events of that event class. Recent claims of observations relating to 'the original Big Bang event' have to be taken as uncertain, in being of indirect effects probably subjected to indeterminable modifications.

Unobservables needed by a science theory make that theory less fully provable, and are at best supported or not supported by observation. A science theory that is more fully provable is stronger than a science theory that is less provable, so a science theory needing less unobservables is more fully provable than a science theory needing more unobservables.

For any piece of science, experimental evidence may seem to support some event description like 'A=B+C' being true for some aspect of the universe. The main issues for science regarding that event description are then ;
1. Is this event description exactly accurate and complete, or is this event description just an approximation, or is this event description just one of multiple possible event descriptions for that event ?
2. Is this event description accurate and complete or approximate or one of multiple possible event descriptions for all of that aspect of the universe, for

just part of that aspect of the universe, or for more than only that aspect of the universe ?
Of course scientists may then actually work on only some of these issues, and not address all of these issues.

Contradiction in modern physics is commonly justified (and comparing different theories dismissed), as 'only being the use of different descriptions'. See http://www.forbes.com/sites/chadorzel/2015/11/19/physics-demands-many-kinds-of-literacy/ Of course different descriptions are different theories and to use more than one must require demonstrable compatibility. Religions have often allowed of incompatibilities and miracles and have included Gods, purpose and ethics which seem to not exist in science though almost all scientists have supported ethics at least. But by now many scientists have studied much of the universe and to date seem to have found no strong evidence of any God existing. Physics seems to work without a God, Chemistry seems to work without a God, Biology seems to work without a God and the heavenly bodies seems to work without a God. So you getting to a heaven looks similar to you winning a big lottery jackpot, possible but very unlikely. In which case it may well be OK to put a little into it but a mistake to put a lot into it ?

To many who consider themselves scientists the chief principle of science philosophy is the requirement that there can be only one correct science theory of the universe which must disprove all others, and that the chief goal of science is to fully define that one correct theory. Of course that claim was strongly challenged by Isaac Newton with his Blackbox Theory that limited the possible scope of science knowledge and so allowed of multiple possible alternative beyond-science theories of the universe. However there is certainly a case that Newton's blackbox science argument though good was just contingent despite its later unintended non-Newtonian backing by eg Heisenberg's Uncertainty Principle, but a much stronger argument against the one-theory science principle is put on this site in the General Image Theory of Science Theories. Nature cannot have actual contradictions so a well-defined valid science theory that can be true cannot include contradictory statements. So if science experiments can yield different contradictory valid interpretations, then they can yield valid different contradictory well-defined theories any of which can be true. This is the only real way that contradiction can have a place in science, and leads to our General Image Theory. And any unproven theories really remain a part of a science along with any proven theories, till disproven, so that science now perhaps chiefly really needs much improved disproving - and to date, in physics especially, science disproving has commonly been really poor. Those claiming to be 'philosophers of science' or 'science historians' would maybe be better advised to draw much more attention to the failings common in science disproving, because that is an area in which they often fail badly. Many physics 'theorists' today push what is really a crazy new 'requirement' for a valid physics, not that it must support experiment but that it must support some existing theories like Relativity and Quantum Mechanics !

For explaining basic physical actions science must require that there be at least one true theory and possibly several compatible true theories. Now most classical physics theories seemed to fall into one of two broad classes of physics theories as being either types of contact-push physics as Galileo and Descartes or types of action-at-distance physics as William Gilbert and Newton. This did seem a requirement of a classical physics theory but not for some modern physics theories unless they are not defined sufficiently accurately to determine such. Hence in the continuum physics of Einstein all physical action seems to derive from contact with his continuum, and all logic says his continuum contact should involve pushings though Einstein himself insisted that it could not involve pushings but must involve 'some other unknown mechanism'. Most who support it take it as really a contact-push physics anyway but avoid discussing that. But some other modern physics theories like those based on probability do also avoid supporting specific mechanisms, so that overall physics theories seem to fall into one of three classes - contact-push physics, action-at-distance physics or unknown-mechanism physics. If a true physics must have a specified mechanism then we are back to the two classical physics theory classes only, and a disproving of one of these theory classes would seem to also be a proving of the alternative theory class. Excepting only if some member of one theory class can be shown to be compatible with some member of the other theory class, in which case both might be valid image theories. If everything that there is in the universe works coherently, as basically seems the case, then it should all have some singular explanation or unified theory of everything. Of course with many things being involved deciding what such a physics is, even approximately, may be extremely difficult. But the modern physics position of accepting many different conflicting theories at the same time does certainly not seem acceptable science. So maybe some non-mainstream physics or some wrongly-dismissed earlier physics could usefully be considered still ? But from both Newton dodging commiting to any gravity mechanism and Einstein dodging commiting to any spacetim-continuum mechanism, it seems that the best of scientists can support views that are not really scientific - commonly thinking to protect themselves or promote themselves ?

Support for 'against-the-mainstream' Gilbert-Newton attraction physics

Gilbert-Newton 'attraction physics' was supported by some other physicists, and also by some notable people outside physics like the chemist-physicist Priestley and the philosopher Kant. Joseph Priestley rejected solidity and saw 'contact-collision' as just repulsion and he saw a strength of attraction theory involving robot atoms responding to signals, rather than involving dead atoms, as its better allowing science to explain animal and human brains thinking processes. (History of Optics 1772, Disquisitions 1777)

And to Immanuel Kant for any physics theory attempting to replace attraction with push impacts, the very existence of spatially extended configurations of matter (as objects of above-zero radius) seems to need some sort of binding force to hold the extended parts of the object together when hit by other objects. Such a force cannot be explained by pushing from other particles, because those particles too must hold together in the same way, so to Kant circular reasoning in physics is avoidable only if there exists at least one fundamental non-push attractive force. (see Metaphysics of Science 1786 at http://philosophiebuch.de/metannat.htm)

Most scientists have supported the theory that there can only be one right theory concerning anything in nature, though this theory could itself be a wrong theory. Isaac Newton did allow that multiple alternative theories might be options in an area of science with limits to observation and experiment. And as different people can interpret and describe the same observation or experiment differently, a scientific realist science can allow of multiple valid theories if they meet the requirements of a valid science as in General Image Theory.

Now to consider the significant general issue of proofs. In mathematics and logic there can be propositions that can be definitively proven in terms of logical necessity in line with Euclid's QEDs. But science rests on observational or experimental evidence which for some things may be such strong evidence as to lead them to being widely taken as proven facts. Like that Earth has revolved daily for many years, and that for Earth's oceans their tides are chiefly caused by the moon. The probability of these facts being wrong are close to zero. However several different science theories have been proposed about how exactly the moon affects Earth. These theories basically rest on different scientists interpretations of some evidence, maybe evidence of planets apparent effects on the motions of other planets or of the apparent effects of planetary bodies on light beams or of the apparent effects of remote controls on TVs or radio-control boats. Some might reasonably conclude that the evidence supports one theory strongly while others might conclude that it gives some support to an alternative theory or just gives no definitive proof of any of the theories.

Though limited support for attraction theory from some like Newton and Kant had little effect on many physicists, it remains the case that there are some very strong arguments in favour of attraction physics that Einstein and others have certainly failed to address. And early Catholic physicists like Galileo and Descartes and some early catholic church Jesuits had improperly dismissed attraction physics though failing to offer any convincing disproofs. Physics now, long being a mess of conflicting theories, needs to reconsider the physics basics starting from the long suppressed action-at-distance physics of William Gilbert, free to read at On The Magnet !

And listen to this video below (may need Right-click Play) ;

Two websites on what physicists and astronomers are up to lately are http://physicsworld.com and www.universetoday.com
And for free online Latin translation (though not very good) see Latin .

You are welcome to link to any page on this site, eg www.new-science-theory.com/albert-einstein.php

IF you like this site then you could maybe make a donation ;

PayPal Donate .

It will help with site development, and just possibly with some key basic physics experiments long planned but never afforded.
[PS. and you may perhaps help make history for science ?]
(The fictional time-travel and multi-universe type ideas of modern physics theory have long totally discouraged certain lines of physics experiment despite there being strong reasons to believe them to be very promising if not essential lines of experiment. Some such lines of experiment considered here identified as early as the 1960s seem still to have had no work done on them and there is maybe not much more time here for this. Science funding both government and private unfortunately now all goes to basically safe standard mainstream science, and no money at all goes to any really innovative risky science though that might pay a thousand times greater.)

otherwise, if you have any view or suggestion on the content of this site, please contact :- New Science Theory
Vincent Wilmot 166 Freeman Street Grimsby Lincolnshire DN32 7AT.

© new-science-theory.com, 2021 - taking care with your privacy, see Sitemap.

William Gilbert - *robot universe signal theory*

Homepage . Rene Descartes . Isaac Newton . Albert Einstein .... Gilbert's De Magnete . De Magnete + ... General Image Theory
- Site Search at bottom v -

From the physics published in his lifetime, William Gilbert or William Gilberd (1544-1603) seemed like Galileo to be largely concerned with experimental physics and less with theory. The one work he published under some peer constraint, in 1600 in Latin, was "De magnete, magneticisque corporibus et de magno magnete tellure; Physiologia nova, plurimis et argumentis et experimentis demonstrata." ('On the magnet, magnetic bodies and the great magnet the earth; A new science, with many both argument and experiment proofs.'). Published in London under Queen Elizabeth I, it was first translated into English nearly 300 years after Gilbert's death, in two versions that both failed to at all clearly present his science theory. He logically examined some of the basics of science as no other scientist ever has, and backed some key science that became widely wrongly disapproved though never disproved. To study this suppressed action-at-distance physics of Gilbert, see the free online new improved English translation De Magnete. Or get the A4 paperback book 'On The Magnet', or ebook 'On The Magnet ebook'.

As a physician to England's rulers and Royal College of Physicians president, Gilbert was eminent in medicine and a hobby or amateur physicist who put much into his physics. De Magnete was a new science work mostly on magnetism with much polemic against mere theorising and for the new experimental science, and was condemned by the Catholic Inquisition. Its new physics theory was patchily put, buried in pro-experiment polemic and views on the baseless musings or mere theorisings of others. Its new experiments and unique theory were clouded by the use of new Latin terminology that made its science difficult to really understand. Gilberts physics theory ideas became a bit more widely known 50 years after he died when his 'De Mundo' was first published still only in Latin. Even now 2022 still sees little backing for a first translation of this to English or other modern language.

William Gilbert's science theory.

(OR here you could listen to our video at the bottom of this page, "William Gilbert, the King of Physics")

Gilbert for his 1600 De Magnete in its preface notably claimed "and the causes are made known of those things which, either through the ignorance of the ancients or the neglect of moderns, have remained unrecognized and overlooked." For when experimental science proper was first developing in Europe with Gilbert and Galileo, the prevailing scholarly philosophy of nature was based on mere thought and was that of Aristotle backed by governments and religion. In Aristotle's divine universe every thing was to some extent self-acting (or 'animate') and thinking, with divinely set motivations and knowledge - so that objects fell to the ground because they 'sought to move themselves to their natural place'. Gilbert and Galileo saw this as involving too much irrational supposition and unable to describe the complex realities of actual natural phenomena shown in experiments to accord with invariant laws of behaviour. More than any other physicist Gilbert had widely studied all the proto-science thinkings of ancient Greek, Arabic, Chinese and other writings being basically everone in the field all of which he referenced and which no other scientist emulated as fully. While Galileo and some others chose to just study and follow simple ancient-Greek Atomist push-physics 'dead matter with law determinism' theory, Gilbert chiefly promoted experiment and initially gave little prominence to the new physics theory that he had developed from his experiments, though promoting this more in his 'De Mundo' published unfortunately long after his death. William Gilbert's early physics ideas published in 1600 were really more considered and more interesting in both their science and their philosophy than those of any other early physicist. But his 'De Magnete' really presented a new big TOE physics, while seemingly hoping to minimise religious opposition to it by presenting as being only a small specialised magnetism work.

It seems that a very wary Gilbert maybe wanted to bring out his science somewhat quietly under-the-radar, and it surely was a rather low-impact publishing. So his failing to include his key theory term 'effluvium' in De Magnete's 'Interpretation of Certain Words' may well have been part of his intentional playing-down of his theory relative to his experiments. Gilbert's new 'magnetical physics' or 'attraction physics' did retain a type of Aristotlian self-action for bodies but, very contrary to Aristotle, only as combined with Democritus Greek-atomist style 'non-free-will-thinking' deterministic automatic invariant law responses to his postulated emitted signals or 'effluvia' (involving no pushing, no free-will-thinking and no god) - so stones fell to the ground only with a specific acceleration in response to a specific strength and direction of gravity signals from the Earth. Where both Galileo and Descartes basicly saw only the mechanical push as determinist, Gilbert saw that at least some of the non-mechanical could also be deterministic. In this Gilbert was maybe somewhat akin to Kepler's pre-1600 maths-universe 'Mysterium Cosmographicum' position of Kepler unlike most others seemingly seeing mathematics as somehow 'physical'. Gilbert hence postulated a new signal-response information-transfer universe of self-acting robot objects basically, and allowed that physical causations might mostly involve either material or non-material (energy) signals and that only for some minor phenomena might physical pushings be actually involved. Cartesian-style science dualism had what is basically an undefined 'free-will thinking' that is like requiring a TV to change channels with no remote-control signal or other action ? Gilbert saw his experiments as experimentally proving that all physical causation involves determinate automatic responses to signals or 'simple thinking', so that actual thinking might then also seem proven as being physical and science as basically offering a potentially complete explanation of the universe - though Gilbert himself did not publicly conclude that and did allow of a God. Gilbert really believed in science and he really preached science with passion, at a time when people only preached religion. But the physics of Gilbert was potentially the most complete physics, though its 1600 Latin did not explain this at all clearly.

Gilbert by 1600 had concluded that the evidence then proved at least that Earth is a big spherical magnet with daily rotation, and that physical forces like magnetism and gravity involve action response to signals. None of these were widely supported views then, being certainly opposed by most churches and governments, though seemingly he won the backing at least to some extent of British queen Elizabeth. Gilbert did basically conclude that his 1600-published experiments fully proved his action-at-distance signal-response physics theory, though he convinced only a few at the time of that as some of what he published was very guarded and unclear. And this was long before remotes and computers became common though such ideas were clearly anticipated in Gilbert's physics and its part-development by Newton and sixteenth century words were maybe not really up to explaining this new signal-response physics theory. Nor was his attempt at updating Latin and his writing aimed at the most intelligent discerning reader including much useless entertaining chat, so his theory really requires some substantial elucidating. While Gilbert produced his action-at-distance physics himself from his experiments, Galileo and later Descartes just lifted their push-physics theory from the ancient-greeks based on no experiments at all. But with 1600 Catholic Europe far outnumbering Protestant Europe, their actually less scientific Catholic physics did prevail. So the greatest English scientist was probably one of the world's first, father-of-Newton, William Gilbert who in addition to his magnetism and eletricity developed his unique action-at-distance signal-response remote-control physics against Galileo-backed simple push-physics. His main physics ideas have long been unreasonably discounted, leaving only minor unavoidable acknowledgement of his magnetism and electricity experiments. Queen Elizabeth the First did give a little helpful encouragement and support to Gilbert's early groundbreaking hobby science, before Galileo, but no modern government has followed her example including not Queen Elizabeth the Second. Of course all historians of Queen Elizabeth the First have totally ignored her helping early

English science, though admittedly she did not do so very strongly or very openly as it was not really safe for even her to do that then, though later another British queen knighted Isaac Newton in 1705 - Queen Anne. And if QE2 made some less intelligent decisions than QE1, it may well be partly due to QE2 having a somewhat less talented and more bossy set of advisors and assistants.

Of course Gilbert had grasped the nettle of 'action at distance' or 'remote-control' - the then most difficult theoretical problem for science to explain. How could bodies, separated and seemingly with nothing in-between them, influence each other ? Besides magnetism, his De Magnete did examine to some extent other 'action at distance' phenomena including especially static electricity and to an extent gravity. It did back the astronomy of Copernicus with an almost final statement of De Magnete acclaiming Copernicus as 'The Restorer of Astronomy', but Gilbert also concluded that his own magnetism experiments added further proof of planetary bodies spinning though not of their orbiting. He argued at some length that observations of the Sun daily rising in the East and setting in the West was not proof that the Sun orbits Earth daily, and that other observations or experiments supported the different explanation of Earth being a big sphere that revolves daily so that things that observations or experiments strongly suggest are true can be entirely wrong. And at the time Copernicus did not actually have a strong case for all of his astronomy and he knew that. Robert Hooke's 1674 'An attempt to prove the motion of the Earth' said Copernicus could not prove Earth orbits the Sun, only that it was probable, but claimed that he could prove it with the evidence of a more powerful telescope and claimed such orbiting to be due to the combination of attraction and inertia in space both put by William Gilbert though not their combination but not mentioning Gilbert. (Newton later showed that such attraction physics did prove planetary orbitings if given the right mathematics though without significantly acknowledging Hooke or Gilbert.) Gilbert somewhat developed this view more fully in his later De Mundo, adding the proposition that the Earth's 24 hour spin was probably initiated by magnetism (though a steady magnetism that initiates the rotation of a body also brings it to a halt unlike a temporary magnetism, and it was later also discovered that some spins can generate magnetism). In 'De Magnete', the Preface of Edward Wright more clearly supports Earth's rotation but not its orbiting. But while Copernicus and most early scientists had not sought to develop any theory explaining the why / how of planetary motion or of any 'action at a distance', Gilbert did. His experiments really disproved simple Greek-Atomist push-physics, as he firmly claimed, though most physicists of the time failed to accept his disproofs which most of them had undoubtedly not understood. Some saw action-at-distance as too similar to how Magic was claimed to work, and some saw it as too similar to how God was claimed to work, but Gilbert was only concerned with how his actual experiments proved physical forces actually work. And his experiments proved the intricacies of magnetic force with its attraction, repulsion and orientation motions that could not really be explained by any simple push-physics. Another more recent study of Gilbert's physics that concluded that his physics was widely misunderstood was the 1983 study by Gad Freudenthal in his "Theory of Matter and Cosmology in William Gilbert's De Magnete." at Isis, vol.74 no.1, pp.22–37 JSTOR

In fact Gilbert's new Magnetical Science was an automatic-response-to-emitted-signals physics, involving different types of attraction and/or repulsion 'magnetical' signals to which different bodies responded - including for him at least 4 'magnetical' signal types being magnetic, electric charge, terrestrial gravity and inter-planetary attraction [Newton later concluded that the last two were the same gravity]. Gilbert termed such signal emissions generally as 'effluvia' or 'emissions' and Newton generally called gravity signal emissions 'spirits emitted' or 'energy emissions'. Gilbert noted that magnetic and electric signal effects decreased with distance from their sources and had some signal range limits and for magnetism at least also some qualitatively different effects at different distances. Though such emissions may be claimed to not be currently directly detectable, many clearly detectable emissions show a decrease in intensity with the square of the distance from their source similar to some of the major action-at-distance forces.

Many early scientists were concerned with deriving improved description of natural phenomena, and afraid of or not at all concerned with trying to explain why nature acted as it did. Thousands of years of mere clever thinking had achieved little real, before the experimental science method emerged and produced quite different ideas on the universe. Gilbert's 1600 De Magnete was mostly just taken as being the most expert scientific work using experiments to describe magnetism and how it works, and only a few like Newton saw the significance of its physics theory. Gilbert saw action at a distance as based on signals that bodies emit (effluvia), and to which signals other bodies reacted automatically and invariantly as robots when signals reached them. Despite Gilbert producing the strongest disproofs of many of Aristotle's ideas and methods, his robotic response theory was commonly misinterpreted as an Aristotle animate universe theory, though it was really more an information-handling robot universe theory perhaps more advanced than the simpler mechanical universe theory that Descartes later produced and which won wide support. And Gilbert's universe had less requirement of gods or of humans than Aristotle's or Descartes'.

Gilbert himself did many experiments, as did Galileo and Newton though not Kepler, Descartes or Einstein. Gilbert centrally claimed that his many experiments on many materials (including even diamond and other gemstones) proved that no inactive matter existed :- "Aristotle's 'simple element' - and that most vain terrestrial phantasm of the Peripatetics, - formless, inert, cold, dry, simple matter, the substratum of all things, having no activity, - never appeared to anyone even in dreams, and if it did appear would be of no effect in nature." ('De Magnete....' Mottelay, Book 1.17 pp.69). Peripatetic or Aristotlian science had a mixed dead-matter and active-matter science with Gods, while earlier Democritus science seems to have been an only-dead-matter fully determinist push-physics no-God atomism (though the later atomism of Epicurus seemingly had a push-physics theory for magnetic attraction but had non-determinist free-will and Gods). Gilbert's 'no dead matter' physics was somewhat in line with the later 'no matter' philosophy of George Berkeley and opposed by the 'no mind' mechanical physics of Rene Descartes. As against Descartes' 'I think, therefor I am - with a universe of non-thinking things' George Berkeley is almost 'I think, therefor thinking things exist - with signals that prompt their thinking' (his 'Esse est percipi' means 'To be is to be perceived' or 'To be is to be observed')- so Berkeley was almost as William Gilbert's physics until the philosophers decided they had to invoke God ? Where Descartes mechanical physics required absolute properties of bodies in them occupying absolute space and not being able to occupy the same absolute space at the same time so that body motion must push, Gilbert physics had only relative requirements. Anything corporeal or non-corporeal (particle or energy) might be a signal, relative to some observer that can respond to it - and anything might be an observer, relative to some signal to which it can respond. The theory also basically required that physical observers, unlike intelligent observers, always respond to signals in fixed predictable science-law ways. George Berkeley's philosophy came close to backing William Gilbert's signal-response physics that was opposed to 'useless dead matter', though seemingly Berkeley never directly studied Gilbert and Berkeley's philosophy ended depending entirely on God quite unlike Gilbert's physics.

Gilbert's basic physics theory reasoning was very soundly based as explained by him in De Magnete book 2 chapter 2. He saw action between 2 bodies as needing some type of 'contact', and so concluded that at-a-distance action must involve something being emitted by one body and reaching the other body. But he saw signal 'contact' as not needing to involve pushing and concluded that the attraction, repulsion and other motions of magnetism could not be due to any type of simple pushing. Gilbert like Newton saw pushing as indiscriminate, so that light things like air should be moved to a visible extent by any push-magnetism, push-electricity or push-gravity that moved heavier things substantially, but Gilbert's experiments proved that was not the case. So he deduced that these forces at least were not indiscriminate push forces, but must be discriminating signal response forces - responses when signals touched bodies without pushing them. Gilbert was said to be very interested and expert in materials and chemistry (as later was Newton). He proved that magnetism did not affect all bodies the same, but could attract some or repel some and/or re-orientate some - depending on the body and on the strength of the magnetic signal. Very unlike a pushing and not explained by Cartesian, Einsteinian or other non-signal theories. Aristotle had said that action-at-distance needed contact but opponents of Gilbert claimed it needed contact AND pushing or that somehow contact necessarily involved pushing though this did not fit experimental facts.

Physics objections to objects touching-pushing has also come from some supporters of 'field' forces as well as from supporters of Gilbert signal forces, though some supporters of field forces did see 'fields' as themselves basically touching-pushing things if not objects - while others have avoided specifying what their 'fields' actually are at all. The zero distance required by Cartesian contact-physics is actually unmeasurable and so unprovable, while the finite distances of 'at-a-distance' physics are measurable and provable. And the differing abilities of Neutrons and Photons to penetrate bodies now suggests maybe that, unlike macroscopic objects, the ability of microscopic objects to penetrate other objects is less affected by 'pushability' or 'massness' properties of the objects than by some 'reactivity' properties. And this maybe backs doubts on pushability existing, though smaller Neutrinos having better penetration than larger Neutrons might give some little support to pushability ? In the 'Newtons Cradle' toy, something may appear to penetrate a series of solid balls though clearly nothing actually penetrates. So the apparent penetration ability of gravity and magnetism might possibly involve some actual penetrations or not. Modern claimed differences in space-occupancy and other properties of 'matter' and 'energy packets' at the microscopic level are maybe doubtful. And for another modern physics argument that things generally do not touch or contact, based on evidence that the outside of atoms generally have electrons which electrostatically repel each other, see No Touching at www.worsleyschool.net/science/files/touch/touch.html. Of course there is more to atoms and to matter generally.

Physicists from Galileo to Einstein wrongly ridiculed action-at-distance or remote-control physics and some saw this Gilbert 'animate' motion as Aristotlian, especially as he of course often used the scholar Aristotlian words of the time he was writing in - though often with new scientific meanings intended. This has been noted by some like Gad Freudenthal, ISIS 1983 at www.jstor.org/pss/232278 and Stephen Pumfrey, CUP 2002. Some strangely saw Gilbert as in line with Jean Buridan (1300-1358), though Gilbert's motion is distinctly his own in concluding that his experiments proved his new theory of active bodies responding automatically in proportion to different emission signals they receive. Gilbert had studied widely and referenced all technologies and ideas of any relevance to his science - mainstream and non-mainstream both current and from early Chinese, Arab, Greek and other societies. He noted their contributions to science knowledge while disproving many of their errors and producing his new theory. Other physicists largely followed greek Atomist 'dead matter' theory as did catholic church Jesuits (though the 1620s did see some Jesuit attacks on ancient-Greek atomism as supported by Galileo in his 1623 'The Assayer'). And later still physicists were to largely confine their studies narrowly to only what were current local science-journal issues.

While he supposed different reasons for why some bodies are magnetic, some are electric and some are neither - Gilbert did not see the different physical forces as needing basically different mechanisms of operation. All forces involved emitted signals and response to such signals. Gilbert basically took all bodies as being simple robots that emitted signals and responded to signals, and this was understood at least by Newton who developed it for gravity especially, but religion saw this as their thinking-spiritual arena needing to be dismissed as occult or alchemist. The key to Gilbert's theory was bodies automatically responding to whatever signals, but Kepler concluded that the heavenly machine is a kind of mechanical clockwork whose motions are caused by magnetic push force threads. Kepler claimed in Epitome of Copernican Astronomy (1618-21) to have built his astronomy "on the Copernicus hypotheses, Tycho Brahe's observations, and the Magnetical Science of William Gilbert" - with Gilbert's magnetical science misunderstood or misrepresented as a push-forcefield threads science. Of course in 1600 Gilbert's ideas were alien and generally not understood correctly as there were no signal response robots or remote-controls built then - the most advanced machines being maybe the mechanical clock and the compass. And it was the 1890's before Nikola Tesla started working on wireless communication and the 1960's before remote-control technology became common. Of course Tesla was chiefly concerned with action-at-distance technology and not with action-at-distance theory. Gilbert publishing the basic theory in 1600 was maybe just too far ahead of his time.

Gilbert claimed to have proven Earth daily rotation as Earth is a magnetic body and all magnetic bodies initially rotate towards any dominant magnetic pole, and space would offer no drag to stop an initial rotating. He argued that Earth's daily tides were chiefly due to lunar attraction (eg De Magnete book 2 chapter 16 etc plus De Mundo) and that they hence confirmed Earth's daily rotation, which he also saw as supported by the best astronomical evidence. But as magnetism produces no orbital type motions and the Sun had only a small then undetected annual effect on tides from Earth orbiting, Gilbert seems to have considered evidence then for Earth orbiting to be weaker and his De Magnete seemed to take no public position on heliocentric vs geocentric astronomy - as later Newton took no public position on attraction vs push physics though Newton advanced a more sophisticated science reason for that position. But Gilbert did certainly oppose geocentric or anthropocentric science generally, in opposition to much religious belief of the time.

From soon after Gilbert's death it seems the only copy in England of Gilbert's second work De Mundo was held by Sir Francis Bacon (1561-1626) who suppressed it instead of helping to get it published. And he in his 1605 Advancement of Learning, his 1620 Novum Organum and other works repeatedly attacked Gilbert's physics theory ideas while ignoring Copernicus and Galileo and at the same time advancing himself as generally promoting new experimental science. Bacon claimed that Gilbert's physics theory was too big to be proved by his experiments - which Bacon unreasonably claimed were not evidence of the Earth being a magnet or rotating or otherwise moving and were not evidence of how magnetic, electric or gravity attractions worked. Park Benjamin Jr noted, in his 1895 'The Intellectual Rise in Electricity, a History' p.320, that Bacon "attacked and condemned over and over again the opinions of a man who could neither speak for himself being in his grave, nor be spoken for by his De Mundo wherein he had set them forth and which in the cabinet of Bacon was effectually silenced and entombed". (see http://ia600200.us.archive.org/6/items/historyofelectri00benjrich/historyofelectri00benjrich.pdf) Some thought that Francis Bacon had improperly used and abused William Gilbert for his own self-promotion as a supposed 'Champion of Science' despite himself having some notable talent. Some other attacks were made regarding Bacon's character but they may not all have been really merited. Bacon was a maybe very tricky religious lawyer, politician and philosopher who somewhat helped with the British colonising of America. He chiefly prospered only after the deaths of Queen Elizabeth and Gilbert in 1603 and though he had struggled somewhat under Queen Elizabeth, Bacon prospered under a new King James and was quickly knighted by him in 1603. But when Bacon died in 1626 he was heavily in debt. Bacon's main philosophy (or claimed 'science') work was his Latin 1620 'Novum Organum' after Aristotle's main 'science' work so he basically saw himself as 'The New Aristotle' among other things. Though this work basically stole with a little rewrite from William Gilbert, Bacon did win favour and turn many in England and elsewhere wrongly against Gilbert. He himself promoted an 'action-at-a-distance' physics that, unlike William Gilbert's sound 'signal-emission action-at-distance' physics, involved spooky action-at-a-distance with nothing causal spanning a distance but it gained little support and much opposition which was often wrongly applied also to Gilbert's physics.

William Gilbert took his degree as Doctor of Medicine in 1569 at St. John's College, Cambridge, and settled in London in 1573 after four years of foreign travel. He was then admitted to the Royal College of Physicians, of which he became first Censor, then Treasurer and finally in 1599 President. In 1601 Gilbert was appointed personal Physician to Queen Elizabeth I, whom he attended in her final fatal illness in 1603. De Magnete's opening 'Address by Edward Wright' stated that Gilbert had delayed publishing for almost 18 years, not disputed by Gilbert, which was till late in life and partly like Copernicus who delayed publishing till on his deathbed to try to avoid persecution or attack by peers, though Gilbert did apparently also tell some people about his science ideas long before publishing and it was believed that Gilbert had for many years been one of several physicians to Queen Elizabeth I, but she liked Gilbert's science work and when he published made him her royal physician or physician-in-chief and gave him a royal allowance or pension from the royal purse. (Note that such 'pension' could present as 'reward for nobler loyalties', as against 'salary' or 'wage' more as 'reward for lowly labour'.) So in 1600 Queen Elizabeth 1 rewarded the publishing of the world's first real science book though previous monarchs had favoured books that were artistic and/or religious, but it was more a partial shift than a major change and there had maybe always been occasional monarchs that favoured reading some philosophy or mathematics and maybe even in Greek or Latin ? Yet this perhaps does make Queen Elizabeth the First the first national leader to champion real science ? Yet this perhaps does make Queen Elizabeth the First the first national leader to champion real science, if done rather late in her life ? Elizabeth was intelligent and fluent in English, Latin and some other languages and was considered open to many ideas, as including even those of astrologer and alchemist John Dee, though generally forming her own conclusions which she could enforce ruthlessly. Rodrigo Lopez who she had in 1581 made her royal physician or physician-in-chief, with a 'life pension' of £50 per year, Elizabeth had executed for claimed treason in 1594. And of course Giordano Bruno, basically a supporter of Copernican astronomy and of other worlds, was burnt at the stake in Rome on the 17th Febuary 1600 aged 51 by the roman catholic inquisition, which included a Cardinal later made Pope, for heracy or basically disagreeing with the church in a trial begun by them in 1593. Though Bruno was no scientist nor even really any astronomer, Gilbert's 1600 'De Magnete' basically included some support for these ideas that Bruno and some others supported, and the catholic inquisition did soon turn on Galileo. Like some other new-thinkers of the time Bruno tried escaping a more restrictive catholic Europe to protestant Europe and lived some of the 1580s in London and published in London then. But like the Roman Catholic christian church then, some anti-Catholic christian churches of the time could also be brutally narrow-minded and kill anti-church 'witches' and 'heretics' as in 1615 protestant Germany Kepler's mother was on non-science grounds charged with witchcraft though acquitted - and Gilbert considered the majority of his peers to be useless or dishonest and insisted on proving so repeatedly.

His delay in publishing his science seems to have largely been due to the times being dangerous for publishing science, but maybe also partly due to him juggling different advice and 'help' being offered to him on its publishing. Interestingly, the year after Gilbert published De Magnete, Queen Elisabeth enacted a Poor Law to help England's poor. Three years after he published De Magnete, and eight months after Queen Elizabeth died aged 69, Gilbert died aged 59 apparently of the Black Death. Queen Elizabeth I died on 24.3.1603 and Gilbert finalised his will on 24.5.1603 leaving all his library books, instruments and minerals to The Royal College of Physicians (who unfortunately lost them). He left most of his assets to his brothers and sisters, and some to friends and employees as well as £10 each to the poor of Colchester and the City of London. Gilbert did leave £10 to his local church where he asked to be buried, and left his soul to God, suggesting a man somewhat religious but not over-convinced about churches. It was a rather normal will for the time, leaving nothing to medicine or to science and so giving little indication of it being of a prominent man of medicine and man of science. But not until the 20th century did wills in England begin to often include medical charities and pet charities and still rarely included anything for basic science. Technological and economic progress can commonly rest on basic physics and other science experiment, so that the greatest advance for medicine was basics physics experiment producing the microscope. Maybe for Earth now the Climate Change danger, the Superbugs danger and maybe the Asteroid danger all need big new basic science advance to handle them and as that is not being done in physics today needs some people to write wills leaving something to some basic science experiment programs ?

William Gilbert died on 30.11.1603. (He did well at making money and with 1603 manual wages about £6 a year, £10 was a reasonable amount then, and Gilbert was said to have spent over £7,000 of his earnings on his physics or at 2022 value over £2,000,000) He may have correctly felt that his science could have a tough time without Queen Elizabeth, and asked his younger brother to support his science and publish his final De Mundo manuscripts. A major drawback to publishing shortly before death is that when dead you cannot defend your ideas from unscrupulous attacks then - as with Francis Bacon's attacks on the dead William Gilbert's science while stealing much of it to claim as his own pretentious science ideas. Bacon attacked Gilbert chiefly for presenting as dealing with the small subject of Magnetism while really dealing with all of science - saying 'Gilbert hath made a philosophy out of observations of the magnet.. he has ascribed too many things to that force and built a ship out of a shell' Yet Bacon basically used Gilbert's wider science himself as his own, with no evidence of his own at all for it, and taking Gilbert's 'orbs of virtue' as the distances to which bodies magnetic, gravity or other perceptions extend and not per Gilbert as being 'effluvia signal ranges', but discussing eight different types of 'effluvia' (including light and sound, and some he calls 'emissions of spirits'), as signals prompting perception for magnetism, gravity and some other types of action. [To be fair, any signal-emitter's signal range is as some signal-detector's detection range for that type of signal - and may differ for some different type of signal-detector, and this generally assumes that signals weaken with distance from their source.] In England for many years after Gilbert's death John Wilkins, Robert Hooke and others promoted Gilbertian 'magnetical science' as also covering gravity and maybe other forces against the Cartesian push-physics gaining ground especially in mainland Europe. Isaac Newton and Robert Hooke essentially backed Gilbert in like him taking the basics of Magnetism as really applying also to gravity and maybe any other physical forces. Of course most non-English commenting on Gilbert took it that he dealt only with magnetism, though many of them had probably never actually studied Gilbert at all.

Gilbert's younger brother took on responsibility for publishing the seemingly partly-English preliminary draft manuscripts for Gilbert's second book De Mundo putting his wider Magnetical Science or Attraction Science. His brother had it translated fully into Latin and soon after 1603 he did it seems manage to provide a few people with manuscript copies, certainly at least Prince Henry and maybe also Francis Bacon, Galileo and Kepler. Gilbert's younger brother dedicated De Mundo to Prince Henry who was the young, popular and Protestant son of King James the First of England, who succeeded Queen Elizabeth I, and who died aged 18 in 1612. He sent several MS copies of De Mundo to the young Prince Henry by 1608, believed to have got from there to the Kings Library and to Francis Bacon (referred to by him in a 1612 publication and passed to Sir William Boswell who published De Mundo) and to Thomas Harriot who mentioned it to Kepler in a 1608 letter. But he could not then get De Mundo published, seemingly due partly to its suppression by Sir Francis Bacon, until in 1651 long after Gilbert's death it did get published though only in Latin - but even Gilbert's Colchester grave inscription is in Latin (though he did write his own will in English, any other of his writing in English being seemingly lost or destroyed).

Gilbert, like Galileo and Newton, held a low opinion of the majority of his peers and just trusted that his own proofs and experiments would sufficiently demonstrate the correctness of his theory whatever most of his prejudiced peers concluded. But before Gilbert reached that conclusion (as did Newton later) Thomas More publishing his 'Utopia' in Latin in 1516 in Holland, and in 1551 in English in his England after his execution there for being Catholic, noted that as Plato had concluded basically rulers often have prejudices to which they have long been accustomed. Gilbert's theory of Earth tides eg De Magnete book 2 chapter 16 etc and more fully detailed in De Mundo was sound and easily beat Galileo's wrong mechanical theory of tides wrongly backed also by Francis Bacon, though probably Kepler and certainly Newton later were also to find that planet orbitings were best explained by attraction physics though neither fully backed such physics which was strongly opposed by the Catholic church especially. But unlike Galileo with his telescope Gilbert prioritized advancing physics over advancing astronomy, which seemingly helped boost the industrial revolution in England more.

Gilbert seemed to basically merge the active-matter of Greek Aristotelianism wih the determinism of a Greek dead-matter Atomism by adding 'effluvia signals' to connect the two in his signal-response physics. However, Descartes and others instead chose to entirely reject Greek Aristotelianism and support a Greek Atomism as a dead-matter push physics. But, as early Christianity burnt most of it, we do not know to what extent ancient Greek determinist atomism was really a push-atomism or an attraction-atomism or a mix ? (see https://aeon.co/essays/is-atomic-theory-the-most-important-idea-in-human-history) It may be that Gilbert was actually a more accurate development of ancient Greek atomism than Galileo and Descartes ? Gilbert's attraction science or remote-control science did gain some limited backing especially among protestants, but catholic church Jesuits and catholic Galileo and Descartes supporters were quick to join Bacon in discrediting Gilbert's theory without any disproof of it, and the later 1651 publication of Gilbert's 'De Mundo' was too little too late. While he published no comment on Galileo, Gilbert concluded that his own experiments disproved the push-physics theory that Galileo backed, though privately he did strongly support Galileo's experiments. And Galileo basically published without any relevant evidence his disapproval of the physics theory that Gilbert's experiments backed, though he did strongly support Gilbert's experiments.

The very many and wide-ranging references in De Magnete and De Mundo showed that Gilbert has studied exceptionally widely but little else is known of that. All of Gilbert's manuscripts and library were seemingly destroyed in The Great Fire of London 1666 or otherwise lost, and the only known 17th century university teaching of Gilbert seems to have been at Gresham College London and at Clare College Cambridge between 1658 and 1678 and that may well have been little (Gilbert had studied at St John's College Cambridge, and Newton was at Trinity College Cambridge from 1661). Robert Hooke was at Gresham and supported Gilbert's attraction physics, including putting attraction physics ideas to Newton, but Descartes 'dead-matter' Catholic-backed push-physics based on ancient greek Atomism generally prevailed over Gilbert's new 'robot-matter' theory largely by name-calling. Newton at first followed the then-prevailing push-physics till seeing Gilbertian attraction physics as better fitting the mathematics of planet orbits and other physical phenomena. Newton's disgust at Gilbert attraction theory being dismissed by merely calling it occult based on 'prejudice' was shown in him saying that in that case all theories involving unseens should be called occult including Galileo-Descartes push-physics (and logically also including Einstein's theory since nobody has directly seen a spacetime continuum).

Gilbert's physical universe had two types of fundamental things ;
1. Various types of robot observer particles that emitted and responded to effluvia force signal emissions, which might mean atoms or parts of atoms and maybe photons etcetera. The internal structure if any of these 'blackbox' things mattered little in Gilbert's theory, only their emission and response to effluvia signals.
2. Various types of effluvia force signal emissions, causing eg electrical, gravitational and proximity responses in some or all of the above particles. The latter seem currently less easily directly detectable than the former.

Two main conclusions of Gilbert were that different types and strengths of signal had different ranges - which for magnetism could be less than an inch for a weak magnet to some miles for the Earth's magnetism - and that signal strength diminished with distance. He showed that for magnetism at least effects also had qualitative difference with distance, attraction-repulsion effects over shorter distances only and orientation-magnetization effects over greater distances. It may be however that a magnetic field's turning ability seems to extend to a further distance than its attraction ability only because its signal strength decreases with distance and a turning response requires less signal strength than does an attraction response ? So maybe a magnetic field really extends to an infinite distance though at decreasing signal strength ? But this does not really seem explained by any other non-signal physics theory so that magnetism field theories seem to explain attraction only but not compass-turning. And further Gilbert also deduced from experiments comparing magnetism and static electricity that different types of effluvia signal emissions also had different abilities to penetrate matter, seeing low-penetration electric charge signals as more likely material particles and high-penetration magnetic signals as more likely non-material energies or 'spirits' - so his effluvia signal emissions were perhaps in modern physics terms 'quanta' that could be mass or could be energy. (of course higher penetration might also be due to a much smaller size like Neutrinos or due to absorption-and-retransmission of some quantities in some directions.) But some interpreted Gilbert's signal range in terms of a 'force field', though the idea of force fields is a quite different idea requiring all space to be filled with something like an energy version of Descartes material ether. From our atmosphere attenuating with altitude, Gilbert concluded that just a few miles above the Earth was empty space containing nothing - but through which his signals including gravitation effluvia 'gravitons' could pass. Planet orbits not having drag made Newton support Gilbert's empty space, though Descartes like Aristotle and perhaps Einstein thought empty space was not possible largely on other theoretical grounds.

It is to be noted that Gilbert did not conclude that magnets or magnetic signals contained contrary properties because they attracted iron and did not attract ice. Gilbert like Newton taking science as not allowing actual contradictions, saw the difference as being in iron and ice having different responses, without any contradiction, to the same unitary thing. Einstein and others unfortunately later made what is maybe an anti-science mistake of taking light (and particles) as both being wave and being not-wave, and adopted the self-contradictory self-disproved 'Duality Theory' instead of accepting that different responses as to light do not imply different source properties as of light. And Gilbert basically held that theoretical ideas are scientific only if they are directly intuitive, as being deduced only from direct observation or experiment, though many other physicists since have wrongly accepted 'counter-intuitive' ridiculous conjectural theory ideas.

One substantial problem for Gilbert's theorising came from magnetism being one the most complex of the physical forces, so his many measured experiments could not yield him the simple mathematical laws that Newton was to later develop in applying Gilbert's theory to gravity. While the other physical forces are simpler central attraction or repulsion forces, magnetism involves poles and includes turning or partial-rotation responses and magnetization responses. And these different magnetic responses to the same signal operate at different ranges, so that apparent signal range is clearly less a property of the signal than an indicator of response sensitivities. Gilbert often refers to electric force as 'a simple attraction' and to gravity as 'that simple straight motion to a centre' and clearly considered more complex magnetism as a force of more interest to physics. Of course a simple attraction can help with more complex motions like orbits if acting in conjunction with body momentum or other force sources. And Newton noted, in his Principia Book 3 Proposition 6 Cor 5, that 'the power of magnetism in receding from the magnet decreases not in the duplicate but almost in the triplicate proportion of distance' as far as Newton's crude by his word magnetism experiments showed assumedly regarding magnetic attraction. (Some magnetic effects are said to work at greater distance than others, and some to work slower, but maybe not all of the magnetic effects have been well studied. If signals go to the same distance and are the same speed but responses and response times differ then that would seem to prove that a signal-response effect is indeed involved and William Gilbert concluded that he had proven that. Of course magnetism experiments could be done today with greater accuracy using electromagnets though it seems no physicist has done much on that.)

Though Gilbert had been a Cambridge university examiner in mathematics, he somewhat distrusted mathematical deduction as being mere logical philosophising as against being experimental proof science, and so stood by minimal logic and minimal mathematics. He perhaps foresaw the mathematics-only-physics and probability-science problems that emerged later. But maybe the idea of an experimental philosophy against conjectural philosophy needed to be combined with the idea of an experimental mathematics against conjectural mathematics ? And in Europe early mathematics was largely used in astrology, as being in 16th Centuary England promoted substantially by astrologer John Dee. Some pre-1500 exploration, developing technology and later emerging experimental science became growing more substantial uses for mathematics. In early Europe astrology was basically a 'more-rational' alternative to religion that did not directly challenge religion, and early alchemy was somewhat similar. Lutheran Europe began developing mathematics more in the early 1500s, with Catholic Europe and Protestant England somewhat behind till the mid 1500s but English science and increasingly mathematics were leading by the mid 1600s. But a bigger problem than mathematics to Gilbert developing his physics theory further was the fact that his knowledge of mechanics and motion being pre-Newton and indeed pre-Galileo was limited. A couple of bits of Gilbert were disproved by later experiments, but were entirely inessential to his theory. Kepler unintentionally showed that good mathematics could be successful even within a poor explanation physics, but not until Newton was Gilbert's 'attraction theory' properly mathematised. And to be fair pro-signal Gilbert did seem to maybe also allow some push physics somehow, while perhaps Newton really entirely rejected push-physics for Gilbertian 'spirits emitted' signal physics though not publicly admitting that. But still there is the question of are philosophy, thought and mathematics more aligned with religion, and really contrary to experimental science and technology as William Gilbert seemed to chiefly argue ? Of course Gilbert did not entirely argue that, but he did strongly favour the latter as better proving truths - what can be done as against what can be thought. In the 1720's, Voltaire thought that the English favoured (Gilbert-)Newton attraction physics while the French favoured Descartes push physics - Voltaire - but really perhaps only the English public and not most English physicists.

The old legal joke "There are three types of unreliable witnesses : simple liars, damned liars, and experts.", was made a statistics joke as "There are three kinds of lies : lies, damned lies, and statistics." But some supporters of experimental 'real' science might prefer "There are three types of doubtful science : hypotheses, science fiction, and mathematics." And while at the same time as Gilbert studied magnetism, Galileo studied gravity. But gravity produces only one simple unselective attraction effect. Magnetism produces selective attraction, repulsion, orientation and induction effects. And still Gilbert produced a rather better explanation for magnetism than Galileo did for gravity. Gilbert who argued strongly against over-theorising was undoubtedly the better at theorising.

While Gilbert produced his useful working mini-magnetic-planet models ('Terrellas'), nobody has made useful working mini gravitational planet models as gravity seems insignificant for normal small bodies and atomic repulsions prevent substantial object compression. (Strangely perhaps it has not yet proved possible to use the fact that groups of neutrons or neutrinos should be more easily compressible.) But with todays radio-control technology it is possible to produce robots or drones that mimic gravitational behaviours. Designing such with spin and a single receptor and single engine with delays would add significant constraints that might maybe give Einsteinian maths to basically Newtonian response, though spin in space itself involves no effort. A range of different alternatives for mimicking gravitation would certainly be possible as discussed in our Gravity section, but it would significantly expand the possibilities for actual astronomical experiments and more.

Gilbert saw the term 'attraction' being chiefly used in England to 1600 as meaning push-physics 'pulling', but not too long after his De Magnete was published Newton and others could see the term 'attraction' being more used in later-1600s England as meaning attraction-physics 'responding to signals'. This did not really happen elsewhere and the first detailed published opposition to Gilbert's physics theory was by Jesuit catholic Niccolo Cabeo in his 1629 'Philosophia Magnetica' which though generally Aristotlean adopted a poor greek-Atomist matter-pushing-matter explanation of magnetism, electricity and gravity that Descartes later supported but Newton disproved. The catholic christian church founded the male Jesuit Order, or Society of Jesus, in 1540 'to fight to defend and propagate the catholic church' basically by opposing protestant ideas and science and to promote catholic alternatives, though a Catholic science was to be a tricky option. From generally opposing Protestant Christianity the Jesuits were soon prioritising opposing emerging Non-Aristotelian Science and in the early 1620s that included some attacks on greek atomist science. But that position soon crystallised turning the catholic church to supporting greek-Atomist physics against protestant attraction physics.

Despite Gilbert disproving much of Aristotle many times in his works and his physics having no apparent place for gods, Gilbert's theory became labeled by many physicists as 'Aristotelian' god-derived - and was rejected in favour of the god-separate Descartes mechanical-push science (fully published by 1644) but maybe akin to throwing out the baby with the bath water ? Information handling robots and remote-control are a more modern technology than mechanical robots, and modern information theory is now doing much work that is basically along Gilbert signal theory lines, though without any great impact on physics theory as yet. Despite the almost universal use now of television remotes and mobile phones all acting in response to remote emitted signals, which perhaps at least partly confirms Newton's view that Gilbert signal theory was at least plausible ? But the majority of physicists still claim that action at a distance is impossible - when most people know it IS possible and works by SIGNALS emitted and responded to as Gilbert concluded that magnetism, electric charge and gravity work. Gilbert termed those natural emitted signals 'effluvia' - from Latin at the time generally taken as meaning 'non-visible characteristic emissions from bodies such as their smells'. But in his preface to De Magnete did clearly state that his use of words often involved new scientific meanings for them. While his natural signals emitted by objects causing magnetism, gravity etcetera were termed 'effluvia' by Gilbert, they were generally referred to by Newton as 'spirits emitted'. But Gilbert saw the evidence as indicating that some 'effluvia' natural signals emitted were probably corporeal particles and that only some were probably non-corporeal non-particle energies or 'spirits'.

The actual observed difference between magnetic behaviour and gravitation behaviour is substantial, so that producing one simple theory to cover both is a substantial problem to any physics. Hence magnetism involves attraction, repulsion, orientation and magnetization affects, while gravity involves only one simple attraction affect. Gravity being basically simple could easily seem to suit a simple Descartes mechanical push theory, which was very difficult to apply to magnetism. But magnetism being more complex perhaps more suits a Gilbert signal response theory, which also was easy to also apply to gravity as attraction theory as Newton showed. And notably gravitational and electromagnetic forces have some common aspects that Gilbert signal response theory handles well. They both have directionality though it may be only directionality relative to another object, and their action also seems to involve a mutuality relative to another object. In fact these forces may well have no objective existence for one object alone, in line with William Gilbert's signal-response theory of forces ? And signal physics has some other interesting aspects, see our Light.

Some like Einstein who followed Descartes basically and took gravity as being a fundamental, took magnetism as being an inessential of less importance to physics theory. Though in science all well confirmed facts are basically equal, Newton did somewhat little to oppose the magnetism-does-not-matter position. But the fact that magnetic and electric charge forces give BOTH attraction AND repulsion behaviour AND other responses AND only on selective bodies, does strongly suggest that the 'force' of these forces at least is NOT in the force itself but in bodies responses - as in Gilbert's signal response theory. A big problem for any push-physics explaining is that pushing is basically indiscriminate but the actual universe includes different attractings including some discrimination attractings as well as different repulsions including some discriminating repulsions and other discriminating behaviours more consistent with an attraction physics than with a push physics. And Newton could get his workable physics maths from attraction physics but not from push physics.

For comparison with other physics theories, Gilbert's three laws of motion would be ;

1. Every observer body will remain at rest, or in a uniform state of motion unless effluvia signal emissions act upon it.

2. When effluvia signal emissions act upon an observer body, it accelerates itself proportional to the signal strength and inversely proportional to the mass of the body and in the direction required by the signals.

3. Every effluvia signal action evokes an equal and opposite effluvia signal emission reaction.

Gilbert's theory might maybe be strengthened with a few additions that would basically make it a gauge bosons particle exchange theory such as some modern particle physicists favour ;
1. Observer bodies emit various effluvia with speed of light velocity in response to various effluvia being received by them.
2. All motion and other natural phenomena are caused by this process (including seemingly causeless radioactive decay).
3. Effluvia are conserved.
4. All observer bodies are aggregates of effluvia.
Then we might have the basis of a relativistic quantum mechanics physics without fields or continua, or of a no-ether Descartes particle push physics without fanciful corkscrew-particle-push or boomerang-particle-push attractions ? Maybe a high-reaction graviton causing the emission in the same direction of a particle pair of a similar low-reaction graviton plus another high-reaction particle (normally multi-directional) giving the gravity momentum effect ? Whatever it would mean, that physics would be about how many different types of effluvia exist and their properties, how many different types of

effluvia-aggregates exist and their properties including what influences effluvia aggregation and dis-aggregation ? And maybe unlike mass, charge and spin could be just signal response properties ?

One apparent difference between Gilbert-Newton signal attraction theory and Descartes push theory is on 'action and reaction are equal and opposite'. Though Gilbert and Newton proved that this did hold for attractions, it may seem that a push must give an equal effect while a small signal might give a big effect. But this apparent problem can also occur in mechanics and can be fully resolved with lever, trigger and conversion (eg $E=mc^2$) effects. And if particles like neutrons are themselves complex systems, then a graviton signal might trigger a series of events including eg neutrino emission that actually produce motion responses ?

The supposedly separate two processes of force-production and acceleration-by-force, may actually be basically one process - ie. bodies in Gilbert theory terms maybe basically respond to external forces by accelerating themselves by producing their own forces as maybe by emitting small particles in response to received signals like remote-controlled rocket engines. This could give a natural 'equivalence' of forces and acceleration having a wider cover and making more sense than Einstein's little Equivalence Principle applying only to gravity. Supporting this is Gilbert and Newton often positing the mutuality of forces between multiple bodies, and as we are in a multiple body universe there can maybe be no proof that one single isolated body would have any force of gravity or any other force ? This mutuality seems clearly related to the 'entanglement' property in quantum information theory (from experiments suggesting that atoms can split one photon into two mirror 'entangled' photons of eg opposite spin polarisation and half the energy with some claiming that these remain somehow actually linked or 'entangled' even when distantly separated). A signal physics can more naturally handle multiple-signal emissions having related information and/or separate mutual signal emissions having related information, without requiring any mystical 'entanglement'. In fact the first published discussion of 'corpuscles being entangled' was in Gilbert's 1600 De Magnete. But the question is are 'entangled particles' connected by magic or by signals - or not connected and just a related-creation giving related properties ? And yet most weirdly no modern physicist at all tries to relate distance-entanglement to action-at-distance as originally put by William Gilbert and backed by Newton but was widely attacked by most physicists completely unreasonably without proper study ?

Gilbert's later De Mundo published long after his death put his physics as a Theory Of Everything, where for any body there is some unstated but implied Universe Equation of the type A+B = A'+B' where ;
A, for gravitational force, the attraction on the body by all other bodies summated, PLUS
B, for any other forces like magnetism applying, the attraction on the body by all other bodies summated
being equal and opposite to the attraction by the body on all other bodies summated for forces applying.

Gilbert's physics was a complete explanation physics whose description was somewhat complicated with its inclusion of (signal) emissions and self-moving responses to them that he considered needed for complete explanation which he saw as required for a science to be really scientific. Some who supported Gilbert's physics tried to simplify their description of it at times by either omitting its responses or omitting both its emissions and its responses. This did not always mean that they believed anything different to Gilbert, but did readily lead to such misunderstandings. And such omissions seemed to allow some possible very un-Gilbertian Galileo/Descartes type push or pull mechanisms being involved, so possibly reducing some objections from that quarter but presenting an incomplete science that could be called mysterious. But Newton was happy to justify in his Principia such omission. The main supporters of Gilbert, in whole or part, were Johannes Kepler and Francis Bacon and later Robert Hooke and Isaac Newton though they did not at all acknowledge Gilbert and Newton insisted that complete explanation was not needed in science.

Hence in his Novum Organum, Francis Bacon compared gravity with magnetism and wrote in line with Gilbert that 'whatever produces an effect at a distance may be truly said to emit rays' - though, unlike Gilbert, Bacon commits to nothing specific about such rays or how they work. And Robert Hooke in a communication to the Royal Society in 1666, wrote "I will explain a system of the world very different from any yet received. It is founded on the following positions. 1. That all the heavenly bodies have not only a gravitation of their parts to their own proper centre, but that they also mutually attract each other within their spheres of action...." This is pure William Gilbert signal-response physics almost word for word, where 'spheres of action' are his signal ranges, though here Hooke does actually omit Gilberts signals and response mechanism and claims to be stating his own new idea. Yet that year Hooke also wrote that Gilbert began to imagine gravity a magnetical attractive power, inherent in the parts of the terrestrial globe and in his 1678 "Cometa" he wrote that the planets "may be said to attract the Sun in the same manner as the Loadstone hath to Iron." Newton also wrote of significant comparison between gravity and magnetism, so that ideas of unification of gravity and electromagnetism clearly much pre-dated Einstein. Gilbertian signal-response physics also more firmly excluded time-reversal in nature than the Cartesian reversible-push physics that was commonly to be wrongly ascribed to Newton. If physical actions are all responses to signals, then response must always follow signal, giving time and making it one-directional. So a bar-magnet will attract a piece of iron but there is no reversing that, and though an electro-magnet can have its poles switched or 'reversed' that is not time reversal and evidence also does not support any reversing of gravity. And surely a physics where an effect does not always follow a cause, cannot really be a scientific physics ?

Hooke and Newton clearly did correctly understand Gilbert's physics basically as a signal-response physics though not all of them fully agreed with all of it, and many have studied Gilbert and not at all understood his theory being instead mesmerised by his declared unusual use of language - as seen in the unfortunate modern study 'Francis Bacon and Magnetical Cosmology' by Xiaona Wang, University of Edinburgh in Isis, Volume 107, Number 4, December 2016. Somewhat better is Mary B Hesse in 1960 in The British Journal for the Philosophy of Science noting that earlier translations of Gilbert were poor, and that both Francis Bacon and Robert Hooke misinterpreted key ideas of Gilbert. And a second strong female supporter of William Gilbert was Sister Suzanne Kelly who in 1965 published an excellent facsimile of Gilbert's 1651 De Mundo (of which I have a nice copy), though neither her nor Mary Hesse might fully support his action-at-distance physics being a signal-response physics as it undoubtedly actually is. Gilbert's attraction physics theory got some marginalised support mainly in England to Newton's time, against the growing influence of Galileo and Descartes supported push-physics. But William Gilbert and his magnetism and electricity did significantly impact early science mainly in England and at least into Newton's time, though it was to be many years later before that particular area of physics was to itself give any further real advances. Much later others like Volta, Faraday, Maxwell and Tesla made some further advances in electromagnetism to great effect with both materials chemistry and biology being significantly involved, but failed to deal with major issues, and of course Einstein failed entirely to deal with any magnetism issues. But of course Gilbert's signal-response physics was basically an advanced information-handling physics way ahead of its time. While physicists like Faraday and Maxwell did very good work, it really only poorly half-explained the simpler aspects of magnetism. Nobody did any real theory follow-up work on the actual full experimental magnetism of Gilbert and Tesla. And it is a bit of a mystery why later no theoretical physicist, including Einstein, thought to work with Tesla ?

A static electric charge stationary on the moving Earth, produces NO magnetic response from a detector compass that is also stationary on the Earth - but it DOES produce a magnetic response from a detector compass that is moving relative to the Earth. And a still or uniform velocity charged particle generates a static electric field, but an accelerating charged particle generates a fluctuating electromagnetic wave. This fits poorly with most types of field physics theory, and better fits a Gilbert style theory where no 'magnetic fields' as such exist but electric charge signals are emitted and detectors simply respond electrically and/or magnetically to such signals motion relative to themselves. A stationary permanent magnet does produce magnetic responses from another stationary permanent magnet, but the absence of macroscopic relative motion can still allow of their microscopic atoms or elementary particles having some relative spin or other motions. A 'stationary' electric wire can carry a moving electron current, and generally 'rest' can include motion and vice versa. Moving any normal macroscopic magnet slowly will emit very low frequency long wavelength (millions of metres, 1.24-peV) E/M waves, and needs to move extremely fast to emit higher frequency shorter wavelength (around 1 metre, 1.24+ μeV) E/M waves like standard radio waves. Eg rotating a magnet at frequency f implies E/M emission with a wavelength λ=c/f relative to some E/M detector ? This does give the problem of infinitely slow motion, if not also of infinitely fast motion, and also of absolute motions versus relative motions ? Of course it also relates to Newton's physics relating accellerations to forces.

Magnets only produce induced electric currents in bodies at a distance from them if there is relative motion between them, as studied by eg Faraday and Maxwell, but otherwise of course a static magnet still produces magnetic induction or resonance in suitable magnetisable bodies ! With no relative motion a magnet does, by induced magnetic resonance, orientate the motion of charged particles and magnetisable bodies near it, which is affected chiefly by the extent to which electrons and/or protons are in bodies paired or not paired. More recent magnetism experiments seem to show that a + or - electric charge in (relative) motion produces a magnetic field that can affect at least some other bodies, and at least some + or - electric-charged bodies in (relative) motion can respond to a magnetic field. So magnetism perhaps concerns only the (relative) motion of two or more + or - electric-charged bodies ? But such 'motions' can just involve claimed sub-atomic 'spins' that are not visible motions or may actually be non-motions ?! And standard experiments with naturally-magnetic magnetite ore or with bar-magnets do seem to show at least some magnetism apparently needing no motion and maybe no + or - electric charges ? So maybe there are really two different kinds of magnetism, one maybe from + or - electric charge relative motion and one from static +/- charge displacement in some bodies having both + and - electric charges ? Study of both of these motion and static magnetism phenomena have still not clarified the issue of exactly how magnetism works, as between the two basic options noted by Newton of either a Maxwellian medium/continuum pushing bodies or Gilbertian information signals that bodies respond to. Einstein supported the former while failing to prove it and Newton favoured the latter and now there seems some growing support for the latter from information signal physics.

'De Magnete' 1600 London edition title page :-  - Click image to enlarge, or to get click-enlarging image.

'De Magnete' 1628 pirate edition title page :-  - Click image to enlarge, or to get click-enlarging image.

GILBERT initially mainly saw himself as a chief advocate of new experimental natural science examining everything, as against the mere dogmatism and philosophising of old natural philosophy, religion and mathematics. He experimented and he talked with miners, sailors and others before writing his De Magnete somewhat in the polemic anti-establishment style that Karl Marx was to later write his Das Kapital and which Galileo was banned from using. While he attacked 'mere theorising', the full title of his De Magnete indicated that scientific proof needs both experiment and theory argument and he did further see himself as the originator of a new signal response physics theory covering magnetism, electricity, gravitation and mechanics - which he sought to prove chiefly through his experiments on magnetism. Gilbert implied perhaps wrongly that an experiment could be interpreted wrongly or differently only if you considered things far removed from the experiment, but that is perhaps a mistake that many scientists even today make. Park Benjamin Junior noted in his 'The Intellectual Rise in Electricity, a History' (1895), p327, that Gilbert's aim was not primarily the making of electrical and magnetic discoveries, but the establishment through such means of a great theory of the physical structure of the universe. Gilbert did prove by replicable experiment both action-at-distance and action-through-matter against any kind of push-physics, and both apply at least to both magnetism and gravity. And finally Gilbert saw his lesser role as establishing some of the real facts of magnetism and electricity - though commonly only this role got properly acknowledged. In his later writing, maybe seeing experiment as becoming more accepted, Gilbert promoted a more fully positive view of science theory.

Gilbert's publication of De Magnete in 1600 Protestant England was chiefly poorly received by the career-scholars that he saw as more concerned with their careers than with the truth, but its concentrating on Magnetism made it somewhat obscure and of little threat to religion though parts of it did make strong claims on sensitive issues like motion of the Earth. But it was chiefly after his 1603 death that Gilbert's physics was largely buried mostly with mere name calling. The early Catholic Church and others often competing with William Gilbert, effectively conspired to deploy a big 'fake physics proof' against his action-at-distance signal-response physics, though still today no valid disproof has yet been produced. And in his Philosophia Magnetica of 1628, Niccolo Cabeo like some other Catholic church Jesuits opposed Gilbert's signal physics with a Cartesian style push-physics. But when the London 1600 De Magnete edition finished, supporters produced two pirate editions in Protestant Holland. However some major occult or religious philosophers of the time, like Robert Fludd (Utriusque Cosmi 1617-24) and Athanasius Kircher (Ars Magnesia 1631), misused Gilbert signal physics as supposedly supporting natural-magic or miracle [as charlatans today can often misuse relativity theory]. But Gilbert's later De Mundo though not published till 1651 did take his action-at-distance or remote-control science far beyond just magnetism and to more of a challenge to religion. Still 2022 sees the Wikipedia section on 'action at a distance' somehow totally ignore Gilbert and rather skip over Newton and ignore the mass of remote-control technology in actual use to 'expertly' claim that action-at-distance does not exist despite the mass of experiments and technologies supporting it and most observing being from some distance. But Gilbert's experimental proofs of action-at-distance will not go away and their backing by remote-control technologies can only increase, so his basic physics still carries some real weight as Isaac Newton also correctly concluded. But still today is taught the fake-Gilbert physics that very wrongly pretends that his action-at-distance physics was not the signal-response physics that it clearly was ? Proving an action-at-distance physics theory seems to be not impossible, though involving some big difficulties, basically needing the active signals to be identified and preferably to be subject to manipulation for experiment. (And on top of that we are also stuck with the commonly taught fake-Newton very wrongly claimed to back Cartesian push-physics though he really backed William Gilbert's action-at-distance signal-response physics ?)

William Gilbert demonstrating a science experiment to Queen Elizabeth I in 1598 ;

While there should be freely available some English version or at least a translatable online version of Gilbert's two major publications, his 1600 'De Magnete' and his 1651 "De Mundo Nostro Sublunari Philosophia Nova" ('A Philosophy of our Sub-lunar World, or A New Science of everything under the moon'), somehow this seems not fully the case. We will try to put more of it online on this website soon, but for now our Gilbert sections have only English extracts of old translations, the full new English De Magnete and full online Latin versions or links - including good translatable Latin online De Magnete and De Mundo.

(Gilbert's 1600 'De Magnete' was published in Queen Elizabeth I's reign in Latin, and two English translations long after in Queen Victoria's reign in 1893 New York and 1900 London, and its newer improved English translation published in Queen Elizabeth II's reign in 2015 London. The most interesting scientist to read is almost certainly William Gilbert, as he really believed in science and really preached science with passion at a time when people only preached religion. Many find reading Einstein tricky, and also reading William Gilbert's 1600 'On The Magnet...' ?! You should maybe try reading Gilbert at least three times, as the three English translations in date order, noting what you thought you chiefly understood each time.) William Gilbert's chief work and certainly the world's first real science book, his Latin 1600 'De Magnete ...' has the best English translation as 'On the Magnet ...' It is a must for any Phisophy of Science, or for any History of Science, and indeed anyone preparing for any science degree should first study this.

See 'William Gilbert forgotten genius' 2003 by David Tilley and Stephen Pumfrey. And for a summary of a 'Gilbert-Newton' view of gravity and like forces see Attraction Theory of Physics. And see an interesting English translation of the 1556 Latin 'De Re Metallica' by Georgius Agricola, On the Subject of Metals, referenced by Gilbert.

In nature atomic fusion in stars relies on gravity and maybe heat, but as a related technology would probably have to rely on magnetism and maybe heat - but heat being causal here is actually unproven, though atomic fusion certainly produces heat, and a cold fusion technology may be possible using maybe some attraction only ? While atomic fission seems to chiefly concern the atom nucleus and so be nuclear, there seems little reason to think that the same applies to atomic fusion. And perhaps of some interest Gilbert's strongly anti-philosophising/reasoning and pro-experiment/experience position was maybe reflected around 1670 in 'Satire Against Reason And Mankind' by John Wilmot Earl of Rochester though that interesting work was perhaps just anti-religion, anti-government, anti-science and anti-social ?

Tell a friend about this website simply,
and they will thank you for showing them the best on the important ba_sics of the science of William Gilbert;
Type friends email address here   ...   Then click to tell your friend
NOTE : You can use this with confidence as we do not share and do not store this information at all.

OR if you like this site you could maybe make a donation ;

It will help with site development, and just possibly with some key basic physics experiments long planned but never afforded.
[PS. and you may perhaps help make history for science ?]
(The fictional time-travel and multi-universe type ideas of modern physics theory have long totally discouraged certain lines of physics experiment despite there being strong reasons to believe them to be very promising if not essential lines of experiment. Some such lines of experiment considered here identified as early as the 1960s seem still to have had no work done on them and there is maybe not much more time here for this. Science funding both government and private unfortunately now all goes to basically safe standard mainstream science, and no money at all goes to any really innovative risky science though that might pay a thousand times greater.)

You can do a good search of this website below ;

[Search] on this site www.new-science-theory.com, with Google.

Or do a search of the web better with DuckDuckGo - Type web search then Enter

If you have any view or suggestion on the content of this site, please contact :- New Science Theory
Vincent Wilmot 166 Freeman Street Grimsby Lincolnshire DN32 7AT.

And do listen to this video (may need Right-click Play) ;

© new-science-theory.com, 2022 - taking care with your privacy, see New Science Theory HOME.

William Gilbert - De Magnete

Homepage . William Gilbert . Rene Descartes . Isaac Newton . Albert Einstein .... De Magnete + ... General Image Theory

The two English translations of William Gilbert's 'De Magnete', produced hundreds of years later in Queen Victoria's reign in 1893 New York and 1900 London, are poor science translations of a rather poorly written masterpiece, but 2015 saw a better new version currently free online On The Magnet, or an A4 book or ebook at ebook or with New Science Theory - already commended by a physics professor as the best translation of De Magnete. Below are key extracts from the USA 1893 P.F.Mottelay translation and more here. Machine translations offered for convenience, also give poor science translation, but Lancaster University UK does have a good online translatable version of the full original Latin 1600 De Magnete. Minor bits of physics were published earlier but this the first substantial physics book was banned or had Book 6 removed in many catholic countries at the time.
And as a sequel to this amazing book, we hopefully await an English translation by Dr Stephen Pumfrey and Dr Ian Stewart of Gilbert's other posthumously published 1651 Latin work "De Mundo Nostro Sublunari Philosophia Nova" (A New Sublunar World Philosophy, or A New Theory of Everything Under the Moon, or A New TOE) with Gilbert's apparently intended title "Physiologiae Nova Contra Aristotelem" (A New Science Against Aristotle). To quote Steve Pumfrey, Lancaster University science historian, "Gilbert's uniqueness in both natural philosophy and cosmology stems from his conviction that he had empirical proof of his theory of active matter." in 'Cambridge Scientific Minds' CUP 2002. (their still keenly awaited translation was initially planned for 2005, but they did inspire the new 2015 translation of De Magnete)

De Magnete basically says that it is new science written chiefly for the more intelligent discerning reader, and it does include much relatively useless entertaining chat not intended to be taken seriously. So it is like a university lecture that is really intended for the benefit of the very best students with good science, while trying to hold the attention of all of the students with some chat. The new experiments in it are certainly intended to be noted and studied, but so also are some of its basic physic theory ideas. Here we are chiefly concerned with the theory.

William Gilbert's 'De Magnete' English translation by P.F.Mottelay New York 1893.

From 'De Magnete' Book 1, Chapter III :

"But inasmuch as the spherical form, which, too, is the most perfect, agrees best with the earth, which is a globe, and also is the form best suited for experimental uses, therefore we purpose to give our principal demonstrations with the aid of a globe-shaped loadstone, as being the best and the most fitting. Take then a strong loadstone, solid, of convenient size, uniform, hard, without flaw; on a lathe, such as is used in turning crystals and some precious stones, or on any like instrument (as the nature and toughness of the stone may require, for often it is worked only with difficulty), give the loadstone the form of a ball. The stone thus prepared is a true homogenous off-spring of the earth and is of the same shape, having got from art the orbicular form that nature in the beginning gave to the earth, the common mother; and it is a natural little body endowed with a multitude of properties whereby many abstruse and unheeded truths of philosophy, hid in deplorable darkness, may be more readily brought to the knowledge of mankind. To this round stone we give the name microge, or Terrella (earthkin, little earth)."

and,

"The terrella sends its force abroad in all directions, according to its energy and its quality. But whenever iron or other magnetic body of suitable size happens within its sphere of influence it is attracted; yet the nearer it is to the loadstone the greater the force with which it is borne toward it."

Of course Gilbert does discuss his theory ideas in various parts of his works often using different terms capable of different interpretation and translation - physics did not yet have an accepted technical jargon then, so that eg Gilbert himself had to invent some terms like 'electricity'. In another bit of Latin innovation, he coined a term for mutually-attracting bodies coming together as 'coition' instead of 'attraction' - but, unlike his new 'electricity', that term did not catch-on in physics.

The latin term 'effluvia', meaning approximately 'emissions', was used by many before and after Gilbert but often in quite different and in some cases very unscientific theories. Hence some supporters of ancient greek Atomism used 'effluvia' as proposed emissions of particles said to push bodies about - including an early theory of magnetism in which magnetic particle effluvia from magnets were supposed to push away the air between a magnet and a piece of iron so that the resulting vacuum sucked iron to magnets. Descartes' physics involved such particle effluvia, and gasses and smells were also often called effluvia. Others have used 'effluvia' with a different sense, as either emissions of energy or of 'soul' or 'spirit' that left one body and if entering another body energised, enlivened or motivated it.

In most of these uses, the proposed 'effluvia' directly caused actions in bodies. Gilbert's physics theory was quite different in involving a variety of effluvia some of which he reasoned were probably particles and some not - and his effluvia signal emissions did not directly cause any actions but acted as signals to bodies receiving them and bodies themselves responded automatically as information response robots. Later such gravity signals were called 'emitted spirits' by Newton and Gilbert maybe should have invented a new term for his effluvia signals, but a term that covered a thing being both an automatic emission and acting as a received automatic signal did not exist then (and in English now might be something more like 'natural emission signals' ?) - making the understanding and translating of Gilbert physics with its robot-matter difficult. He is clear about the working of his magnetic effluvia and electric effluvia but is less clear about gravity and somewhat confusingly also uses the term effluvia for gasses with no such action. Uniquely his physics theory's ultimate atomic particles are basically nanorobots as the basis of all physics - including electricity, magnetism and gravity.

NOTE. Gilbert's effluvia signal emission explains gravitational and electric charge attraction decreasing as the square of the distance from a body, as his effluvia signal emissions spread and dilute evenly and the surface of spheres increases as the square of their radius. Inverse square force necessarily follows from any theory involving emissions of particles or of waves, excepting possibly when traveling through a medium (eg gas, liquid or solid) when losses might be expected to involve actual attenuation being somewhat greater than the square of the distance. Hence such forces, like light, following the inverse square law over astronomical distances would seem to involve either 0% interaction, 100% propagation and/or no medium ? (magnetism is a somewhat more complex effect that does not simply follow the inverse square rule anyway).

Non-emission physics theories, like Maxwell's field theory and Einstein's continuum theory, include inverse square action perhaps non-necessarily and even arbitrarily ? Also in a Gilbert type theory a constant signal-response time, a signal saturation level and/or a maximum response level might replace Einstein's perhaps anomalous constant velocity of light ?

Collision push-theories of forces like gravity are assumed to work something like 'billiards averaged' - where the typical collision is glancing-collision where a ball from one direction collides causing another ball to move away at some angle, but the average being exactly head-on causing the other ball to move away in the same direction though happening much less often. However, signal response systems may always respond precisely to the directionality of incoming signals - as some plants and animals respond to a light source, moving directly towards or directly away or eg spiraling towards like moths. Of course individual 'force events' may perhaps never be detectable, only average responses ?

When a beam of light hits a sheet of glass, a wave theory or a particle theory may seem to require that the light be entirely reflected or entirely refracted - but in fact at least normally some of BOTH happens. While either light theory can be elaborated to explain such double-happening, it seems maybe simpler to take it as not being down to either type of mechanical contact but as down to marginal attraction/repulsion responses Gilbert-Newton theory fashion ? Of course Gilbert, Descartes, Newton and Einstein all supported determinist theories where if you know the full details then any event will involve single determinate outcomes though a multi-event event might have multiple single determinate outcomes. They all rejected probabilistic or indeterminate events in physics as being 'uninformed' or 'inadequately experimented' events only. Yet for some kinds of 'probabilistic' events mathematical laws have been produced that some see as giving an alternative type of. or elaboration of, physics theory.

'De Magnete' page 155 :-  - Click image to enlarge, or to get click-enlarging image.

The London 1900 S.P.Thompson english translation of De Magnete was a very impressive book, a giant red hardback measuring about 18 inches by 12 inches and 6 inches thick and a great weight, limited to only 250 copies. Its fine illustrations and artwork also made that 'De Magnete' a work of art as well as an outstanding work of science. The times have no doubt gone when young students could study such a very impressive science book to maybe more than match religions best holy books, but regrettably it was still a rather poor science translation of Gilbert's odd 1600 Latin original which was maybe a little less impressive. Eg Gilbert's use of 'we' can seem to oddly vary from like the 'royal we' as replacing 'I' or 'God and I' or 'my government and I', to the 'generic we' as replacing 'you and I' or 'all people' in eg 'we must hence conclude', and maybe intended to be more humble or less self-promoting than frequent use of 'I' ? For Thompson's english translation of De Magnete see Thompson's De Magnete. De Magnete seems to show that Gilbert believed in a God that does not interfere in the normal working of the universe and so not impact the scientific study of the universe. For more on translating Gilbert's Latin see Translating Gilbert.

Gilbert's 'De Mundo....'

De Magnete was published in non-Catholic 1600 England in some limited number of copies before the death in 1603 of both Gilbert and a somewhat sympathetic Queen Elizabeth. (Two further pirate editions were printed in Gdansk in Poland in 1626 and 1633.) Yet prior to 1600 it seems Gilbert was rightly afraid to publish his ideas on science, astronomy and gravity and that, apparently aided by suppression by Sir Francis Bacon, ensured that it was nearly 50 years after his death before they were somewhat more published in his De Mundo in Latin in a non-Catholic Holland. It is not clear if the 'De Mundo' that we have is a complete or accurate reflection of the writing that Gilbert left on his death, but nothing else seems to be available. See eg De Mundo.

De Mundo showed among other things that Gilbert concluded that there must be some force natural to planetary bodies, which was proportional to their mass, mutually attractive and decreased with distance. An attractive force that was emitted from the sun making planets orbit it, that was emitted from the Earth making the moon orbit it, and that was emitted from the moon making Earth tides. Basically just what astronomy needed.

He did not link that planetary force specifically either with magnetism or with earths gravity, though assuming several different types of forces and saying objects weight consists only in their responding to attractions from another body like the earth or other planetary body. Gilbert assumed that non-iron matter is unaffected by magnetic attraction, but it does produce and respond to the other gravitational and electrical 'magnetical' attractions. So in assigning planetary bodies attractions proportional to their masses he was postulating not planetary magnetism effects but a planetary gravity, though without specifically linking that to terrestrial gravity as Newton later did so successfully. Of course, having also studied magnetic and electric forces, Gilbert was well aware that multiple kinds of forces existed so that it was reasonable to think that planetary attraction not amenable to experiment may be a different kind of long-range attraction force. For his ideas that forces are needed to change bodies motions and do so in inverse proportion to their mass, see De Magnete Book 5 Chapter 12, Book 6 Chapter 3 and Book 6 Chapter 5.

The version of De Mundo published was not specifically approved by Gilbert and included some sections that may have been mere 'musings'. It came from preliminary draft manuscripts and gave his signal attraction physics as applying much more fully and widely than De Magnete indicated, to include stuff like planet and universe motions, Earth tides and weather effects and probable chemistry and medicine effects. And physics does undoubtedly actually have such wide effects. Gilbert's attraction physics necessarily includes signal emission, signal transmission, signal reception and signal response, possibly subject to some affects by the environs giving variation in some physical signal forces, but the published De Mundo did not go further into his physics effluvium signal mechanism details than De Magnete.

Kepler certainly did learn of these astronomy ideas of Gilbert, as least in general from De Magnete and possibly something of the then still unpublished

De Mundo. He did acknowledge Gilbert but developed an unworkable greek Atomism based mechanical-field push modification as his own theory (akin to the later Descartes fluid-ether vortex theory) which he wrongly thought better than Gilbert's theory. Newton later disproved Kepler's theory and proved that planetary attractions were the same attraction force as Earth gravity, though modern physics does still assume that there are different types of attraction forces including some short-range atomic or nuclear forces.

You are welcome to **link** to any page on this site, eg www.new-science-theory.com/william-gilbert-de-magnete.php

OR maybe make a small donation ;

PayPal Donate.

(it will help with site development, and just possibly with some experiments long planned but never afforded.)

If you have any view or suggestion on the content of this site, please contact :- New Science Theory
Vincent Wilmot 166 Freeman Street Grimsby Lincolnshire DN32 7AT.

© **new-science-theory.com, 2021** - taking care with your privacy, see New Science Theory HOME.

William Gilbert - *De Magnete, De Mundo and selected extracts*

Homepage . William Gilbert . Rene Descartes . Isaac Newton . Albert Einstein .... Gilbert's De Magnete .... General Image Theory

William Gilbert's 'De Magnete' was written about 1583 when there was little if any science and anything half scientific risked imprisonment or execution, so its publication was delayed 'almost 18 years' till 1600 and many copies were sold with Book 6 backing Earth rotation and Copernicus cut out. Its two standard english translations were not done till about 300 years later, but even the translators admitted were very problematic. Some well selected extracts from the P.Fleury Mottelay translation can be read below, and the full S.P.Thompson translation at De Magnete. Or see the better improved 2015 English translation currently free online On The Magnet also in print as A4 book or ebook or combined with New Science Theory.

Gilbert's 'active matter' physics with its robot atoms emitting and responding to signals is very unlike other physics theories, and Newton very successfully used its 'attraction physics' basics so it may still merit some consideration. It was certainly bare-bones theory, needing some addition, but Gilbert certainly believed that it could provably explain magnetism, electricity, gravity and all the basics of the physical universe. And Newton seems to have privately agreed. Gilbert's second draft work De Mundo was not published till 1651 but tried to deal with the theory more in expanding on its astronomy, tides, weather and chemistry implications. Showing how his Magnetical or Attraction Physics could be a successful signal-response Theory Of Everything.

William Gilbert's 'De Magnete' - P.Fleury Mottelay translation extracts.

That electric and other attractions are responses to signals and are not any kind of pushing.
(Book 2.2 pp.89-92 on rubbed-amber static electricity)

And that amber does not attract the air is thus proved : take a very slender wax candle giving a very small clear flame ; bring a broad flat piece of amber or jet, carefully prepared and rubbed thoroughly, within a couple of fingers' distance from it ; now an amber that will attract bodies from a considerable radius will cause no motion in the flame, though such motion would be inevitable if the air were moving, for the flame would follow the current of air. The amber attracts from as far as the effluvia are sent out; but as the body comes nearer the amber its motion is quickened, the forces pulling it being stronger, as is the case also in magnetic bodies and in all natural motion ; and the motion is not due to rarefaction of the air or to an action of the air impelling the body to take the vacated place ; for in that case the body would be pulled but not held, since, at first, approaching bodies would even be repelled just as the air itself would be: yet in fact the air is not in the least repelled even at the instant that the rubbed amber is brought near after very rapid friction. An effluvium is exhaled by the amber.......... A breath, then, proceeding from a body that is a concretion of moisture or aqueous fluid, reaches the body that is to be attracted, and as soon as it is reached it is united to the attracting electric; and a body in touch with another body by the peculiar radiation of effluvia makes of the two one: united, the two come into most intimate harmony, and that is what is meant by attraction. This unity is, according to Pythagoras, the principle, through participation, in which a thing is said to be one. For as no action can be performed by matter save by contact, these electric bodies do not appear to touch, but of necessity something is given out from the one to the other to come into close contact therewith, and be a cause of incitation to it.

and later (Book 2.2 pp.96-97)

The effluvia spread in all directions........hold and take up straws, chaff, twigs, till their force is spent or vanishes; and then these small bodies, being set free again, are attracted by the earth itself and fall to the ground. The difference (distinction) between electric and magnetic bodies is this: all magnetic bodies come together by their joint forces (mutual strength); electric bodies attract the electric only, and the body attracted undergoes no modification through its own native force, but is drawn freely under impulsion in the ratio of its matter (composition). Bodies are attracted to electrics in a right line toward the centre of electricity: a loadstone approaches another loadstone on a line perpendicular to the circumference only at the poles, elsewhere obliquely and transversely, and adheres at the same angles. The electric motion is the motion of coacervation of matter ; the magnetic is that of arrangement and order. The matter of the earth's globe is brought together and held together by itself electrically. The earth's globe is directed and revolves magnetically; it both coheres and, to the end it may be solid, is in its interior fast joined.

Magnetism is by speed of light or faster signals with some effective signal range or distance. (Book 2.7 pp.123-124)

The magnetic force is given out in all directions around the body; around the terrella it is given out spherically; around loadstones of other shapes unevenly and less regularly. But the sphere of influence does not persist, nor is the force that is diffused through the air permanent or essential; the loadstone simply excites magnetic bodies situate at convenient distance. And as light - so opticians tell us - arrives instantly in the same way, with far greater instantaneousness, the magnetic energy is present within the limits of its forces; and because its act is far more subtle than light, and it does not accord with non-magnetic bodies, it has no relations with air, water, or other non-magnetic body; neither does it act on magnetic bodies by means of forces that rush upon them with any motion whatever, but being present solicits bodies that are in amicable relations to itself. And as a light impinges on whatever confronts it, so does the loadstone impinge upon a magnetic body and excites it. And as light does not remain in the atmosphere above the vapors and effluvia nor is reflected back by those spaces, so the magnetic ray is caught neither in air nor in water. The forms of things are in an instant taken in by the eye or by glasses; so does the magnetic force seize magnetic bodies. In the absence of light bodies and reflecting bodies, the forms of objects are neither apprehended nor reflected ; so, too, in the absence of magnetic objects neither is the magnetic force imbibed nor is it again given back to the magnetic body. But herein does the magnetic energy surpass light, - that it is not hindered by any dense or opaque body, but goes out freely and diffuses its force every whither.

Magnetism and gravity involve control signals that are more penetrating than electric charge signals. (Book 2.16 pp.135-136)

On the other hand, in all the bodies that have a material cause of attraction (eg. amber, jet, sulphur) action is hindered by interposition of a body (as paper, leaves, glass etc.). and the way is obstructed and blocked so that that which is exhaled cannot reach the light body that is to be attracted. But coition and movement of the earth and the loadstone, though corporeal hinderances be interposed, are shown also in the efficiencies of other chief bodies that possess the primary form. The moon, more than the rest of the heavenly bodies, is in accord with the inner parts of the earth because of her nearness and her likeness of form. The moon causes the movement of the waters and the tides of ocean ; makes the seashore to be covered and again exposed twice between the time she passes a given point of the heavens and reaches it again in the earth's daily rotation : this movement of the waters is produced and the seas rise and fall no less when the moon is below the horizon and in the nethermost heavens, than when she is high above the horizon. Thus the whole mass of the earth, when the moon is beneath the earth, does not prevent the action of the moon; and thus in certain positions of the heavens, when the moon is beneath the horizon, the seas nearest to our countries are moved, and, being stirred by the lunar power (though not struck by rays nor illumined by light), they rise, approach with great impetus, and recede. Of the reason of this we will treat elsewhere : suffice it here just to have touched the threshold of the question. Hence, here on earth, naught can be held aloof from the magnetic control of the earth and the loadstone, and all magnetic bodies are brought into orderly array by the supreme terrene form, and loadstone and iron sympathize with loadstone though solid bodies stand between.

Magnetism involves signals similar to light. (Book 5.11 pp307)

As in many other demonstrations, so in this most indisputable diagram of the forces magnetical effused by the form, we grasp the true efficient cause. And this (the form), though it is subject to none of our senses and is therefore less perceptible to the intellect, now appears manifest and visible before our very eyes through this formal act, which proceeds from it as light proceeds from a source of light.

Bodies respond to magnetic signals automatically and not by temperament (Book 2.3 pp.102)

For of what use can temperament be in magnetic movements that are calculable, definite, constant, comparable to the movements of the stars

Bodies need no senses or thoughts to respond to magnetic signals (Book 5.12 pp.311-312)

The human soul uses reason, sees many things, investigates many more ; but, however well equipped, it gets light and the beginnings of knowledge from the outer senses, as from beyond a barrier - hence the very many ignorances and foolishnesses whereby our judgments and our life-actions are confused, so that few or none do rightly and duly order their acts. But the earth's magnetic force and the formate soul or animate form of the globes, that are without senses, but without error and without the injuries of ills and diseases, exert an unending action, quick, definite, constant, directive, motive, imperant, harmonious, through the whole mass of matter ......Yet these movements in nature's founts are not produced by thoughts or reasonings or conjectures, like human acts, which are contingent, imperfect, and indeterminate, but connate in them are reason, knowledge, science, judgement, whence proceed acts positive and definite from the very foundations and beginnings of the world

Planets rotate and orbit in response to signals from the Sun (Book 6.4 pp.333-334)

The earth therefore rotates, and by a certain law of necessity, and by an energy that is innate, manifest, conspicuous, revolves in a circle toward the sun; through this motion it shares in the solar energies and influences; and its verticity holds it in this motion lest it stray into every region of the sky. The sun (chief inciter of action in nature), as he causes the planets to advance in their courses, so, too, doth bring about this revolution of the globe by sending forth the energies of his spheres - his light being effused..... So the earth seeks and seeks the sun again, turns from him, follows him, by her wondrous magnetical energy. ..... And such are the movements in the rest of the planets, the motion and light of other bodies especially urging ....Thus each of the moving globes has circular motion, either in a great circular orbit or on its own axis or in both ways.

Bodies mutually attract in proportion to their mass (De Mundo ...)

"The force which emanates from the moon reaches to the earth, and, in like manner, the 'magnetical virtue' of the earth pervades the region of the moon: both correspond and conspire by the joint action of both, according to a proportion and conformity of motions, but the earth has more effect in consequence of its superior mass ; the earth attracts and repels the moon, and the moon, within certain limits, the earth ; NOT so as to make the bodies come together as magnetic bodies do, but so that they may go on in a continuous course."

(There is now a good Lancaster University translateable online version of the original Latin 'De Mundo' at - De Mundo)

Translating Gilbert's 'De Magnete' and 'De Mundo'.

At school myself having English as a first language and moving to concentrating on science, the other languages that I was taught were Gaelic, French, Scientific German, Scientific Russian, Scientific Greek and Scientific Latin.

Translating fiction literature must prioritise an attempt to conserve writing style as well as general meaning, and for old literature this will often involve conserving the flavour of the period in which that fiction literature was written. But translating original scientific work has to prioritise conserving its science meaning, so that writing style and period flavour must then be very much a secondary concern.

The two late translations of Gilbert's Latin De Magnete were unfortunately done more as translations of fiction literature, losing much of the science meaning. Even the title is poorly translated as in De Magnete's 'physiologia' being translated as 'philosophy' or 'physiology' when it should translate more accurately as 'natural science' or 'science'. Gilbert noted in its preface that he was assigning new specific scientific meanings to some words, and one of the chief words of his science is the word 'effluvium' or plural 'effluvia'. The only use of the word effluvium in science today is as meaning 'waste emission', but Gilbert certainly never used it with that meaning but used it with either the general meaning as 'emission' or with a science meaning in his 'magnetical' science as 'signal emission'. This term did have a range of uses in Gilbert's time, but is simply not translated in either the Mottelay or Thompson translations, though in translating Gilbert's science for today 'effluvium' certainly needs to be translated appropriately as maybe 'emission' or 'emission (signal emission)' since postulated natural magnetic, electric and gravitational signal emissions is clearly what Gilbert uses 'effluvium' to signify in his physics. But he was fully aware of and rather confusingly sometimes also used 'effluvia' with its other more general meaning, as in 'grosser effluvia' versus his physical forces 'rarer effluvia' though by that he was undoubtedly as elsewhere more dishinguishing 'corporeal' versus 'non-corporeal' emissions or 'particulate' versus 'energy' emissions.

In his De Magnete 'Definitions of Terms' Gilbert does not include his term 'effluvia' nor some of his other basic physics terms, taking their contextual uses as sufficiently explaining them. That Gilbert's is a non-push signal physics is clearly shown in his referring to magnetic action avoiding the more popular saying that magnets 'magnetisare' (or magnetize) for saying they 'excitatum' (or elicit response) - and Newton later did the same. Gilbert's physics likewise favours mutual action and more general actionlike or causelike terms over the single-actor Cartesian pushlike or forcelike terms. And when Gilbert's science refers to corporeal effluvia and non-corporeal effluvia it clearly means particle signal emissions and non-particle energy signal emissions. Again the two standard translations done to date fail to give clearly the intended science meanings.

It is generally possible for someone to use different words to have the same meaning, as 'big' and 'large', though this is generally less likely in science terminology. It is also generally possible for someone to use the same word to have different meanings in different settings, as in Optics 'light or dark' and in Mechanics 'light or heavy'. Gilbert's De Magnete used several Latin terms that could translate as 'magnetic' or 'magnetical', but it seems clear that it is better translated simply with 'magnetic' when involving actual magnets or magnetism and as 'magnetical' for a broader meaning of 'magnetism-related' or 'magnetic-like' as 'magnetic or electric or gravitational' - as to meaning 'attraction-physics-related' or 'remote-control-physics-related' or 'signal-physics-related'. Hence the only recorded English known of Gilbert is one letter to William Barlow that included calling mathematician Giovanni Francesco Sagredo (1571-1620) "a great magnetical man" though Sagredo seemingly at most aided Galileo and Sarpi in their replicating Gilbert's magnetic experiments but had studied De Magnete, and chiefly was a friend and maybe patron of the young Galileo for a time. (and whom Galileo made one of his characters in his 'Dialogues concerning Two New Sciences',1632) And my copy of Sister Suzanne Kelly's facsimile Latin later Gilbert 'De Mundo' shows continued likewise Gilbert terminology use. One clear example of its use as 'magnetic-like' is seen in the last De Mundo quote given above here, which many wrongly took as Gilbert claiming that Earth's tides and planetary orbits were caused by magnetism. Gilbert stated that he disapproved of using the term 'attraction' in physics as at the time it was commonly taken as signifying a simple pulling or pushing while his experiments showed more at work. The two earlier English translations of Gilbert's 'De Magnete' translate almost all of Gilbert's different Latin terms akin to 'magnetic' or 'magnetical' unfortunately the same, Mottelay almost all as 'magnetic' and Thompson almost all as 'magnetick'.

Gilbert coined the science term Electricity which stuck in physics and so needs no translation, but some of his terms like 'coition' did not stick and so had no physics meaning but still were not translated. Gilbert's 'coacervationis' got translated as the meaningless 'coacervation' when it should better be 'aggregation', and his 'coition' should maybe be 'coition (mutual attraction aggregation)'. De Magnete at times shortens the phrase translated as 'orbe of virtue' to just 'orbe' perhaps better translated as 'sphere of action (signal range)' and just 'range'. And he also confusingly used the term 'versorium' both for the magnetised Compass (a magnetism indicator) and also for his own invention the non-magnetised Electroscope (an electricity indicator) not clearly distinguished in translations. Gilbert used various Latin terms for Philosophy, Natural Philosophy and Science but too often his 'Natural Philosophy' is translated as 'Philosophy' and at times his 'Philosophy' really means 'Science' or 'Theory' as in 'Magnetic Philosophy'.

Gilbert's science is much concerned with forces and their effects on the motion of masses, and especially on remote-action forces - Magnetic, Electric and Gravitational in that order. He used 'moles' to mean 'mass' and 'mole gravata' to mean 'inertial mass' or 'what resists motion change' as did Kepler, but this is sometimes wrongly translated as eg 'bulk' or 'volume'. But certainly Gilbert did use some other words of his time problematically, like maybe form, anima and spirit, sometimes maybe as new science terms but sometimes maybe with their philosophy meanings or common meanings, so translation of some terms in Gilbert's works must remain uncertain. Hence the word 'form' in Gilbert often means simply 'shape', and in noting that physical forces are spherical he is making a real scientific point (in some respects perhaps mistakenly). But he did at times use 'form' as some at the time used 'spirit' as basically meaning energy and for his science he used 'form' more like modern electricity uses 'charge' as what determines the type of force (positive or negative) that a body has but for Gilbert distinguishing the magnetic, electric and gravitational forces. So bodies can have a magnetic form, an electric form or/and a gravitational form allowing them to emit and respond to the appropriate force signals - but Gilbert himself saw this applying to electricity only in some lesser way. This use is perhaps more in line with Aristotle usage, where an objects 'Form' was basically how its matter is organised or shaped to make it the kind of object that it is and its typical activity. (and a persons 'Soul' to Aristotle was basically the 'Form' of the person). Hence in his 1605 Advancement of Learning, Francis Bacon wrote "When we speak of forms, we mean nothing else but those laws and determinations of the pure act which sets in order and constitutes a simple nature. The form of heat and the law of heat are the same thing."

If early science Latin could be tricky, so also could early science English. Hence in England in William Gilbert's time the English term 'attraction' to most implied a push-physics action but by Isaac Newton's time the English term 'attraction' to most implied a Gilbert-physics action and Newton himself said that he used it to cover either type of action. Because some in his day used the term 'attract' to basically mean 'pull' as he noted, Gilbert also additionally used other terms besides 'attract' like 'allure', 'incite', 'excite' and his own 'coition' to clarify that his physics involved bodies responses to emitted signals and did not involve any mechanical pullings (or pushings). But this maybe did not so greatly clarify for every reader.

While the major physical forces seem to act spherically and produce only rectilinear motion towards or away from their centre of force, Gilbert noted that magnetism has opposite poles or verticity and produces orientation or rotation motions also which in a no-drag vacuum might give a persistent spin. It may be of some interest that only the sphere and the disc can have motion without it changing the space location that they occupy, and that uniform rotation/spin motion cannot itself be distinguished from rest by an external observer if the parts of the sphere or disc are not distinguishable. But Descartes and other physics theory largely proceeded to ignore non-rectilinear motion, though any snooker player could see ignoring spin as a big mistake. Gilbert has to date been almost impossible to study for any modern physics student or physicist, and does really need some much improved science translation.

Isaac Newton's Principia also suffered some similar Latin translation problems, especially in many places where he refers to Gilbertian attraction physics. Hence his Principia use of the term 'virtus' in Definition V11 was translated reasonably by Andrew Motte in 1729 as the science term 'force'. But Newton's use of the same term 'virtus' in Book 2 Section V Scholium was translated less reasonably for 1729 by Motte as the term 'virtue' which in science was later displaced by the term 'power', both strictly meaning the ability to generate a force. But both Newton and Gilbert did at times stretch their use of Latin. Where Gilbert's term 'effluvia' has to date always been untranslated remaining 'effluvia', Newton's use of a Latin equivalent in relation to gravity has always been translated but as the non-science term 'spirits emitted' rather than a more scientifically meaningful 'energy emissions' or 'signal emissions'. Yet it is very clear that Newton actually meant by it non-particle emissions that give power to or elicit responses from other bodies.

Gilbert's 'De Mundo'.

Galileo owned and studied a copy of Gilbert's De Magnete, though possibly without its Book 6, received from an Italian philosophy professor. But before its publication, something at least of some of his De Mundo was also known to at least Thomas Harriot and Francis Bacon - perhaps directly or indirectly through two manuscripts added to Prince Henry's library between 1603 and 1608 with one seemingly later transferred to the King's Library of the British Museum. It is unclear if the two manuscripts were similar or were both the full De Mundo like the current King's Library version. It seems that Francis Bacon (favoured by king James the First above Gilbert from 1603) must have got one of the Prince Henry's library copies, partial or complete, and only on his death in 1626 it passed to Sir William Boswell who (maybe using also some other manuscript) got it published in Holland in 1651. Between 1603 and 1651 England was politically much less settled than it was under Elizabeth. It is known that Thomas Harriot told Johannes Kepler about Gilbert's as then unpublished De Mundo manuscript in a 1608 letter and that Kepler requested a copy of it - though it is not known if he got a copy.

In 1965 Sister Suzanne Kelly got published in Amsterdam a facsimile of Gilbert's 1651-published De Mundo, being a good facsimile of an original copy of that held by Amsterdam University. An unusual case of a Catholic Nun helping keep alive a good bit of physics theory science, after hobby geneticist Gregor Mendel a Catholic Monk unusually helped found the science of genetics that was to underpin evolution theory science - early scientists of course suffered extreme oppression especially from the Catholic church and governments that it controlled. Suzanne Kelly's study of the De Mundo unpublished manuscript held by the King's Library of the British Museum led her to conclude that its Latin differs substantially from the published book Latin - but she could not establish the significance of the differences and the British Museum version remains unpublished still. The differences in Latin noted by Suzanne Kelly maybe suggest different translators into Latin of the same writing by Gilbert in English ? She also published a separate 142 page commentary on her facimile Gilbert's De Mundo, sometimes called Volume 2 of her De Mundo, and see her De Mundo Origins. (Much of the writings left by early English scientists William Gilbert and Isaac Newton are still not 'open science', either still being entirely unpublished or being text only published in image form and not published as more useable text.)

Gilbert's posthumous 1651 De Mundo has in 2022 still been published in Latin only, though Dr Stephen Pumfrey and Dr Ian Stewart have for some time planned translating it into English. They have now put a good translateable online version of a London University copy of the 1965 De Mundo facsimile at - De Mundo.

Some energy questions.

When some assign eg 'gravitational potential energy' to a body, it may be asked what is the gravitational potential energy of a stationary body half way between two planets of equal mass ? Or what happens to the energy of the body if it is moved closer to one of those planets ?

In any gravity theory where bodies are accelerated by external particle momentum being added to them (as from Descartes ether motion) or by external field energy being added to them (as from Maxwell/Einstein fields) them assigning 'gravitational potential energy' to bodies seems purely notional and not actually existing in the body but only existing notionally. An external energy source might not then add such energy as kinetic energy, not itself lose any energy to the body - but do so only when actual acceleration work is done on the body.

But in a gravity theory where bodies accelerate themselves as by emitting or converting part of their own mass in response to signals (as a William Gilbert style active-matter theory) then gravitational potential energy would actually does exist in the body itself ? Then actual gravitational acceleration would involve a body losing some energy to its environs, unless signals also triggered some endothermic reaction drawing energy from the environs ?

So if total energy is given by $E=mc^2$ then body mass when a body is gravitationally accelerated, should maybe slightly increase with an external energy theory but slightly decrease with an active-matter theory ?

What exactly are signals ?

Basically, physical signals are any physical properties of (or physical emissions from) an entity that some other entity can in any manner respond to and so can act as forces - and originating entities can be termed emitters and affected entities can be termed receivers or detectors. An attraction signal physics necessarily includes signal emission, signal transmission, signal reception and signal response, possibly subject to some affects by the environs giving variation in some physical signal forces. Emitters that have mental abilities may send intentional signals that may be termed messages, and receivers that have mental abilities can view signals as being intentional messages or as being unintentional information or data.

Any signal emitter or receiver that has mental abilities may also be able to produce and respond to non-physical or mental signals (as 'ideas'), and produce or respond to physical signal representations of such (as ie 'words').

William Gilbert's signal physics theory is concerned with only physical emission signals, chiefly physical force signals, though it could maybe be extended to deal with more than that. If light was a physical force signal then Gilbert-Newton physics would be as concerned with light's emission and detector responses to light as with light itself. If physical objects can respond to gravity signals then they are physical observers of gravity, but only if an object has mental ability can it be a mental observer. Einstein's relativity physics of course failed to consider physical observers at all, maybe making it only a mental relativity physics.

It can be taken that a Gilbert-Newton physical observer uses itself as its only reference frame.
A programmed mental observer may be programmed to use any one reference frame, or be programmed to conditionally choose from some set of reference frames.
A free-will mental Einstein observer can itself choose to use any alternative reference frames.
(But it may be difficult to distinguish some conditional choosing from free-will, and so some programmed observers from free-will observers.)

NOTE. Electrically charged bodies, in addition to producing charge signals or fields, seem to produce electromagnetic radiation when absolutely accelerated but not when only relatively accelerated. Hence charged-body electromagnetic radiation seems to be produced only by acceleration in the reference frame of the particle itself - which seems predicted properly only in a Gilbert-Newton signal physics theory or a valid image-theory of such theory. If some radiation emission events are caused by prior radiation reception events then their preferred frames of reference may be linked. It seems necessary to conclude that every physical signal radiation event has a unique reference frame, and this could be called the Reference Frame Exclusion Principle ?

You are welcome to link to any page on this site, eg www.new-science-theory.com/william-gilbert-de-magnete-2.php

IF you like this site then . Bookmark

If you have any view or suggestion on the content of this site, please contact :- New Science Theory
Vincent Wilmot 166 Freeman Street Grimsby Lincolnshire DN32 7AT.

© new-science-theory.com, 2022 - taking care with your privacy, see New Science Theory HOME.

Galileo Galilei - *and his mechanics and motion*

HOME .... William Gilbert . Rene Descartes . Isaac Newton . Albert Einstein .... Nikola Tesla .... General Image Theory

Galileo Galilei (1564-1642), was a good early astronomer and in physics mainly experimented in mechanics and gravity and basically invented the science use of the telescope. Some of his publications were banned and he was put under strong religious and legal requirement to restrict how his later publications presented, and from his Catholic Inquisition 1633 trial was put under house arrest till his death aged 77. Galileo did numbers of publishings or semi-publishings, often to attract rich patrons, but no significant science till after 1600. (The rich in Italy then enhanced themselves by funding promising artists, alchemists or scientists if now fund pet-homes, institutions or big pharma.) Galileo's main published works were his 1632 astronomy 'Dialogue concerning the Two Chief World Systems' and his 1638 mechanics 'Discourses(or Dialogues) concerning Two New Sciences' both in ancient-Greek philosophy dialogue style, the latter with some experimental proofs and some sections more a Euclid/Newton style. Galileo published little theory and at times took a Newton black-box position as in saying the cause of gravity was not of urgent importance, maybe from fear of persecution or from failure to produce any satisfying gravity theory from the simple greek-Atomist push-physics theory he adopted - but he like many early scientists saw experiment as more important. Galileo was an early outstanding experimental scientist and an early astronomer, yet studying chiefly gravity he wrongly rejected William Gilbert's magnetism-electricity based attraction physics theory and moon-attraction tides theory as his own very wrong mechanical-push tides theory showed Galileo was really worse than useless on physics theory - but his big Catholic Inquisition trial did give him much more publicity.

'Dialogues concerning Two New Sciences' (1638), Translated by H Crew and A De Salvio

Galileo stated that he himself had not sought the publishing of this his major work on mechanics and motion. It largely deals with mechanical strength under gravity, motion generally and especially motion under gravity including motion on inclined planes, projectile motion and pendulum motion. Its dialogue style (between Salviati, Sagredo and Simplicio) made the presentation of its science somewhat more difficult for readers, but partly suited the nasty religious and legal requirement that Galileo was under to publish his ideas as 'only ideas'. Some key extracts follow;

So on motion, Page 154 ;

"*Uniform Motion.*
In dealing with steady or uniform motion, we need a single definition which I give as follows.
DEFINITION. By steady or uniform motion, I mean one in which the distances traversed by the moving particle during any equal intervals of time, are themselves equal."

And Page 169 ;

"*Sagrado.*
... it would seem that up to the present we have established the definition of uniformly accelerated motion, which is expressed as follows ;
A motion is said to be equally or uniformly accelerated when, starting from rest, its momentum receives equal increments in equal times."

And Page 215 ;

"... Furthermore we may remark that any velocity once imparted to a moving body will be rigidly maintained as long as the external causes of acceleration or retardation are removed..."

And expressing a Newton-like blackbox view on different theories of gravity, Pages 166-167 ;

"*Salviati.*
The present does not seem to be the proper time to investigate the cause of the acceleration of natural motion concerning which various opinions have been expressed by various philosophers, some explaining it by attraction to the centre, others to repulsion between the very small parts of the body, while still others attribute it to a certain stress in the surrounding medium which closes in behind the body and drives it from one of its positions to another. Now all these fantasies, and others too, ought to be examined ; but it is not really worth while. At present it is the purpose of our author merely to demonstrate and to investigate some of the properties of accelerated motion (whatever the cause of this acceleration may be) - meaning thereby a motion, such that the speed goes on increasing after departure from rest, in simple proportionality to the time...."

Of course Galileo basically took Earth's gravity as constant since its strength varies very little over moderate distances from Earth's surface, despite Gilbert having repeatedly noted that the strength of forces including magnetic and electric forces decreased with distance from their source.

And supporting the existence of a vacuum, Page 81 ;

"... in the previous experiment we weighed the air in vacuum and not in air or other medium."

Galileo claimed to have a Universal Law of Gravitation covering both terrestrial gravity and the motion of planets which he was afraid to discuss. But this looks more an aspiration than a reality, as he seems not to have considered gravitational force as decreasing with distance from its source, which was central to Newton's later Universal Law of Gravitation and had been considered earlier by others and was demonstrated earlier by Gilbert for magnetic and electrical forces at least. But Earth's gravitational force does not decrease much if the highest you test from is the top of a Pisa tower, and worse was his testing gravity by the acceleration it produced on bodies using a gravity clock to measure it. He used a water version of the sand hour-glass or egg-timer - but if gravity was weaker and actually produced less acceleration, then his gravity clock would run proportionately slower so that the acceleration, and gravity strength, would appear constant. Clearly a clock must be independent of the event it is measuring, so Galileo should maybe have used iron filings and a magnet horizontally for a magnetic clock - he is known to have certainly have been acquainted with magnetism although the copy of Gilbert's De Magnete that Galileo studied may have excluded Book 6, where Gilbert included a basically correct theory of Earth's tides (detailed later in his De Mundo), as Galileo then put much effort into producing his own quite incorrect mechanical theory of tides. And many kinds of clock are of course possible as using astronomical, physical, chemical, biological or possibly even mental processes of determinate duration, and time measurement was perhaps also an issue even later for Einstein ?

Yet on Pages 261-262 ;

"*Sagrado.*
(if each planet had started from rest at particular heights under gravity and) fall with naturally accelerated motion along a straight line, and were later to change the speed thus acquired into uniform motion, the size of its orbit and its period of revolution would be those actually observed.
Salviati.
I think I remember him having told me that he once made the computation and found a satisfactory correspondence with observation. But he did not wish to speak of it ..."

It seems that at least in two 1615 private letters, Galileo supported the basics of Gilbert's magnetical or attraction physics but feared that if he publicly supported such the Catholic church might burn him to death as it had done to Bruno in 1600. He guardedly admitted that he believed that 'the Sun can be described as the soul of the world and transmits a spirit all around that gives life and movement to all things'. (see Alberto Martinez, http://notevenpast.org/giordano-bruno-and-the-spirit-that-moves-the-earth/) But later in his polemical 'Il saggiatore' (The Assayer) 1623, Galileo was seemingly supporting greek Atomist mechanical push physics against Gilbert attraction physics in claiming that science should concern itself only with the size, shape and relative motion of objects - a clearly unreasonable narrow view but supported by some Jesuits and Rene Descartes and many others. Galileo praised Gilbert while opposing his attraction physics with no actual disproofs or even discussion being provided, and seemed to use bits of trickery at times especially to try to avoid the scary anti-heresy and anti-science Catholic Inquisition. Some may have felt that attacking a Protestant scientist might help placate the then very scary Catholic church. But early science was competitive as with Galileo refusing to help Kepler obtain a telescope (see Philip Ball https://aeon.co/essays/science-is-becoming-a-cult-of-hi-tech-instruments) and also unreasonably rejected Kepler's proof that planet orbits are elliptical and are not circles. He may possibly also have pressured Kepler to drop his early support of William Gilbert's action-at-distance signal-response physics. And Galileo in his 1623 'The Assayer' also publicly condemned the German astronomer Simon Marius wrongly as being a plagiarist, but also as being a Protestant and not a good Catholic ! And he did also note that 'He who slings the most mud wins !'.

While the Catholic church strongly opposed Galileo's post-1609 moving-Earth astronomy, they did back Galileo's push-physics against William Gilbert's action-at-distance physics. But that may have been chiefly rather because of Gilbert's published 1600 support of a moving-Earth astronomy. Galileo's 'invention' of the telescope (really more its improvement and use in astronomy) maybe led him to seeing astronomy as more important than physics, and to wanting to advance that quickly despite astronomical evidence at the time being to many less convincing than experiment evidence. But beginning his telescope studies Galileo basically in 1610 abandoned Experimental Physics for Observational Astronomy. He produced an awful wrong mechanical push theory of Earth tides as later did Descartes, and Kepler produced a mechanical-field push theory of Earth tides - but all were easily disproved later by Newton who correctly developed Gilbert's better attraction theory of Earth tides (useable field or continuum mechanical push explanations of tides may well be possible but seem hard to find). Kepler, but not Galileo, correctly had gravity decrease as the square of the distance from its source and a better mathematics of planet motion within his own pseudo-Gilbert physics. But Galileo's motion under gravity experiments did basically show how planet ellipse type motion in nature could derive from linear motion. He was just not himself very strong on such theory. And 2022 still sees some of Galileo's writings not yet translated from their Italian to English, though he did publish some of his work in Latin like his 1610 Sidereus Nuncius.

Science and churches.

Early science in Europe faced sometimes fierce opposition from churches that often dominated governments, with Bruno being burnt at the stake in catholic Rome in 1600 (the year that Gilbert after much hesitation finally published his work in a then only slightly less intolerant protestant England under Queen Elizabeth who however died just months before Gilbert's death in 1603). But churches generally preferred to control dissent and science more often by reasoning and by nasty threats and less often by extreme nasty action. The catholic church executing Bruno in 1600 acted as a strong threat to all dissidents and emerging scientists, and its Jesuit Order which was founded to counter non-Catholic ideas soon began pushing an acceptable greek-Atomist physics that was basically taken up as a self-perpetuating mainstream physics that Newton considered 'prejudice'. The catholic church also ensured that Galileo faced legal restriction and also pressured Descartes strongly, maybe encouraging them both to attack protestant Gilbert's physics. In Galileo's 1633 Catholic Inquisition trial the one book condemned besides his 1632 'Dialogue Concerning the Two Chief World Systems' was Gilbert's De Magnete, though the Catholic Church had already banned that and some other books including Copernicus and some Galileo. When Copernicus published his astronomy it was a fairly good theory backed by only a little evidence and, though Galileo and others did then add some further supporting evidence, it was only later that Newton was to tie together such evidence with a stronger theory which really proved it. The catholic Jesuit 'scholars' who had strongly opposed William Gilbert were also involved in Galileo's catholic inquisition trial really directed perhaps more at Copernicus.

It has been noted that "Soon after 1600, when William Gilbert published his famous book on magnetism, a copy was given to Galileo by an Italian professor of natural philosophy at Padua, probably Cesare Cremonini ... Galileo remarked that the professor seemed afraid that Gilbert's work might infect the other books on his shelves (or that Galileo believed he wanted to free his library of its contagion)." Galileo was probably joking in a somewhat insulting manner, but the more theory inclined Cremonini may well have thought De Magnete of more interest to the experimentalist Galileo and may have given him friendly advise that the catholic inquisition might investigate anyone caught owning it. Cremonini himself was an atheist Aristotlean and had survived several investigations by the catholic inquisition and he did have a healthy fear of them and considered limited accommodation necessary. Galileo generally thought likewise. see 'Galileo at Work, His Scientific Biography' Stillman Drake 2003 p.62-63 - or - http://www-spof.gsfc.nasa.gov/earthmag/demagadd.htm Around 1602, on studying De Magnete, Galileo did some magnetic experiments with Sarpi and Sagredo but produced nothing new on magnetism beyond what Gilbert had, as later Newton also did with basicly similar result. Hence basically Galileo did Gilbert's magnetic experiments but did not commit to interpreting the results as Gilbert had, while later Newton did Gilbert's magnetic experiments and accepted the Gilbertian interpretation of the results as Gilbert had but also allowed of some unspecified Galileo-Descartes push-physics explanation being maybe possible. Experiments can be interpreted differently, and Galileo was probably less well acquainted with Gilbert's interpretation than Newton. It seems that Galileo's copy of De Magnete contained just one note by Galileo but many underlinings relating to Gilbert's experiments, indicating that he closely studied Gilbert's experiments but not Gilbert's theory as indeed many at the time probably did. He certainly was not the 'honest reader' to which Gilbert's preface addressed De Magnete, which was written chiefly as new science for the more intelligent reader but to also entertain and hold all readers and so included a deal of entertaining chat not intended to be taken seriously, so it was easy to take his experiments as his only serious science though Gilbert undoubtedly hoped that much of his basic physics theory would also be taken seriously. It is not clear if Galileo and friends also replicated Gilbert's electrical experiments though some others did, but it was to be only after the 1820 Oersted discovery that electric currents also produce magnetism that real further progress was made in magnetism or electromagnetism, and certainly Galileo entirely failed to incorporate any of it into his physics as later Einstein was also to fail to incorporate electromagnetism into his relativity physics.

Churches being inclined to the view of God as the cause of everything, led many early scientists to omitting causal theory from their science. Yet churches generally really saw God as at least largely an imaginary unseen that science would never be able to fully prove or disprove. And the churches were in fact less concerned about what caused day to day events, than with science contradicting some particular words in their holy books. So their real opposition was to science claiming that the Earth is not the centre of the universe but is just one planet of several orbiting the Sun, and to science claiming that humans were not specially created but evolved from apes.

So early scientists even claiming that almost everything was caused by God, could still be in trouble. Of course Descartes basically did just get away with claiming that everything was caused by God AND that nothing was caused by God. But in the end physics survived, in a maybe highly prejudiced form, chiefly because the power of religion in Europe gradually weakened and science was increasingly seen as being of practical use - especially for war weapon development.

You are welcome to **link** to any page on this site, eg www.new-science-theory.com/johannes-kepler.php

IF you like this site then you could maybe make a donation ;
.
It will help with site development, and just possibly with some key basic physics experiments long planned but never afforded.
[PS. and you may perhaps help make history for science ?]
(The fictional time-travel and multi-universe type ideas of modern physics theory have long totally discouraged certain lines of physics experiment despite there being strong reasons to believe them to be very promising if not essential lines of experiment. Some such lines of experiment considered here identified as early as the 1960s seem still to have had no work done on them and there is maybe not much more time here for this. Science funding both government and private unfortunately now all goes to basically safe standard mainstream science, and no money at all goes to any really innovative risky science though that might pay a thousand times greater.)

© **new-science-theory.com, 2022** - taking care with your privacy, see New Science Theory HOME.

Johannes Kepler - his astronomy and physics

Homepage . William Gilbert . Rene Descartes . Isaac Newton . Albert Einstein .... Galileo Galilei .... General Image Theory
- Site Search at bottom v -

Johannes Kepler (1571-1630) was one of the best early astronomers and a very able mathematician. His published works, in Latin, included his 1596 'Precursor of Cosmographic Dissertations', his 1609 'New Aetiological Astronomy', his 1611 'Dioptrics', his 1619 'The Harmonies of the World' and his 1618-21 'Epitome of Copernican Astronomy' naming his 'giants' in the preface to Book 4 writing that he built his astronomy physics 'from the hypotheses of Copernicus, the observations of Tycho Brahe and the magnetical science of William Gilbert' - and he was also somewhat of a friend of Galileo. And nicely distinguishing hypotheses, observations and science.
Kepler's 1627 'Rudolphine Tables' allowed the positions of planets to be approximately computed and most importantly predicted, making Kepler the foremost astronomer of his time. His optics work was also useful. However, here we consider Kepler's theories for explaining his astronomy - first two mathematical fictions he never fully abandoned then (after briefly supporting Gilbert attraction physics) a third weak push-physics causal theory based on ancient greek Atomism that did not really explain it and was later actually entirely disproved by Newton though without him claiming that and anyway most physicists of the time rejected Newton's physics till much later after textbooks had converted it into a Cartesian Newtonian physics that is still taught wrongly as being Newton's physics.

'Epitome of Copernican Astronomy' - Book 4 (1619), Translated by Charles Glenn Wallis

Kepler stated that this work of his was designed to serve as a supplement to Aristotle's 'On The Heavens', for these were times when Aristotle had commanding support from Christian churches, many governments, and most scholars. Though in 1600 England William Gilbert had been somewhat braver in dismissing Aristotle as irrelevant to science, it was a time when scientific thinking risked imprisonment or even execution. In reality both Gilbert and Kepler rejected substantial if different parts of Aristotle as some others were also doing at the time.

Kepler's early attempts at an 'astronomy science' were based on a view of the universe having been created by a God having chosen to create a musical universe or a mathematical universe along the lines of ancient-Greek Pythagorean ideas but based on 5 notes rather than 7 and on the 5 Platonic regular solid polyhedra. His early creationist physics were both basically mathematical logic attempts at a logical-universe astronomy. Early-Kepler physics involved a view of some godly mathematics being primary in the universe, so he tried producing a Geometry Mathematics Physics and a Music Mathematics Physics. Others like Euler supported similar physics theories, though attempts to produce a physics from mathematics seems to have given many dud theories. (Einstein and post-Einstein physics also seem to have produced mathematics-derived theories, with Wave Mathematics Physics and Quantum Mathematics Physics ?) But Kepler's pre-1600 mathematical-universe 'Mysterium Cosmographicum' showed that, unlike most others, Kepler seemingly saw mathematics as somehow 'physical'. This maybe reflected William Gilbert, unlike most others with their deterministic push-physics, seeing that at least some of the non-mechanical could also be deterministic ? Of course in 1600 Kepler studied William Gilbert and for a time at least concluded that the universe though maybe conceptually mathematical is NOT physically mathematical but is 'Experienceal', though Kepler's physics then went from temporarily a Gilbert 'signal response' experience to a Galileo greek-Atomist Descartes 'touch' experience possibly on some pressure from Galileo. For a time Kepler like Galileo supported the astronomy of Copernicus only as a push-physics astronomy, though Gilbert and Newton realised that only an attraction physics could actually explain planetary motions. But between Kepler's different published astronomy theories he did seem to have had a brief period of favouring Gilbert's attraction physics which may possibly have really helped him develop his astronomy maths, as it later helped Newton develop his. This he indicated in a June/July 1600 essay dedicated to Archduke Ferdinand in which he supported a somewhat vague force-based hypothesis of lunar motion - 'In Terra inest Virtus, quae Lunam ciet' ('There is an influence in the Earth that triggers the Moon to move') per Max Caspar's Kepler, p.110. Einstein's later physics however involved another form of 'push-physics without push' that was really without any explanation though with useful maths.

Kepler's actual discovery of his Three Laws of Planetary Motion seems to have developed chiefly from his study of the orbit of Mars under Tycho Brahe, continued beyond Brahe's 1601 death still using Brahe's detailed observational data. He seemed to have developed his elliptical planetary orbits and his first two laws by 1605 though publication was delayed till 1609, and he seems to have developed his 3rd law later in 1618 and published that in 1619, see Kepler. But early science could be very competitive as with Galileo refusing to help Kepler obtain a telescope (see Philip Ball) and also unreasonably rejecting Kepler's proof that planet orbits are elliptical and not circles. Kepler was however somewhat helped by one or two catholic Jesuits (see Thonyc). But it is known that Kepler was also studying William Gilbert's 1600 'De Magnete' early in this period, and it claiming proof of causal action-at-distance forces operating between solar system bodies that might give non-circular planetary orbits. And Kepler did himself publicly acknowledge the major contributions of both Brahe and Gilbert to his astronomy ideas. His first two laws of planetary motion were developed from 1600 to 1605 as (1.) The orbits of the planets are ellipses with the Sun at a focus and (2.) Planets sweep out equal areas in equal times in orbiting the sun. They were published in his 1609 Astronomia Nova (A New Astronomy) with the third law added for his 1618-21 Epitome of Copernican Astronomy as (3.) The square of the orbit time for a planet is proportional to the cube of its distance from the sun. But these were only good to an approximation and perhaps without real explanation though all involved planet motion differences depending on distances from the Sun.

Kepler's 1618-21 'Epitome of Copernican Astronomy' used many of Gilbert's magnetism phenomena and illustrations (but not always correctly) for an astronomy that explained the motion of planets and moons. Gilbert had developed a non-magnetism signal response attractive force astronomy theory in outline, of which Kepler had at least some general knowledge that he did not acknowledge and he maybe did not include this non-magnetism force, but Kepler instead presented (as though it was Gilbert's theory) a theory of his own involving the claim that planets, moons and stars were rotating magnets and their magnetism maintained planet orbits with forcefield-thread vortexes acting mechanically. He also seems to have failed to inform Galileo of Gilbert's basically correct theory of Earth's tides when Galileo was working on a quite wrong mechanical theory, and he included a useless but popular argument against a 'mind' parody version of Gilbert's actual signal-response or attraction theory. He had studied Gilbert's De Magnete whose Book 6 gave only a basic statement of Gilbert's conclusion that tides are chiefly caused by the gravitational attraction of the Moon, and may or may not have also studied a manuscript De Mundo where Gilbert detailed his tides theory. However, Kepler never fully abandoned his God-mathematic astronomy physics, in 1621 publishing an expanded second edition of his 1596 Mysterium Cosmographicum.

Yet in Kepler's 'Epitome of Copernican Astronomy' Part 3 section 3 ;

"But isn't it unbelievable that the celestial bodies should be certain huge magnets ?

Then read the philosophy of magnetism of the Englishman William Gilbert ; for in **that book**, although the author did not believe that the Earth moved among the stars, nevertheless he attributes a magnetic nature to it, by very many arguments, and he teaches that its magnetic threads or filaments extend in straight lines from south to north. Therefore it is by no means absurd that any one of the primary planets should be what one of the primary planets, namely the Earth, is."

Gilbert used his magnetic lines as metric aids only, and the nearest Kepler comes to putting any actual argument against Gilbert's signal theory (though maybe more appropriately directed against others who supported planets having minds) is in Part 2 section 3 ;

"the material globe would have no faculty of obeying or of moving itself." and "it is asked by what means the mind knows where the centre is, around which the orbit of the planet should be organised ; and how great the distance of the mind and its globe from that point is."

Of course magnets move in response to other magnets without having minds, eyes, or legs - and magnetic or gravity signals would have directionalities and strengths needing simple response to these only, making Kepler's argument useless against Gilbert's theory. Though clearly aware of inertia, Kepler imagined that the Sun must push the planets around their orbits and not just attract the planets. Hence Kepler puts his own perhaps much more problematic ether forcefield mechanical energy threads theory in Part 2 section 3 ;

"*Then does the Sun by the rotation of its body make the planets revolve ? And how can this be, since the Sun is without hands with which it may lay hold of the planet, which is such a great distance away, and by rotating may make the planet revolve with itself ?*
Instead of hands there is the virtue of the body, which is emitted in straight lines throughout the whole amplitude of the world, and which - because it is a form of the body - rotates along with the solar body like a very rapid vortex ; moving through the total amplitude of the circuit whatever magnitude it reaches to with equal speed ; and the Sun revolves in the narrowest space at the centre.

Can you make the thing clearer by some example ?
Indeed there comes to our assistance the attraction between the loadstone and the iron pointer, which has been magnetised by the loadstone and which gets magnetic force by rubbing. Turn the loadstone in the neighbourhood of the pointer ; the pointer will turn at the same time. Although the laying hold is of a different kind, nevertheless you see that not even here is there any bodily contact.

Then what takes place now by the Sun's rotating around its axis ?
Indubitably by the turning of the solar body the virtue too is turned, just as by the turning of a loadstone the attractive force of one part is transferred to different regions of the world. And since by means of that virtue of its body the Sun has laid hold of the planet, either attracting it or repelling it, or hesitating between the two, it makes the planet also revolve with it and together with the planet perhaps all the surrounding ether. Indeed, it retains them by attraction and repulsion ; and by retention it makes them revolve."

And in Part 2 section 2 ;

"*Whence do you prove that the matter of the celestial bodies resists its movers, and is overcome by them, as in a balance the weights are overcome by the motor faculty ?*
This is proved in the first place by the periodic times of the rotation of the single globes around their axes, as the terrestrial time of one day and the solar time of approximately twenty-five days. For if there were no inertia in the matter of the celestial globe - and this inertia is as it were a weight in the globe - there would be no need of a virtue in order to move the globe ; and if the least virtue for moving the globe were postulated , then there would be no reason why the globe should not revolve in an instant. But the revolutions of the globes take place in a fixed time, which is longer for one planet and shorter for another : hence it is apparent that the inertia of matter is not to the motor virtue in the ratio in which nothing is to something. Therefore the inertia is not nil, and thus there is some resistance of celestial matter.
Secondly, this same thing is proved by the revolution of the globes around the Sun - considering them generally. For one mover by one revolution of its own globe moves six globes, as we shall hear below. Wherefore if the globes did not have a natural resistance of a fixed proportion, there would be no reason why they should not follow exactly the whirling movement of their mover, and thus they would revolve with it in one and the same time. Now indeed all the globes go in the same direction as the mover with its whirling movement, nevertheless no globe fully attains the speed of its mover, and one follows another more slowly. Therefore they mingle the inertia of matter with the speed of the mover in a fixed proportion."

It is to be noted that none of the many magnetic experiments published in Gilbert's De Magnete indicated that a magnet could make others orbit around itself. Kepler's explanation theory did not have the experimental backing, from Gilbert or anybody else, that Kepler clearly considered stronger than the purely mathematical 'most perfect regular solid figures' and 'harmonic octave ratios' which he had earlier tried to use in Aristotelian fashion. Continuing with his magnetic threads theory in Part 2 section 3 ;

"*How is it possible that the virtue flowing from the body of the sun should be weaker in the greater interval at A than near the sun at E ? What weakens the virtue or makes it feeble?*
Because that virtue is corporeal and partakes of quantity: wherefore it can be dispersed and thinned out. Therefore since as much power is diffused throughout the very wide orbital circle of Saturn as is collected in the very narrow orbital circle of Mercury: therefore it is very thin throughout the parts of the orbital circle of Saturn, and hence it is very feeble; but it is most dense at Mercury and hence is very strong.

If it were a question of the body of the sun, I might grant to it this natural power of moving: but you draw out this material power from the body and place it without a subject in the very spacious ether. Doesn't this seem absurd ?
That it should not seem absurd is clear from the example of the loadstone, to which this same objection can be made. But in neither case is this force without a proportional subject. For in this way at the very source the subject of the natural faculty is the body of the sun, or the threads stretching out from the centre to its circumference; thus even in this very emanation, I think a rational distinction should be made between the immaterial form of the solar body, which flows as far as the planets and beyond, and its force or energy which actually lays hold of the planet and moves it - so that the form is the subject of the force, though it is not a body but an immaterial form of a body.

Could you give an example of this thing ?
There is a true example in the light and heat of the sun. There is no doubt but that just as the whole sun is luminous, so it is all on fire, and that on account of the density of its matter it should indeed be compared to a glowing mass of gold, or to anything else which may be denser. Now from that light of the sun there emanates and comes down to us a form which is not corporeal, not material, which we call the illumination or rays of the sun and which however is subject to dimensions and accidents. For it flows on straight lines and may be condensed or rarefied, and many indeed be cut by a mirror and by glass, namely, by reflection and refraction, as we are taught in Optics. Moreover, this form of the sun's light bears its heat with it; and in proportion to the greatness or smallness of the strength whereby it falls upon bodies which can be illuminated, it warms them to a greater or to a lesser extent. Therefore just as that form or illumination - which form we know with certainty to flow down from the light of the sun - is the subject of the heat-

giving faculty, which has similarly been extended from the sun, through a form; so too the solar body's immaterial form, come down as far as the planets, has as its companion the form of that energetic virtue in the solar body; and this form strives to unite like things to itself and to repel unlike.

What is the likeness between the form of light and the form of this prehensive virtue ?
There is a very close likeness in the genesis and conditions of both forms: the descent of each from the luminous body takes place instantaneously; each remains of average greatness and smallness without loss, is not taxed; nothing perishes in the journey from its source, nothing is scattered between the source and the illuminable or movable thing.
Therefore each is an immaterial outflow, not like the outflow of odours, which are conjoined to a decrease of the substance; not like the outflow of heat from a raging furnace, or anything similar, by which the spaces in between are filled. For this form is not anywhere except in the opposite and withstanding body; the form of the light on its opaque surface, but the form from the motor virtue in the total corporeality: but in the intermediate space between the sun and the surface, the form is not but has been. But if they were to meet the concave spherical surface of an opaque body, both solar forms would be scattered in that concavity together with all that abundance with which they have emanated from the body of the sun: in this way as much of the form would be in a wide and farther-away sphere of this sort as is in the narrow and nearer sphere. And since the ratio of convex spheres is the ratio of the squares of their diameters : therefore the form will be made weaker in unequal spheres in the ratio of the square of its distance. And again because circles have the same simple ratio as their diameters: therefore in longitude the form is weaker in the same ratio of its distance from its source."

Though Kepler's own theory differed from Gilbert's it was still often seen as basically an attraction by an unseen emission theory akin to Gilbert's - and some like Galileo and Descartes considered this somehow 'occult' and preferred to look for more simply mechanical if still unseen causation. Of course Kepler correctly followed Gilbert in having the Moon causing Earth tides, and additionally correctly had gravity decrease as the square of the distance from its source though incorrectly assigned by him to magnetism - basically using a Gilbert approach and logic. And while Kepler preferred his own mechanical adaptation of Gilbert's theory, he did at some points use an actual Gilbert signal attraction theory of gravity to aid explanation. He did have gravitational effluvia as sometimes material and sometimes immaterial, though he seemingly chiefly settled on an atomist Galileo-Descartes material push-physics mechanism and view of matter.

Kepler rightly judged Tycho Brahe's improved planet motion accuracy as important chiefly because lower accuracy wrongly supported circular orbits and a push-physics, while better accuracy more correctly supported ellipse orbits and really Gilbertian attraction physics. But Kepler's explanation of planetary motion was only of planets motion around the Sun, not involving the motion of moons around planets - though Gilbert did consider both. Kepler's theory of planet motion involved the Sun being the sole causal agent, and involved no mutual action - though mutual action was central to Gilbert. Kepler saw the Sun basically as having solid pushing magnetism extensions that transferred its own rotation to planet orbitings, and as also having magnetism elastic extensions that somewhat pulled planets towards itself. This was an explanation that could theoretically work for planet orbiting only, and nothing else, and it was far from realistic physics. It seems unlikely that Kepler actually developed his astronomy maths from that, and somewhat more likely that he really did so like Newton from the attraction physics that he sometimes adopted ? (See www-history.mcs.st-andrews.ac.uk/Extras/Keplers_laws.html and www-history.mcs.st-andrews.ac.uk/HistTopics/Kinematic_planetary_motion.html)

But at times Kepler did almost seem to perhaps get nearer to the truth, as in this quote from one of his letters.
"My aim is to say that the machinery of the heavens is not like a divine animal but like a clock, and that in it almost all the variety of motions is from one very simple magnetic force acting on bodies - as in the clock all motions are from a very simple weight." - Letter to J. G. Herwart von Hohenburg, 16 February 1605, KGW 15, 146. (From www-history.mcs.st-andrews.ac.uk/Quotations/Kepler.html) Rather later Einstein was to make claims about the precession of Mercury that disputed Kepler's laws of planetary motion but he put as disproofs of all of Newton and little effort was made to find alternative explanation such as eg the Sun's magnetism with its 11-year cycle.

Like most early scientists Kepler had to do paid work, in his case chiefly as an astrologer - and was a somewhat unconventional Christian, in his case an excommunicated Protestant. And the early Kepler to 1600 clearly strongly believed that the universe was designed by God. But he then studied William Gilbert and seems to have privately come to believe that the design of the universe might be independant of God. But he never publicly acknowledged that belief, as neither did Gilbert or Newton, because religion was just much too powerful throughout Europe then and it was just not safe to publicly doubt God if you did have any doubts.

Like Kepler's magnetic ether forcefield vortex theory of planetary orbits, Descartes' later simpler ether fluid vortex theory of planetary orbits was also disproved by Newton. Both were basically versions of greek-Atomist push-physics. Since Kepler had wrongly presented his explanation theory as being Gilbert's theory, some wrongly took its valid disproofs as being valid disproofs of Gilbert's theory - though Newton knew that he could not disprove the basics of Gilbert's theory and indeed developed that for his universal gravitation theory. (One thing this demonstrates is that a science theory can have a stronger maths but still have a weaker explanation logic.)

Kepler's general forcefield idea was basically reflected in Maxwell's forcefield theory and in Einstein continuum theory, but with these seemingly requiring that something non-mechanical, which cannot be pushed by objects, can be mechanical and push objects. And somehow selectively, when anything that can push should push anything that is pushable ? But 'field' and 'charge' type jargon, and ridiculous rubber-sheet 'analogies' maybe just hide selective push and other problems and avoids giving any actual testable mechanical push explanation ? Kepler claimed his emitted forcefield acted in a simple mechanical push manner on bodies, but maybe just required forms of emitted energy that have only little interaction with matter, but when they do interact can produce motion - maybe something like the photo-electric effect where atoms can emit a massive electron in response to an incoming little photon ? Field theories have involved various strange logics that generally are not clearly specified.

IF you like this site then you could maybe make a donation ;

It will help with site development, and just possibly with some key physics experiments long planned but never afforded.
[PS. and you may perhaps help make history for science ?]
(The fictional time-travel and multi-universe type ideas of modern physics theory have long totally discouraged certain lines of physics experiment despite there being strong reasons to believe them to be very promising if not essential lines of experiment. Some such lines of experiment considered here identified as early as the 1960s seem still to have had no work done on them and there is maybe not much more time here for this. Science funding both government and private unfortunately now all goes to basically safe standard mainstream science, and no money at all goes to any really innovative risky science though that might pay a thousand times greater.)

You can do a good search of this website below ;

Search on this site www.new-science-theory.com, with Google.

Or do a search of the web better with DuckDuckGo - Type web search then Enter

otherwise, if you have any view or suggestion on the content of this site, please contact :- New Science Theory
Vincent Wilmot 166 Freeman Street Grimsby Lincolnshire DN32 7AT.

You are welcome to **link** to any page on this site, eg www.new-science-theory.com/johannes-kepler.php

© **new-science-theory.com, 2022** - taking care with your privacy, see New Science Theory HOME.

Nikola Tesla - his remote-control science technologies

Homepage . William Gilbert . Rene Descartes . Isaac Newton . Albert Einstein .... Johannes Kepler.......... General Image Theory
- Site Search at bottom v -

Serbian-American inventor Nikola Tesla (1856–1943) was best known for his alternating-current electricity supply systems favoured by business and less for his remote-control physics that he rightly saw as more important and which only much later became key to modern life. Most people of the time could not see much further than the light bulb, though maybe some like Einstein claiming to be physicists should have seen the significance of Tesla's remote-physics experiments but it seems not. He did mostly work on electromagnetic technologies and published little on theory, though much of his work basically assumed the theory of William Gilbert for magnetism, electricity and gravity of action-at-distance signal-response remote-communication and remote-control physical forces, which was also assumed though not openly acknowledged in Newton's physics. Tesla was an exceptional inventor and experimental physicist. But though Tesla's time was also somewhat-younger-Einstein time and both worked in the USA at the same time, Einstein never backed Tesla and indeed many may have quite wrongly taken Einstein's physics as somehow disproving Tesla's confirmed remote-electric physics though Einstein certainly never himself attempted any actual disproof of action-at-distance physics. Yet in physics theoretical physicists have the loudest voices, and tend to dismiss actual experimental physicists as 'just engineers' as Tesla has wrongly been dismissed then and mostly still wrongly now.

The largely dismissed science of Nikola Tesla

Tesla was the son of an Eastern Orthodox priest and emigrated to the USA in 1884, where he worked for a short time in telephony and at Continental Edison in the new electric power industry. Then he got private finance to set up his own laboratories and companies in New York to develop a range of electrical and other devices. His alternating current induction motor and related patents, licensed by Westinghouse Electric in 1888, earned him a considerable amount of money for a short time and established him then as the world's leading electricity scientist of the time. While there were some working on simple-science direct-current electricity, Tesla developed his better trickier-science alternating-current electricity which was quickly widely adopted. Tesla conducted a significant range of experiments with different oscillators/generators, electrical discharge tubes, and early X-ray imaging. He also worked as some others did on radio communication and notably experiment on remote-power and built and demonstrated a first wireless remote-control boat, but his remote-control and remote-power were unfortunately rejected until many years after his death.

Throughout the 1890s, Tesla pursued his ideas for wireless remote-power lighting and worldwide wireless remote electric power distribution in his high-voltage, high-frequency power experiments. In 1893, he made pronouncements on the possibility of wireless communication with his devices. By around 1893 Tesla fully realised the greater significance of his remote-physics and was publicly demonstrating radion remote-control by 1898. So in his 1905 U.S. Patent 787,412 Tesla stated "By use of such a generator of stationary waves and receiving apparatus properly placed and adjusted in any other locality, however remote, it is practicable to transmit intelligible signals, or to control or actuate at will any one apparatus for many other important and valuable purposes." (See Tesla's Patent) But his earlier radio patent lost out to Marconi's 1896 London radio patent and it was Marconi who was given a Nobel prize in 1909. It is likely that Tesla's science would have prospered better had he like Marconi moved to a more attraction-physics England rather than to a more cut-throat America, however Edison had lured him to America and then quickly betrayed him ? Of course, like the world generally, England today has perhaps become quite intransigent on much science.

It seemed that Tesla had advanced too far ahead of American business, government and physicists who were basically still chiefly concerned with simpler street lighting. He was also no doubt mistaken in saying that he had received radio signals that might be from off-Earth people, though they were likely from natural atmospheric or natural off-Earth sources. Soon he was being ostracised and not backed or funded. Tesla tried to put some of his ideas to practical use in his unfinished Wardenclyffe Tower project, seemingly intended as an intercontinental wireless communication and power transmitter, but ran out of funding before he could complete it. Tesla's hopes for some of his technology really far exceeded its more limited if still potentially valuable practical uses some of which were only commercially adopted after his death many years later. And after Wardenclyffe he experimented with a number of other inventions in the 1910s and 1920s with varying degrees of success. Having spent most of his money and with much of his science dismissed, Tesla then lived in a series of New York hotels, leaving behind unpaid bills when he died in January 1943.

The early Nikola Tesla

Tesla later wrote that he first became interested in demonstrations of electricity by his high school physics professor, probably Martin Sekulić. Tesla noted that these demonstrations of this "mysterious phenomena" made him want "to know more of this wonderful force". But the young Tesla's being able to perform integral calculus in his head, prompted his teachers to think he was cheating. He then attended a technical university in Austria where he initially did well though clashing with a professor and ending by dropping out.

In 1881, Tesla moved to Hungary to work at the Budapest Telephone Exchange then under construction. But for a first few months he worked as a draftsman in the Central Telegraph Office instead until the Budapest Telephone Exchange became functional. Tesla was allocated its chief electrician position and made many improvements to the Exchange equipment. Then in 1882, Tesla got a job in Paris with the Continental Edison Company, and began working in what was then a new industry installing indoor incandescent electric lighting citywide. The company had a subdivision the Société Electrique Edison that Tesla worked at, in charge of installing the lighting system. There he gained a great deal of practical experience in electrical engineering and was soon designing and building improved versions of generating dynamos and motors. He was also sent to troubleshoot engineering problems at other Edison utilities being built around France and in Germany.

In 1884, Tesla got a move within Edison to the United States at the Edison Machine Works in New York with the task of building the large electric utility in that city. As in Paris, Tesla worked on troubleshooting installations and improving generators. One of the projects given to Tesla was to develop an arc lamp-based street lighting system. Arc lighting was the most popular type of street lighting but it required higher voltages than the then current Edison low-voltage incandescent system and was not put into production.

And within six months Tesla quit Edison and with the help of business investors patented an arc lighting system including an improved DC generator.

But his business investors soon dropped Tesla and he struggled for a while. But in 1886, Tesla found new business investors and developed an induction motor that ran on alternating current which was a power system format that was rapidly expanding in Europe and the United States because of its advantages in long-distance, high-voltage electricity transmission. His motor used polyphase current, which generated a rotating magnetic field to turn the motor, a principle that Tesla claimed to have conceived in 1882. This innovative electric motor was patented in May 1888 and was a simple self-starting design that did not need a commutator, thus avoiding sparking and the high maintenance of constantly servicing and replacing mechanical brushes. The money Tesla made from licensing his AC patents to Westinghouse Electric made him independently wealthy and gave him funds to pursue his own interests for a time.

In 1889, Tesla traveled to the Exposition Universelle in Paris and learned of Heinrich Hertz's experiments on electromagnetic radiation, including radio waves. Tesla found this new discovery "refreshing" and decided to explore it more fully in repeating, and then expanding on, these experiments. Tesla tried modifying a Ruhmkorff coil to create his own Tesla coil with an air gap instead of insulating material between the primary and secondary windings and an iron core that could be moved to different positions in or out of the coil. In 1891, aged 35, Tesla became a naturalized citizen of the United States and patented his Tesla coil.

From then Tesla experimented with transmitting power both by magnetic inductive and electric capacitive coupling using high AC voltages generated with his Tesla coil. He attempted to develop a wireless electric lighting system based on near-field either inductive or capacitive coupling and did a series of public demonstrations where he lit Geissler tubes and even incandescent light bulbs from across a stage. He spent most of the decade working on variations of this new form of remote-lighting, and in 1893 Tesla said that he was sure a system like his could eventually conduct "intelligible signals or perhaps even power to any distance without the use of wires".

An observer of a Tesla remote-power demonstration at the 1893 World's Columbian Exposition in Chicago noted: "two hard-rubber plates covered with tin foil were about fifteen feet apart and were terminals of wires leading from transformers. When the current was turned on, electric lamps or tubes which had no wires connected to them, but could lay on a table between the suspended plates or could be held in the hand in almost any part of the room, were made luminous. These were the same experiments and apparatus shown by Tesla in London about two years previous, "where they produced so much wonder and astonishment". And in 1894, Tesla was experimenting on what he referred to as radiant energy of "invisible" kinds being X-rays. Tesla wanted his science to do good, as in maybe providing free electric power, but rich USA science funders then wanted only to fund science that made them money and so wanted to fund only bad science or gold-making alchemy-science.

Tesla did like some other scientists have real funding problems and his experiment ideas often needed large funds, despite him like Newton and Gilbert not having to support a wife or children. Science theory is generally cheaper than experiment and often draws easier funding from governments or universities of which Tesla experiments struggled to attract. Yet science really rests on experiment and Tesla really believed in and practiced well confirmed and publicized physics experiments. Unfortunately that his science experiments maybe leaned more directly towards technologies may be in part because he felt that might better attract funding than 'undirected' experiment though that may maybe not always give the best science possible. He did try sometimes desperately hard to accommodate potential funders though often they did not have very good intentions.

Tesla and Radio Remote-Control

In 1898, Tesla demonstrated a boat operated by a coherer-based radio remote-control — which he called a "telautomaton" — to the public during an electrical exhibition at Madison Square Garden in New York. Some of the crowd that witnessed the demonstration made outrageous claims about the workings of the boat, such as magic, telepathy, and being piloted by a trained monkey hidden inside. There were others at the same time working on related issues like Marconi on radio remote-communication, though that action-at-distance was remote-communication rather than remote-control. But on remote-control the physics profession really failed to back Tesla, and indeed only reluctantly on remore-communication him or Marconi, though Tesla did try to sell his remote-control to the U.S. military for a radio-controlled torpedo, but they showed little interest and his radio remote-control remained a novelty until World War I and afterward, when a number of countries began using it in military programs. But increasing work on action-at-distance radio etcetera by Tesla and others did not prompt any real physics re-evaluation of action-at-distance physics theory by Einstein or other physicists which somehow very wrongly remained ridiculed by 'theorists' and insubstantially studied. The mass of physics theorists ignored physics experimentalists and ignored a mass of physics experiments that did not suit them, as earlier physics theorists had really ignored magnets and apples falling on heads, but this was early Einstein thought-physics time. Tesla tried promoting radio-remote-communication including claiming to have received radio signals from space, for which some called him crazy though later radio-astronomy became accepted. Tesla did try further demonstrating his "Teleautomatics" or radio remote-control to little effect then and to be largely ignored by the physics majority. But the remote-control science of Tesla did still become very popular many years later with universal remote-control TV, remote-control toys and much other remote-control becoming common from around the 1960s especially. His remote-power was not really implemented till the 2010s in the form of remote-chargers for different electric devices. But popular interest in Tesla has also held up well despite his unwarranted dismissal by a large majority of professional physicists at the time and perhaps his showing how science-funders can severely restrict science progress.

Of course Tesla's remote-science substantial dismissal did follow on from a majority of physicists long failing to properly consider William Gilbert's 1600 action-at-distance remote physics which was basically backed by Newton. And indeed the majority of philosophers doing likewise, other than maybe George Berkeley with his perhaps rather fanciful immaterialist signal-response philosophy?

Tesla from 1884 to 1943 and Einstein from 1933 to 1955 were both working in physics in America, but Einstein never backed Tesla or his physics. So how exactly did physics get away with ignoring Tesla's confirmed and publicized outstanding remote-distance-physics experiments - basically by lying and misinterpreting them as had also been done earlier with Newton's light experiments and with William Gilbert's magnetism experiments ? Really remote-control signal-response physics can be called Gilbert-Newton-Tesla physics, and still stands strongly against Galileo-Descartes-Einstein or Quantum Mechanics/String Theory style push-physics. Tesla's work on remote action-at-distance physics led to remote radio and tv technologies, remote phone and radioastronomy technologies and to other remote technologies. But the work of Einstein on his relativity physics has, even 100 years later, still produced little or no new technology, and certainly not the time-travel it implied. So today almost every person carries a remote-communication device and almost every home has at least one remote-control device - and nobody has time-travel. A science that produces little or nothing is surely little or nothing.

Tesla disputed especially Einstein's claimed speed of light limit and his claimed space-curvature (maybe not both equally validly) and in 1935, on his 79th birthday, he wrote that the Theory of Relativity was just "a mass of error and deceptive ideas violently opposed to the teachings of great men of science of the past and even to common sense. The theory wraps all these errors and fallacies and clothes them in magnificent mathematical garb which fascinates, dazzles and makes people blind to the underlying error. The theory is like a beggar clothed in purple whom ignorant people take for a king. Its exponents are very brilliant men, but they are metaphysicists rather than scientists." (NYT, 7/11/1935, p.23) Most modern 'theoretical physicists' are no doubt really metaphysicists and not physicists, as Tesla viewed Einstein and his 'thought-experiment' theory. Tesla himself seemed inclined to using multiple theories, as juggling bits of attraction-repulsion signal-response theory and force-field and ether theory. Einstein and Tesla

were no friends though when asked or pressed by Time Magazine, on Tesla's 75th birthday in 1931, Einstein did pen an open letter congratulating Tesla on his birthday and commending him as an engineer which no doubt annoyed Tesla. There has been a rumour that Einstein once said something nice about Tesla but there is no evidence for that and nobody in physics is known to have nominated Tesla for the Nobel prize though there was a malicious rumour that he was going to be awarded the 1915 Nobel Prize in Physics.

Einstein got the 1921 Nobel Prize in Physics "for his discovery of the law of the photoelectric effect", though the Photoelectric Effect was discovered in 1887 by Heinrich Hertz and in 1900 Max Planck had suggested that light was quantal in it before Einstein later in 1905 published his small theory paper interpreting the photoelectric effect as involving quantal light. This small work of his really achieved little or nothing. And a logical quantal light theory interpretation of the photoelectric effect might seem to be a particle-set response as requiring receipt of some set of particles within some time period. And key public experiments of Tesla that seemed to support action-at-distance physics were ignored by most physicists then, as still today and as some key experiments of William Gilbert were long ignored also. Most physicists treatment of Tesla was all generally similar to the way William Gilbert and Isaac Newton were also treated by most physicists in their time for backing action-at-distance signal-response physics. And Tesla was subject to some strong public personal criticism. But in technology production it is really action-at-distance physics that has won, though most physicists now still cannot accept the clear fact shown by our actual world and they are holding back physics and the world.

Some of his science technologies Tesla basically gave away, some others were basically stolen while others were ignored for many years till long after his death. And Tesla was certainly much attacked wrongly chiefly by competitors, while disgracefully mainstream science never really gave his science any support maybe in part because one or two of his science ideas then did not seem as useful as he imagined. Of course similar no doubt also applied to many other scientists to at least some extent over the years. And Tesla was quite widely known as wanting his science to do good for the world, maybe more than some other scientists. A conspiracy theory has persisted that one or more United States governments supressed some of Tesla's good science. Of course many conspiracy theories have been shown to be wrong, though a few may have been shown to be right. But people who have done wrong will generally try to hide the fact, so that in some cases the truth cannot be proven though it may be possible to make a reasonable guess at what was the truth as to for example who killed various princes. So it may not be possible to prove some historical theory right, but it may also not be possible to prove it wrong - some reasonable-guess theory may be true. And if Tesla may not have had one wrong done against him, he may have had some different wrong done against him. Wrong doers or course may be governments, churches or others. And some people or even a whole society can do wrong while wrongly believing that they are doing right. The History Channel investigative TV series 'The Tesla Files' being rerun now maybe inspires little confidence in being run rather like various other conspiracy TV series such as 'Ancient Aliens', and maybe deals more with engineering than with physics. So it may produce some interesting bits of information on Tesla, while missing chunks of perhaps more relevant information on him. But his later work did face claims that Tesla was crazy and his science wrong, and perhaps strangely were not opposed by Einstein or other physicists of the time. He followed Gilbert and Newton, in having his science technologies hijacked while his basic science ideas in line with theirs were very wrongly dismissed.

IF you like this site then you could maybe make a donation ;

It will help with site development, and just possibly with some key basic physics experiments long planned but never afforded.
[PS. and you may perhaps help make history for science ?]
(The fictional time-travel and multi-universe type ideas of modern physics theory have long totally discouraged certain lines of physics experiment despite there being strong reasons to believe them to be very promising if not essential lines of experiment. Some such lines of experiment considered here identified as early as the 1960s seem still to have had no work done on them and there is maybe not much more time here for this. Science funding both government and private unfortunately now all goes to basically safe standard mainstream science, and no money at all goes to any really innovative risky science though that might pay a thousand times greater.)

You can do a good search of this website below ;

Search on this site www.new-science-theory.com, with Google.
Or do a search of the web better with DuckDuckGo - Type web search then Enter

otherwise, if you have any view or suggestion on the content of this site, please contact :- New Science Theory
Vincent Wilmot 166 Freeman Street Grimsby Lincolnshire DN32 7AT.

You are welcome to **link** to any page on this site, eg www.new-science-theory.com/nikola-tesla.php

---

© new-science-theory.com, 2022 - taking care with your privacy, see New Science Theory HOME.

Rene Descartes - *mechanical push universe theory*

Homepage . William Gilbert . Isaac Newton . Albert Einstein ...... Descartes' Principles . Descartes' The World ....General Image Theory
- Site Search at bottom v -

Rene Descartes or Rene Des Cartes (1596-1650) was a philosopher and mathematician with a basically simple theory of the universe that many saw as in line with the emerging science of the time. He basically took mechanics and its push-action and made it a complete theory of a determinate-law universe composed of only matter and pieces of moving matter pushing other pieces of matter, and its only energy being the motion property of matter. In 'The World' his Cartesian physics hypothesised a fluid matter ether vortex motion pushing the planets around the sun, with other particle push theories for terrestrial gravity and for some magnetism in his conjecture-physics. (Descartes discussed his magnetism with Princess Elisabeth of Bohemia in the 1640s, with neither mentioning William Gilbert and both confining themselves to conjectures and no actual science.) For religious believers Descartes also posited a second God-determined immaterial spiritual energy universe with no physical connection to the material universe but only to the mind or soul of humans. Descartes did his science as a philosopher, he sat-and-thought and he sold his thinking as being 'logical-thinking' or 'necessarily-right or certain thinking' in line with his famed proposition 'I think, therefore I am'. Much of his certain Cartesian physics soon faced strong disproofs which were denied or ignored. Descartes' basic ideas were perhaps best put in his 1644 Principia Philosophiae (Principles of Philosophy), or an English Google Books version of his 'Discourse....'. His full greek-Atomist push physics theory in 'The World' was not published till 1664, after his death - see The World.

Descartes' science theory.

Descartes was primarily a logician who did much interesting work in philosophy and mathematics. He used logic rather than experiment in developing his new 'science', and his logic is maybe best known for his 'most certain' proposition "I think, therefore I am". He might perhaps logically have taken a Gilbert-like conclusion from that, that the universe certainly contained thinking things and did not certainly contain any non-thinking things or any God. However perversely Descartes' certain-logic went largely with the opposite simple Greek-Atomist push-physics theory conclusion Galileo had adopted - that most things in the universe are non-thinking, and that there was a God that alone with humans could think. He posited a separate spiritual-mental universe distinct from the physical universe and beyond the scope of science, which could suit supporters of religion and churches. And of course this Dualist science conclusion did seem to accord with common views of a stone not thinking, if not with Descartes' own certainty-logic. 'I think' seems to fit better with 'I observe' and with observers existing - and directly from that with things to think upon or observe from, such as signals regarding other objects existing. But 'I think' really gives less support to the view that no-thinking no-signals solid dead push objects exist, which perversely is the basis of Descartes' Cartesian physics theory making him perhaps a mere theoriser such as William Gilbert had railed against. But Descartes seems not to have studied Gilbert who had claimed that his experiments disproved dead-matter push-physics in showing that matter responds to magnetic, electric and gravitational signals somewhat in common with thinking or at least with a deterministic thinking. But unlike Gilbert, Descartes allowed of only free-will thinking and of no deterministic thinking and required all determinism to involve only Greek-Atomist pushings.

The basics of Ancient-Greek Atomism from around 500 BC by Leucippus and Democritus and later Epicurus around 300 BC, was for a random motion push-physics against some orderly thinking-atoms physics akin to Aristotle around 330 BC, as from Lucretius 'De rerum natura' around 50 BC,

Lucretius "De rerum natura" 1.1021-1028 (around 50 BC)	Lucretius "On the nature of things" 1.1021-1028 (translated by A.M. Esolen - 1995)
Nam certe neque consilio primordia rerum	For surely the atoms did not hold council, assigning
ordine se suo quaeque sagaci mente locarunt	order to each, flexing their keen minds with
nec quos quaeque darent motus pepigere profecto	questions of place and motion and who goes where.
sed quia multa modis multis mutata per omne	But shuffled and jumbled in many ways, in the course
ex infinito vexantur percita plagis,	of endless time they are buffeted, driven along,
omne genus motus et coetus experiundo	chancing upon all motions, combinations.
tandem deveniunt in talis dispositutas,	At last they fall into such an arrangement
qualibus haec rerum consistit summa creata	as would create this universe

Of course the Ancient-Greek argument of 'random push' vs 'ordered think' should have become an argument of Galileo-Descartes 'random push' vs Gilbert-Newton 'ordered response'. And actually the chief support for ancient-Greek Atomist push-physics in Europe before Galileo adopted it were some alchemists including Daniel Sennert. While along similar lines, some opponents of William Gilbert's 1600 action-at-distance signal-response physics falsely attacked it as supposedly requiring thinking atoms though it was a good new physics which involved responding atoms but certainly did not involve thinking atoms.

The 'Cartesian' push-physics of Descartes also seems to rest on a very doubtful view of the human senses, taking touch as being the only certain sense as supposedly being unique in not requiring sensory signals. Of course touch may seem less certain for liquids and very uncertain for gasses. Gilbert and Newton took all proper experiment or experience as equally valid for science. Should the assumed sensitivities of any observer be allowed to determine the validity of alternative physics theories anyway ? The issue arises for Descartes' physics, but perhaps applies also to some modern physics theories effectively taking sight as the only sure sense and light the only sure signal ?

George Berkeley's 'to be is to be perceived' philosophy concluding that non-thinking things did not exist, was no great challenge to Descartes matter push physics and no help to Gilbert signal physics since Berkeley somehow additionally concluded that signals informing thinking did not exist either. Berkeley was chiefly concerned with some 'non-causal non-physical thinking', while Gilbert was chiefly concerned with the significance of basic 'causal physical thinking' or natural responses to natural signals as in magnetic attraction etcetera. And a Descartes physics keeping science out of spiritual-mental matters was less a problem to religion than a Gilbert science that looked like allowing science to explain all including the mental and spiritual. Descartes science was confined to the merely technological, leaving religion to lord over the more important human and spiritual arenas. His science also required humans to be unique in the universe, unlike Gilbert's, and so was more acceptable to the Catholic church then.

Descartes basically saw dead as simpler than live, and that the simplest force was the push, leading him to follow the ancient Greek Atomists as had Galileo, requiring the physical universe to be basicaly dead and based only on the one simple push force though he did require an additional separate complex spiritual universe also. William Gilbert had concentrated on doing actual science experiments which to him proved that the physical universe actually was not quite so simple and fitted a signal-response physics with a number of different forces better while not needing it to explain life processes, and Isaac Newton was one of only a few physicists then who seemed to favour that. Later physics theories became perhaps rather more confused. Descartes push-physics basically followed earlier opposition to Gilbert's signal-response physics by Jesuit catholic Niccolo Cabeo in his 1629 'Philosophia Magnetica' which though generally Aristotlean basically followed the simple ancient greek Atomism as in explaining 'action-at-distance' by ether-push. In his supposedly logic-derived material universe theory, Descartes saw objects as mechanical only and animals also as only 'dead' mechanical clockwork robots, and the human body, senses and brain largely likewise - except that humans alone had soul/self-awareness like God. His mechanism for automatic reaction by animals (and the human body largely) to 'signals' was as to direct push forces - so light basically punches eye nerves. Descartes theory viewed all 'signals' (or Gilbert corporeal and non-corporeal 'effluvia' and Newton 'spirits emitted') as corporeal material particles that pushed sense organs mechanically and mechanically caused animal actions deterministically, so that animals reacted more as billiard balls to other billiard balls and less as thinking things or robots responding to information signals.

Descartes physics included a no-empty-space ether theory requiring that parts of a material ether tends to rotate in a vortex tending to rotate bodies in it as planets rotate and whose motion also causes some matter in it to move to its outside and that push some other matter towards its center as in terrestrial gravity. So he basicly had an invisible-vortex push theory for planetary gravity motion, supported by Christiaan Huygens though with a strong disproof by Newton that was not generally accepted, and for terrestrial gravity Descartes had a related but separate theory of celestial-particles moving away from the centre of the earth and so displacing and pushing-down masses in their path. His discriminate-push magnetism 'ethers' were also invisible material particles and physically pulled and pushed magnets, for which he had to postulate left and right handed corkscrew shaped particles working like corkscrews. A somewhat tricky idea needing exact alignments and with much experimental evidence against it. Any way you align a bit of iron it is equally attracted to a magnet, and magnets do not just attract and repel so Descartes could not explain compass-motion orientation at all. Descartes' universe was a mechanical ('wind-up') clockwork robot universe, with energy only as the property of matter being in motion and nothing other than God and human souls being non-material. His material universe was all matter with no empty space and with no separate energy besides the kinetic motion energy of bodies. His 'no empty space' was in line with Aristotle and Huygens but opposed the experimental evidence offered by Gilbert, Newton and others who supported non-corporeal energies or 'spirits' also existing - separately from matter and being maybe not visible but detectable by experimental science.

To Descartes the essential properties of bodies were only the absolute requirements that they must occupy some space and no two bodies could occupy the same space at the same time, so that any body motion contact involved pushing other bodies from the space they occupied. Such contact explains single-force pushings but not really any pullings, for pullings need vacuum or low pressure (which Descartes ether theory like most continuum theories cannot really allow) or corkscrew action push-pullings which are unlikely when multiple forces can do pullings even operating in the same space.

Descartes saw bodies as having different sizes or shapes, and their pushing motions explaining all physical behaviour including gravity, electricity and magnetism. One body could not penetrate another body, though a larger body might contain spaces that a smaller body could enter as especially might a thin fluid. Mass was simply the measure of the size and pushability of bodies. He had no real explanation of how bodies could differ in shape and fluidity if no attraction-type forces were involved - and so also no real explanation of conjoined-bodies pulling each other. Gilbert and Newton correctly saw this theory as always requiring detectable effects like ether drag that could not be confirmed. Gilbert concluded that magnetism cannot work by push since magnets showed no effect on air or on candle flames, and Newton concluded that space had no push-ether or push-continuum since planet orbits show no significant slowing. Both supported space as being empty or non-material and allowed both corporeal matter bodies and non-corporeal force energy or 'spirit' bodies, and saw 'Mass' as a measure of objects gravity production and response.

Descartes' philosophical Logical Mechanical Universe science theory basically followed ancient greek Atomism and influenced many and basically still does. He made a major contribution to philosophy, and his basic science theory ideas have been adopted perhaps wrongly by the majority of physicists to date. Descartes produced 'laws of motion' that read almost the same as Newton's, though his motion examples are often about bodies being pushed by unseen ethers more like Aristotlian motion. Newton published a disproof of the part of Descartes' 'dead-matter' theory that involved ether vortex motion pushing planets around, but seems not to have taken that as essential to Descartes' physics and electro-magnetic field theory based on a modified Descartes ether idea became accepted by most physicists until the Mitchelson-Morley experiment of 1887 indicated that either the ether did not exist or ether motion did not exist, which Einstein agreed, though his spacetime continuum was ether-like if not a full replacement.

There was much support for Descartes ether push physics even after chunks of his theory were firmly disproved. Hence Russian mathematician Leonhard Euler (1707-1783) rejected the planet motion and Earth tide ideas of Descartes. But still Euler supported general Descartes ether push physics, as even in publishing a 'proof' of Descartes push-attraction ether corkscrew magnetism in 1744 - another piece of Descartes theory that was not long in being disproved. Euler was maybe just another example of great mathematics backing bad science - see www.math.dartmouth.edu/~euler/

Directly opposing William Gilbert, Rene Descartes believed in the certainty of rigorous logical reasoning though not merely mathematical reasoning, and that experience and experiment were of less certainty. He largely went with the Catholic Inquisition requirement for Galileo that his science be put as 'just ideas'. Descartes held that his was the best science possible, with logically imagined causation explaining the universe to the best extent possible though it might be impossible to establish the actual causes of phenomena like gravity. Hence on causation he states a neo-blackbox position in his Principles of Philosophy Part 3.CCIV ;

"That, touching the things which our senses do not perceive, it is sufficient to explain how they can be."

"I most freely concede this, and I have done all that was required, if the causes I have assigned are such that their effects accurately correspond to all the phenomena of nature, without determining whether it is by these or by others that they are actually produced. And it will be sufficient for use in life to know the causes thus imagined....."

Hence Descartes himself was maybe less fully committed to his push-physics than most of the physicists who supported it. And of course other physicists were soon producing real evidence that solid objects are not solid but are largely empty space with some perhaps-solid particles. So a billiard ball pushing another billiard ball may well be largely space 'pushing' space - with at most a very few particles contacting. So must the transfer of momentum from all the particles of one ball, to all the particles of the other ball, involve action-at-a-distance and not involve push-contact ? If most apparent contact is not contact, then any push physics has a problem and maybe needs mechanical ethers or particle emissions - and proving their solidity may not really ever be possible ? Contact requires a zero distance that is not measurable and so not provable like finite distances. If Descartes' push physics rested on touch, and Einstein's on vision, then maybe Gilbert/Newton response attraction physics alone having observers and signals within the physics was also the least dependent on particular human senses ? Descartes' physics also included solid push ethers, though some other push-physics theories do not.

Wave theory involving motions of material media became a significant part of Descartes physics, but from Einstein's time waves not involving motions of material media were postulated and were incompatible with Descartes physics. And determinate-law energy or 'spirit' that is not the motion property of matter yet affects the motion of matter, as in Gilbert-Newton attraction-physics forces and in 'field' forces, was also incompatible with Descartes physics.

In Descartes push-physics all physics energy is matter motion, and all matter motion is energy including uniform motion. In Gilbert-Newton attraction-physics all energy is either signal motion which is uniform motion, or is signal response motion which is acceleration motion and signals can be non-corporeal. And while in any Descartes-style physics all energy is basically absolute, any signal-response physics allows of at least some energy being basically relative. Hence all signal emission necessarily involves some motion of material or energy signals from an emitter relative to some receiver or observer that may themselves be in motion, and all responses to such signals as act as physical forces are relative matter accelerations. In both of the two seemingly very different types of physics, energy is basically linked with motion but in basically different ways. There are of course other types of physics, as those that try to replace matter itself with energy often in the form one wave motion of something basically undefined or with some 'wave motion of nothing' or unassociated energy. And it seems that waves of any specified quantal frequency do especially respond to other waves of that quantal frequency only, as in standing-wave interference. While some would take that as only a rare phenomenon of little significance with an almost infinite variety of different frequencies possible, others postulate it as being more fundamental. All these types of physics have energy motion issues including interaction motion issues. The main experimental science concerns have to be trying to determine the validity of similarities or differences in the mathematics and predictions needed by such different physics theories.

In classical Galileo-Descartes push physics, matter chiefly has the contact- push property where amount of push defines 'mass and energy is only the motion property of matter - including waves of such matter. In classical Gilbert-Newton signal attraction physics, matter chiefly has the signal-response property and energy is only matter response - including waves of such matter. But of course many physicists now claim that there are 'non-matter waves' and 'non-matter energy', often omitting firm definitions as of mass, as in theories like the Quanta Physics of Vertner Vergon. And some of such physics theories also claim that non-matter waves and/or energy somehow have a push property as in the ElectroMagnetic Radiation Pressure (EMRP) gravity theory. Descartes matter push has a well defined mechanism, from two pieces of matter being unable to occupy the same space at the same time, but the EMRP 'push' really seems to be more some unexplained moving away that maybe cannot be properly called a push (eg see www.blazelabs.com/f-g-intro.asp). But the basics of Descartes push physics have perhaps still not been firmly disproved, since it remains somewhat doubtful that light or any form of energy has yet been firmly shown to be other than matter response or matter motion ?

Despite clear disproofs of substantial elements of Descartes physics, it has had one perhaps unlikely success area - in gasses, two of which seem able to occupy the same space and do not appear to push each other. Yet today's standard 'Kinetic Theory' of gasses is simple Descartes push physics theory assuming microscopic gas molecules are solid moving balls, though with no Descartes ether, and it seems OK at explaining gas temperatures, pressures and wave motions including sound etcetera. So Descartes push physics maybe lives on for the macroscopic behaviour of gasses, as Gilbert-Newton attraction physics lives on for the common behaviour of gravitational and magnetic bodies. (Of course Descartes physics needs an ether to try to explain at-a-distance forces like gravity, and an attraction physics can undoubtedly also explain gasses and maybe more convincingly.)

But attempts to prove modified Descartes general physics theories still continue, as with Steven Rado's 'Aethro-kinematics' push physics which basically is Descartes physics with ether vortex-motion replaced with ether torus-motion to 'explain' gravitational, magnetic and other forces. It is not clear that its ether torus-motion has any real basis, though it is partly supported by some interesting experimental torus-model evidence - see www.aethro-kinematics.com/. But in any case the known experiment mathematics of these forces does not agree with the known experiment mathematics of vortex/torus motions - so the latter cannot give the actual elliptical orbits of the planets and where 3 gravity forces can add as 1 force from a common center of gravity, 3 vortex/torus motion forces cannot add in that way so a Newton-disproof still holds. (a similar problem also seems to apply in trying to add multiple space-curvature forces, as balloons expanded 1 percent or 3 percent do not exert proportionately more force ?) But the physics or physical chemistry of gasses is still now generally explained in terms of Cartesian push physics usually without mentioning Des Cartes or considering any possible alternative explanations.

Descartes produced a philosophy that could allow the Catholic church to accept science, though much earlier the ancient-Greek Epicurus seems to have done similar in producing a philosophy that could allow religion generally to accept Democritus atomist 'science' though Descartes may have done a more rigorous job of it. But some supporters of Descartes physics, as Einstein for his physics, have and do claim 'compatibility with Newton' falsely. To the extent that they define their mechanisms both Descartes and Einstein seem to require basically similar push mechanisms for planetary motion, which Newton proved are not compatible with his planetary maths or with actual planetary motion. For either Cartesian or Einsteinian theory to be viable they seem to actually need gravity mechanisms different to their claimed mechanisms. But generally Descartes' Cartesian physics is now taught as being 'Newtonian physics' with a small Newton content added on. And the currently best developed Cartesian push physics theories are perhaps Particle Exchange Quantum Mechanics and Lorentz-Fitzgerald Ether Fieldforce Theory, which may well both involve the same mathematics as a properly developed Gilbert-Newton signal response physics. These three may well be valid image theories.

For comparison with other physics theories, Descartes three laws of motion would be ;

1. Every body will remain at rest, or in a uniform state of motion unless pushed or pulled.

2. When a body is pushed or pulled, it accelerates proportional to the force of the push or pull and inversely proportional to the mass of the body and in the direction pushed or pulled.

3. Every push or pull has an equal and opposite reaction.

PS. For some modern Descartes physics, with a fair sprinkling of some other 'non-mainstream physics', see the Natural Philosophy Alliance and the World Science Database - at //thescientificworldview.blogspot.com and //www.worldsci.org.

You should be able to read here Descartes 1644 Principia Philosophiae (Principles of Philosophy) but somehow the original seems not available online anywhere. But an online English translation of part of it is available and discussed.

Tell a friend about this website simply,
and they will thank you for showing them the best on the important basics of the science of Rene Descartes ;
Type friends email address here   ...   Then click to tell your friend
NOTE : You can use this with confidence as we do not share and do not store this information at all.

OR if you like this site you could maybe make a donation ;

.

It will help with site development, and just possibly with some key physics experiments long planned but never afforded.
[PS. and you may perhaps help make history for science ?]
(The fictional time-travel and multi-universe type ideas of modern physics theory have long totally discouraged certain lines of physics experiment despite there being strong reasons to believe them to be very promising if not essential lines of experiment. Some such lines of experiment considered here identified as early as the 1960s seem still to have had no work done on them and there is maybe not much more time here for this. Science funding both government and private unfortunately now all goes to basically safe standard mainstream science, and no money at all goes to any really innovative risky science though that might pay a thousand times greater.)

You can do a good search of this website below ;

Search on this site www.new-science-theory.com, with Google.

Or do a search of the web better with DuckDuckGo -   Type web search then Enter

---

You are welcome to **link** to any page on this site, eg www.new-science-theory.com/rene-descartes.php

© **new-science-theory.com, 2021** - taking care with your privacy, see New Science Theory HOME.

Rene Descartes - 'Principles of Philosophy' and 'Treatise on Light'

Homepage . William Gilbert . Rene Descartes . Isaac Newton . Albert Einstein .... Descartes The World .... General Image Theory

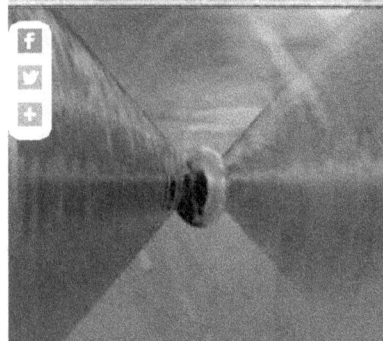

Like Newton and Gilbert, Descartes published his major works in Latin - but he did oversee approved translations into his native French - including his Principia Philosophiae (Principles of Philosophy) which had first been published in 1644. Most of his published works were philosophical but did include some acclaimed 'science'.

Below we will look at some of his 'Principles of Philosophy' and 'Treatise on Light', and his 'The World' published after his death and giving more on his physics is dealt with in a separate section. An English translation of part of Descartes' Principia Philosophiae (Principles of Philosophy), though not approved by him, can be read online at http://www.classicallibrary.org/descartes/principles/ - a good website.

Descartes' Principles of Philosophy translated by John Veitch.

You can read in our section on Descartes 'The World', his argument concluding that the only essential property of a body is extension or space-occupancy, and that all extension assumed some body and so there could not be any empty space only dead matter and the universe must be infinite. He also argues that the only other certain property of a body is motion, and that 'motion can only be produced by other motion' - only by pushings or pullings.

So in Part 4, Descartes argues that "we can perceive no external objects unless some local motion be caused by them in our nerves, and that such motion cannot be caused by the fixed stars, owing to their great distance from us, unless a motion be also produced in them and in the whole heavens lying between them and us (through a continuous material ether)" Descartes like Aristotle basically opposed empty space on theoretical grounds.

(Gilbert saw motion as only a derivative of primary forces associated with bodies, like magnetism and gravity, and saw empty space as generating no forces and so being really empty. Earth's atmosphere attenuating with altitude and planet orbits not suffering drag were seen as evidence for empty space. Newton concluded that the chief property of matter is inertia and any piece of space generating no resistance to motion must be empty space - and perhaps came near to adding a second chief property of matter as gravity implying that any piece of space generating no gravity must be empty space containing no matter ? Of course Newton taking a black-box position did not actually conclude the latter, and allowed there might be some massless bodies with no gravity.)

In his Principles, Descartes repeatedly argued against empty space, to make a material ether central to his physics, hence ;

"XVIII. How the prejudice of an absolute vacuum is to be corrected.
......And accordingly, if it be asked what would happen were God to remove from a vessel all the body contained in it, without permitting another body to occupy its place, the answer must be that the sides of the vessel would thus come into proximity with each other.
For two bodies must touch each other when there is nothing between them, and it is manifestly contradictory for two bodies to be apart, in other words, that there should be a distance between them, and this distance yet be nothing; for all distance is a mode of extension, and cannot therefore exist without an extended substance."

With empty space logically abolished, Descartes' imagining fills his universe with three types of matter, or elements. The first element is matter made up of a non-particle fluid moving so quickly that it shatters any body it hits and produces heat and light, and the sun and stars are composed of this element. The matter of the second element is made up of microscopic spherical particles, making a stable fluid, and this element fills space and propagates light. Finally, the third element of which planets and common objects are formed is made of larger particles least well-suited to motion.

He tackles light as the particles of his second element transmitting motions, somehow in a straight line instantly, "like a stick transmits a push on one end to the other end" - though for sound he used a normal wave theory. And to explain magnetism Descartes claimed that novel emitted effluvia particles of "threaded parts" passed through a network of one-way threaded passages in iron and worked like a corkscrew.

Interestingly Descartes considered light at some length as a signal ;

Descartes' Treatise on Light - translated by George MacDonald Ross

"Chapter I: On the difference between our sensations and the things that produce them

In proposing to write this treatise on Light, the first thing I want to bring to your attention is the fact that there can be a difference between the sensation we have of it (that is, the idea of it formed in our imagination via our eyes), and what there is in the objects which produce this sensation in us (that is what there is in flame or the sun which is called 'light'). For although most people are convinced that the ideas we have in our thinking are entirely similar to the objects they come from, I can see absolutely no reason why we should be certain of this - on the contrary, I am aware of many observations which should make us doubt it.

You know, of course, that words make us form conceptions of the things they signify even though they have no resemblance to them, often even without our paying any attention to the sounds of the words or the syllables of which they are composed. Thus it can happen that, after hearing something said of which we have perfectly understood the sense, we are unable to say what language it was spoken in. But if words, which have meaning only as a human institution, are enough to make us form conceptions of things they bear no resemblance to, why could not Nature too have instituted some sign which would make us have the sensation of light, but without containing in itself anything similar to this sensation? And is this not how she has instituted smiles and tears to make us read joy and sadness on people's faces?

But perhaps you will say that our ears really only make us perceive the sound of the words, and our eyes the countenance of the person who smiles or weeps, and that it is our spirit which, having grasped the meanings of the words and countenance, represents them to us at the same time. I could reply to this that it is likewise our spirit which represents to us the idea of light whenever the action which signifies it comes into contact with our eyes. But rather than wasting time in disputation, I would prefer to give another example.

When we ignore the meanings of words and listen only to their sound, do you think the idea of this sound formed in our thinking bears some resemblance to the object that causes it? Someone opens their mouth, moves their tongue, emits their breath - but I see nothing in all these actions that is not very different from the idea of the sound which they make us form in our imagination. And the majority of philosophers assure us that sound is nothing but a certain vibration of the air which comes and beats against our ears; so that if the sense of hearing brought the true image of its object into our thinking, instead of making us have a conception of sound, it would have to make us have a conception of the motion of the parts of the air then vibrating against our ears. But since not everyone, perhaps, will be prepared to believe what philosophers say, I shall give yet another example.

The sense which is considered the least deceptive and the most certain is that of touch; so, if I show you that even the sense of touch makes us conceive many ideas which have no resemblance at all to the objects that produce them, I do not think you should find it strange if I say that the sense of sight can do the same. There is no one who does not know that the ideas of tickling and of pain which are formed in our thinking on the occasion of our coming into contact with external bodies bear no resemblance to them. You gently pass a feather over the lips of a sleeping child, and it senses that you are tickling it: do you think that the idea of tickling which it conceives resembles in any respect the qualities of the feather? A soldier returns from a battle: during the heat of the action he could have been wounded without noticing it; but now that he is beginning to cool off, he feels some pain, and believes he has been wounded. A surgeon is called, his armour is removed, the surgeon makes a visit, and finally it is found that what he felt was nothing other than a buckle or a strap which had got caught up under his armour and caused the trouble by pressing into him. If his sense of touch, in making him aware of this strap, had impressed the image of it on his thinking, he would have had no need of the surgeon to tell him what he was feeling.

So, I see no reason why we should believe that whatever it is in objects that gives rise to our sensation of light is any more like that sensation than the actions of a feather or a strap are like the sensation of tickling or pain. However, I have certainly not brought up these examples in order to make you believe absolutely that light is different in objects from what it is in our eyes; but only in order to make you reserve judgment about it; and, by keeping you from being prejudiced by the contrary opinion, to enable you to join me now in a more fruitful examination of its nature ...."

Most of Descartes actual science theory being basically 'logical imaginings', based on a weaker knowledge of actual physical phenomena than Gilbert or Newton, perhaps added little of practical use to physics theory at the time, with the exception of his work on light based on a particle theory and adding to knowledge on refraction especially. Newton published a strong disproof of Descartes' material ether and specifically of his ether vortex theory of planetary motion. Descartes had basically better organised ancient greek Atomist theory and incorporated it into his God-based Dualist philosophy so it could better suit religion. Descartes' physics gained wide support, and later Maxwell and Einstein produced alternative 'non-material' ether/continuum Descartes-style theories though they perhaps lacked the relatively clear simple logic of Descartes' material push ether physics.

Though modern science is really still based on Descartes-style dead matter mechanical push-physics theory, many of its statements in fact read like excited active matter theory statements :-
1. A typical modern explanation of part of Brain action - "A neuron accepts signals from other neurons through branchlike structures called dendrites. Whenever enough messages arrive from neighbouring neurons to excite it, a neuron sends an electrical impulse."
2. A typical modern explanation of part of Atomic action - "By absorbing photons of some one wavelength, an atom can be excited to any of various discrete energy levels and then it can emit light of various wavelengths."

These clearly read like active-neurons and active-atoms statements, and not like Descartes mechanical push statements. Even modern declared dead-matter theorists seem often to use active-matter language (as easier or clearer assumedly). From radioactivity and other atomic behaviour, we now KNOW that atoms are not simple small billiard balls as might best suit Descartes-style dead-matter theory - and often at least equally well fit an active-atoms theory akin to Gilbert's.

It was certainly Descartes' advances in mathematics that were of more practical use to progress in physics theory, and the same can probably be said of Newton and then of Einstein also?

Otherwise, if you have any view or suggestion on the content of this site, please contact :- New Science Theory
Vincent Wilmot 166 Freeman Street Grimsby Lincolnshire DN32 7AT.

You are welcome to **link** to any page on this site, eg www.new-science-theory.com/rene-descartes-principles.php

IF you like this site then . Bookmark

© **new-science-theory.com, 2022** - taking care with your privacy, see New Science Theory HOME.

Rene Descartes - 'The World'

Homepage . William Gilbert . Rene Descartes . Isaac Newton . Albert Einstein .... Descartes Principles .... General Image Theory

Below you can read the significant parts of Descartes 'The World', defining the basics of his physics theory which was partly described earlier in his works around 1644. Its publication was apparently planned for 1633, but was abandoned when in 1633 Galileo was charged with heresy by the Roman Catholic Church Inquisition, despite it being Descartes trying to make a Christian science, and it was not actually published until 1664 - after Galileo's 1642 death and Descartes' own 1650 death.

Though this Cartesian physics theory is supposedly entirely deduced by logic from 'the most certain of ideas', it is really just ancient-Greek atomism based on little or no experiment and as formulated by him it had many weaknesses yet seemed for some to plausibly explain a reasonable range of natural phenomena and appeared as though others might be able to build upon it to give a workable physics. However this ancient-Greek atomist push-physics though supported by Galileo and some alchemists had been substantially disproved by William Gilbert in his 1600 'De Magnete' and later further also by Newton.

extracts from Rene Descartes 'The World' translated by Michael S. Mahoney.

CHAPTER FIVE
On the Number of Elements and on their Qualities

I conceive of the first, which one may call the element of fire, as the most subtle and penetrating fluid there is in the world...... Thus, there is never a passage so narrow, nor an angle so small, among the parts of other bodies, where the parts of this element do not penetrate without any difficulty and which they do not fill exactly.

As for the second, which one may take to be the element of air, I conceive of it also as a very subtle fluid in comparison with the third; but in comparison with the first there is need to attribute some size and shape to each of its parts and to imagine them as just about all round and joined together like grains of sand or dust. Thus, they cannot arrange themselves so well, nor so press against one another that there do not always remain around them many small intervals into which it is much easier for the first element to slide than for the parts of the second to change shape expressly in order to fill them.....

Beyond these two elements, I accept only a third, to wit, that of earth. Its parts I judge to be as much larger and to move as much less swiftly in comparison with those of the second as those of the second in comparison with those of the first. Indeed, I believe it is enough to conceive of it as one or more large masses, of which the parts have very little or no motion that might cause them to change position with respect to one another.....

If we consider in general all the bodies of which the universe is composed, we will find among them only three sorts that can be called large and be counted among the principal parts, to wit, the sun and the fixed stars as the first sort, the heavens as the second, and the earth with the planets and the comets as the third. That is why we have good reason to think that the sun and the fixed stars have no other form than that of the wholly pure first element, the heavens that of the second, and the earth with the planets and comets that of the third.....

CHAPTER SIX
Description of a New World, and on the Qualities of the Matter of which it is composed

For a short time, then, allow your thought to wander beyond this world to view another, wholly new one, which I shall cause to unfold before it in imaginary spaces. The philosophers tell us that these spaces are infinite, and they should very well be believed, since it is they themselves who have made the spaces so. Yet, in order that this infinity not impede us and not embarrass us, let us not try to go all the way to the end; let us enter in only so far that we can lose from view all the creatures that God made five or six thousand years ago and, after having stopped there in some fixed place, let us suppose that God creates from anew so much matter all about us that, in whatever direction our imagination can extend itself, it no longer perceives any place that is empty.....

Even though our imagination seems to be able to extend itself to infinity, and this new matter is not assumed to be infinite, we can nonetheless well suppose that it fills spaces much greater than all those we shall have imagined ......Let us not permit our imagination to extend itself as far as it could, but let us purposely restrict it to a determinate space that is no greater, say, than the distance between the earth and the principal stars of the firmament, and let us suppose that the matter that God shall have created extends quite far beyond in all directions, out to an indefinite distance.....

My plan is not to set out (as they do) the things that are in fact in the true world, but only to make up as I please from this matter a universe in which there is nothing that the densest minds are not capable of conceiving, and which nevertheless could be created exactly the way I have made it up.....

CHAPTER SEVEN
On the Laws of Nature of this New World

.....I will set out here two or three of the principal rules according to which one must think God to cause the nature of this new world to act and which will suffice, I believe, for you to know all the others.

The first is that each individual part of matter always continues to remain in the same state unless collision with others constrains it to change that state. That is to say, if the part has some size, it will never become smaller unless others divide it; if it is round or square, it will never change that shape without others forcing it to do so; if it is stopped in some place, it will never depart from that place unless others chase it away; and if it has once begun to move, it will always continue with an equal force until others stop or retard it.....

I suppose as a second rule that, when one of these bodies pushes another, it cannot give the other any motion except by losing as much of its own at the same time; nor can it take away from the other body's motion unless its own is increased by as much.....

I will add as a third rule that, when a body is moving, even if its motion most often takes place along a curved line and........ can never take place along any line that is not in some way circular, nevertheless each of its individual parts tends always to continue its motion along a straight line.....

I could set out here many additional rules for determining in detail when and how and by how much the motion of each body can be diverted and increased or decreased by colliding with others, something that comprises summarily all the effects of nature. But I shall be content with showing you that, besides the three laws that I have explained, I wish to suppose no others but those that most certainly follow from the eternal truths on which the mathematicians are wont to support their most certain and most evident demonstrations; the truths, I say, according to which God Himself has taught us He disposed all things in number, weight, and measure.

The knowledge of those laws is so natural to our souls that we cannot but judge them infallible when we conceive them distinctly, nor doubt that, if God had created many worlds, the laws would be as true in all of them as in this one.....

Nonetheless, in consequence of this, I do not promise you to set out here exact demonstrations of all the things I will say. It will be enough for me to open to you the path by which you will be able to find them yourselves, whenever you take the trouble to look for them. Most minds lose interest when one makes things too easy for them. And to compose here a setting that pleases you, I must employ shadow as well as bright colors. Thus I will be content to pursue the description I have begun, as if having no other design than to tell you a fable.

CHAPTER EIGHT
On the Formation of the Sun and the Stars of the New World

Whatever inequality and confusion we might suppose God put among the parts of matter at the beginning, the parts must, according to the laws He imposed on nature, thereafter almost all have been reduced to one size and to one middling motion and thus have taken the form of the second element as I described it above. For to consider this matter in the state in which it could have been before God began to move it, one should imagine it as the hardest and most solid body in the world. And, since one could not push any part of such a body without pushing or pulling all the other parts by the same means, so one must imagine that the action or the force of moving or dividing, which had first been placed in some of the parts of matter, spread out and distributed itself in all the others in the same instant, as equally as it could.

It is true that this equality could not be totally perfect. First, because there is no void at all in the new world, it was impossible for all the parts of matter to move in a straight line. Rather, all of them being just about equal and as easily divertible, they all had to unite in some circular motions. And yet, because we suppose that God first moved them diversely, we should not imagine that they all came together to turn about a single center, but about many different ones, which we may imagine to be diversely situated with respect to one another.....

Thus, in a short time all the parts were arranged in order, so that each was more or less distant from the center about which it had taken its course, according as it was more or less large and agitated in comparison with the others.......Only one must except some which, having been from the beginning much larger than the others, could not be so easily divided or which, having had very irregular and impeding shapes, joined together severally rather than breaking up and rounding off. Thus, they have retained the form of the third element and have served to compose the planets and the comets, as I shall tell you below.....

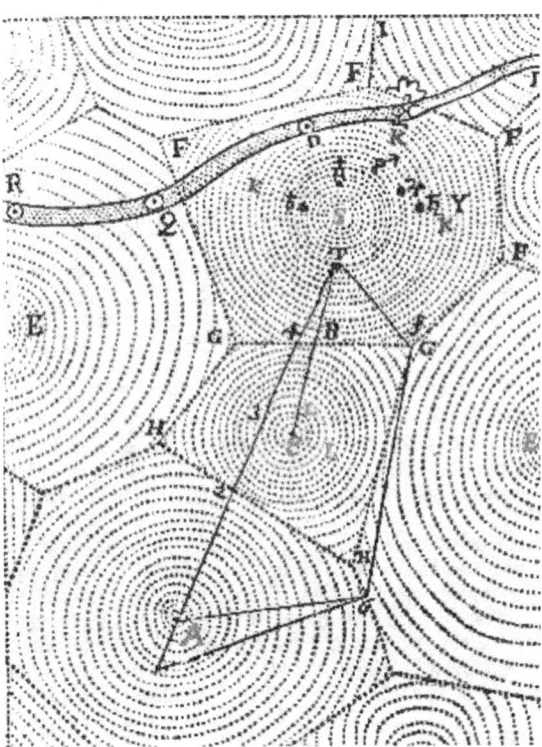

Imagine, for example, that the points S, E, C, and A are the centers of which I speak, that all the matter contained in the space FGGF is a heaven turning about the sun marked S, that all the matter of the space HGGH is another heaven turning about the star marked C, and so on for the others. Thus, there are as many different heavens as there are stars, and, since the number of stars is indefinite, so too is the number of heavens. Thus also the firmament is nothing other than the breadthless surface separating all the heavens from one another.

Imagine also that the ...... speed of matter in each stellar vortex decreases little by little from the outside circumference of each heaven to a certain place (such as, for example, to the sphere KK about the sun [S], and to the sphere LL about the star C) and then increases little by little from there to the centers of the heavens because of the agitation of the stars that are found there.....

Whence you will be able to understand immediately that the highest planets must move more slowly than the lowest (i.e. those closest to the sun), and that all the planets together move more slowly than the comets, which are nonetheless more distant.....

Note finally that, given the manner in which I have said the sun and the other fixed stars were formed, their bodies can be so small with respect to the heavens containing them that even all the circles KK, LL, etc.,........can be considered merely as the points that mark the heavens' center. In the same way, the new astronomers consider the whole sphere of Saturn as but a point in comparison with the firmament.

CHAPTER NINE
On the Origin and the Course of the Planets and Comets in General; and of Comets in Particular

Now, for me to begin to tell you about the planets and comets, consider that, given the diversity of the parts of the matter I have supposed, even though most of them in breaking and dividing by collision with one another have taken the form of the first or second element, there nevertheless does not cease still to be found among them two sorts that had to retain the form of the third element, to wit, those of which the shapes were so extended and so impeding that, when they collided with one another, it was easier for several to join together, and by this means to become larger than to break up and become smaller; and those which, having been from the beginning the largest and most massive of all, could well break and shatter the others in striking them but not in turn be broken or shattered themselves.....

If you imagine two rivers that join with one another at some point and then separate again shortly thereafter before their waters (which one must suppose to be very calm and to have a rather equal force, but also to be very rapid) have a chance to mix, then boats or other rather massive and heavy bodies that are borne by the course of the one river will be easily able to pass into the other river, while the lightest bodies will turn away from it and will be thrown back by the force of the water toward the places where it is the least rapid.

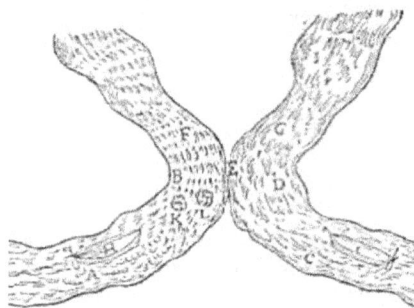

For example, if ABF and CDG are two rivers which, coming from two different directions, meet at E and then turn away from there, AB going toward F and CD toward G, it is certain that boat H following the course of river AB must pass through E toward G, and reciprocally boat I toward F, unless both meet at the passage at the same time, in which case the larger and stronger will break the other. By contrast, scum, leaves of trees, feathers, straw, and other such light bodies that can be floating at A must be pushed by the course of the water containing them, not toward E and toward G, but toward B, where one must imagine that the water is less strong and less rapid than at E, since at B it takes its course along a line that less approaches a straight line.

Moreover, one must consider that not only these light bodies, but also others heavier and more massive can join upon meeting and that, turning then with the water that bears them, several together can compose large balls such as you see at K and L, of which some, such as L go toward E and others, such as K, go toward B, according as each is more or less solid and composed of more or less large and massive parts.....

Know also that we should take those that thus tend to range toward the center of any heaven to be the planets, and we should take those that pass across different heavens to be comets. Now, concerning these comets, one must note first that there must be few of them in this new world in comparison to the number of heavens. For, even if there were many at the beginning, over the course of time in passing across different heavens almost all of them would have to have collided with one another and broken one another up (just as I have said two boats do when they meet), so that now only the largest could remain.

One must also note that, when they pass thus from one heaven into another, they always push before them some small bit of the matter of the heaven they are leaving and remain enveloped by it for some time until they have entered far enough within the limits of the other heaven. Once there, they finally loose themselves from it almost all at once and without taking perhaps more time to do so than does the sun in rising at morning on our horizon. In this way, they move much more slowly when they thus tend to leave some heaven than they do shortly after having entered it.

For example, you see [below] that the comet that takes its course along the line CDQR, having already entered rather far within the limits of the heaven FG, nevertheless when it is at point C still remains enveloped by matter from the heaven FI, from which it comes, and cannot be entirely freed of that matter before it is around point D. But, as soon as it has arrived there, it begins to follow the course of the heaven FG and thus to move much faster than it did before. Then, continuing its course from there toward R, its motion must again slow down little by little in proportion as it approaches point Q, both because of the resistance of the heaven FGH, within the limits of which it is beginning to enter, and because, there being less distance between S and D than between S and Q, all the matter of the heaven between S and D (where the distance is smaller) moves faster there, just as we see that rivers always flow more swiftly in the places where their bed is narrower and more confined than in those where it is wider and more extended.

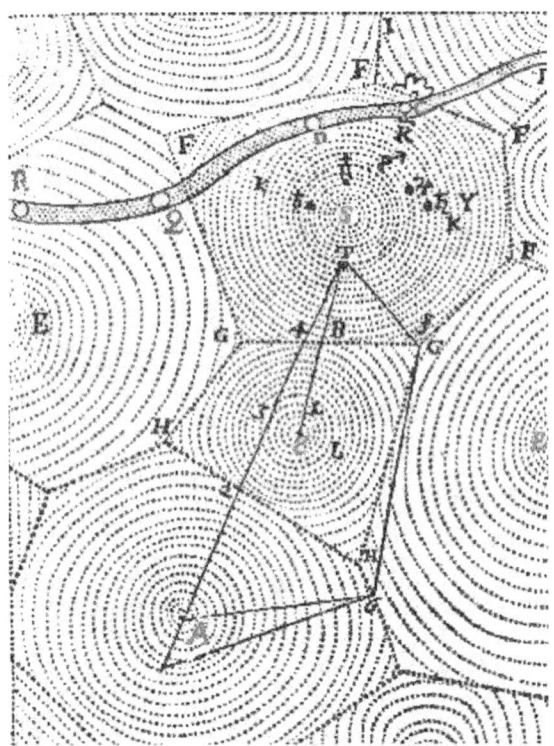

Moreover, one should note that this comet should be visible to those who live at the center of the heaven FG only during the time it takes to pass from D to Q, as you will soon understand more clearly when I have told you what light is. In the same way, you will see that its motion should appear to viewers to be much faster, its body much greater, and its light much brighter, at the beginning of the time they see it than at the end.....

CHAPTER TEN
On the Planets in General, and in Particular on the Earth and Moon

Similarly, there are several things to note concerning the planets. First, even though they all tend toward the center of the heavens containing them, that is not to say thereby that they could ever arrive at those centers. For, as I have already said above, the sun and the other fixed stars occupy them.....

There can be diverse planets, some more and others less distant from the sun, such as here Saturn, Jupiter, Mars, T [Earth], Venus, Mercury. Of these the lowest and least massive can reach to the sun's surface, but the highest never pass beyond circle K.....

It is not simply those that outwardly appear the largest, but those that are the most solid and the most massive in their interior, that should be the most distant.....

The...... matter of the heaven must make the planets turn not only about the sun, but also about their own center (except when there is some particular cause that hinders them from doing so), and consequently that the matter must compose around the planets small heavens that move in the same direction as the greater heaven.....

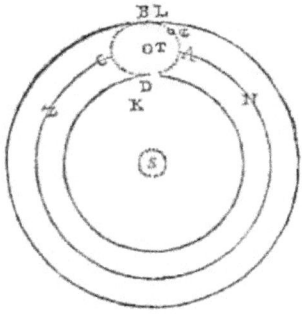

For, since the parts of the heaven that are, say, at A move faster than the planet marked T, which they push toward Z, it is evident that they must be diverted by it and constrained to take their course toward B. I say toward B rather than toward D; for, having inclination to continue their motion in a straight line, they must go toward the outside of the circle ACZN they are describing, rather than toward the center S.

Now, passing thus from A to B, they force the planet T to turn with them about its center. In turn, this planet in so turning gives them occasion to take their course from B to C, then to D and to A, and thus to form about the planet a particular heaven, with which it must thereafter continue to move from the direction one calls the "occident" [west] toward that which one calls the "orient," [east] not only about the sun but also about its own center.

Moreover, knowing that the planet marked Moon is disposed to take its course along the circle NACZ (just as is the planet marked T) and that it must move faster because it is smaller, it is easy to understand that, wherever it might have been in the heavens at the beginning, it shortly had to tend toward the exterior surface of the small heaven ABCD, and that, once having joined that heaven, it must thereafter always follow its course about T along with the parts of the second element that are at that surface.....

I shall not add here how one can find a greater number of planets joined together and taking their course about one another, such as those that the new astronomers have observed about Jupiter and Saturn.....

CHAPTER ELEVEN
On Weight

Now, however, I would like you to consider what the weight of this earth is; that is to say, what the force is that unites all its parts and that makes them all tend toward its center, each more or less according as it is more or less large and solid.

That force is nothing other than, and consists in nothing other than, the fact that, since the parts of the small heaven surrounding it turn much faster than its parts about its center, they also tend to move away with more force from its center and consequently to push the parts of the earth back toward its center.

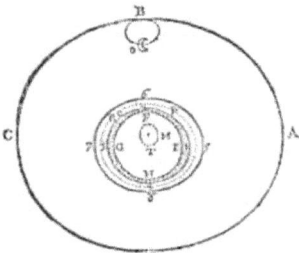

Planet T's "small heaven" (circle ABCD) contains earth (circle EFGH) surrounded with layers of water (circle 1234) and air (circle 5678). The "matter of heaven" fills all the space between the circles 5678 and ABCD. As this matter circulates, it causes T to turn on its axis, and carries the moon around ABCD. Inhabitants of T cannot sense that they are spinning in space because they are moving along with everything else in the swirling vortex.

You may find some difficulty in this, in light of my just saying that the most massive and most solid bodies - such as I have supposed those of the comets to be - tend to move outward toward the circumferences of the heavens and that only those that are less massive and solid are pushed back toward their centers. For it should follow therefrom that only the less solid parts of the earth could be pushed back toward its center and that the others should move away from it. But note that, when I said that the most solid and most massive bodies tended to move away from the center of any heaven, I supposed that they were already previously moving with the same agitation as the matter of that heaven.

For it is certain that, if they have not yet begun to move, or if they are moving less fast than is required to follow the course of this matter, they must at first be pushed by it toward the center about which it is turning. Indeed, it is certain that, to the extent that they are larger and more solid, they will be pushed with more force and speed.....

CHAPTER FIFTEEN
That the Face of the Heaven of that New World must Appear to its Inhabitants completely like that of Our World

Having thus explained the nature and the properties of the action I have taken to be light, I must also explain how, by its means, the inhabitants of the planet I have supposed to be the earth can see the face of their heaven as wholly like that of ours.

First, there is no doubt that they must see the body marked S as completely full of light and like our sun, given that that body sends rays from all points of its surface toward their eyes. And, because it is much closer to them than the stars, it must appear much greater to them.....

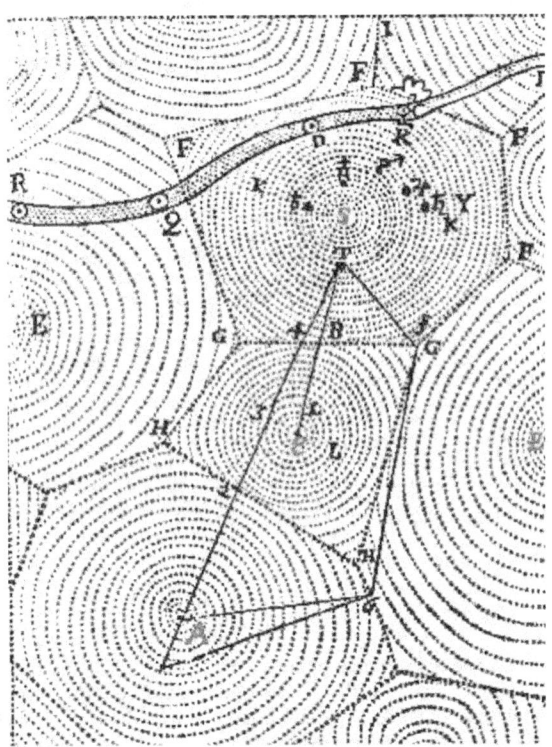

You must...... consider in regard to their arrangement that they can just about never appear in the true place where they are. For example, that marked C appears as if it were in the straight line TB, and the other marked A as if it were in the straight line T4.....

And one must suppose those lines TB, T4, and ones like them to be so extremely long in comparison with the diameter of the circle the earth describes about the sun that, wherever the earth is on that circle, the men on it always see the stars as fixed and attached to the same places in the firmament; that is, to use the terms of the astronomers, they cannot observe parallax in the stars.

Regarding the number of those stars, consider also that the same star can often appear in different places because of the different surfaces that divert its rays toward the earth. Here, for example, that marked A appears in the line T4 by means of the ray A24T and simultaneously in the line Tf by means of the ray A6fT. In the same way are the objects multiplied that one looks at through glasses or other transparent bodies cut along several faces.

Moreover, regarding their size, consider that they must appear much smaller than they are, because of their extreme distance; for this reason the greater part of them must not appear at all, and others appear only insofar as the rays of several joined together render the parts of the firmament through which they pass a bit whiter and similar to certain stars the astronomers call "nebulous," or to that great belt of our heaven that the poets pretend to be whitened by the milk of Juno.....

Moreover, it is very probable that those surfaces, being in a matter that is very fluid and that never ceases to move, should always shake and quiver somewhat, and consequently that the stars one sees through them should appear to scintillate and vibrate, just as ours do, and even, because of their vibration, appear a bit larger. In this way, the image of the moon appears larger when viewed from the bottom of a lake of which the surface is not very stirred up or agitated, but merely a bit rippled by the breath of some wind.

And, finally, it can happen that, over the course of time, those surfaces change a bit, or indeed even that some of them bend rather noticeably in a short time, even if this is only on the occasion of a comet's approaching them. By this means, several stars seem after a long time to change a bit in place without changing in size, or to change a bit in size without changing in place. Indeed, some even begin rather suddenly to appear or to disappear, just as one has seen happen in the real world.

As for the planets and the comets that are in the same heaven as the sun, knowing that the parts of the third element of which they are composed are so large or so joined severally together that they can resist the action of light, it is easy to understand that they must appear by means of the rays that the sun sends toward them and that are reflected from there toward the earth, just as the opaque or obscure objects that are in a room can be seen there by means of the rays that the lamp shining there sends toward them and that return from them toward the eyes of the onlookers.....

The motion those planets have about their center is the reason why they twinkle, though much less strongly and in another way than do the fixed stars; because the moon is deprived of that motion, it does not twinkle at all.

As for the comets that are not in the same heaven as the sun, they are far from being able to send out as many rays toward the earth as they could if they were in the same heaven, not even when they are all ready to enter it. Consequently, they cannot be seen by men, unless perhaps when their size is extraordinary. The reason for this is that most of the rays that the sun sends out toward them are borne away here and there and effectively dissipated by the refraction they undergo in the part of the firmament through which they pass.....

Sir Isaac Newton - *mathematical laws Black Box theory*

Homepage . William Gilbert . Rene Descartes . Albert Einstein ...... Newton's Principia . Against Descartes .... General Image Theory
- Site Search at bottom v -

Isaac Newton (25.12.1642-20.3.1727 or 4.1.1643-31.3.1727) was a great mathematician and a great physicist, and he was probably the most incisive thinker ever known. He chiefly established that natural phenomena generally follow determinate mathematical laws in demonstrating consistent laws of motion, of gravity and of other phenomena. When young he backed Cartesian physics but ended backing Gilbert's action-at-distance attraction physics, however under very strong peer pressure Newton produced his non-committal 'black box' theory of science as allowing only proving how things happen but not necessarily allowing proving or disproving alternative theories of why things happen and seemingly allowing as possible options both Cartesian push-physics and Gilbertian signal-emission physics. But having published disproof of the Descartes option, it must logically be taken that he wanted his physics seen as a signal-response Gilbertian physics though afraid to spell it out clearly. Newton may have seen his blackbox physics as only a more rigorous logical requirement of experimental science, from William Gilbert's earlier requirement that science could not go beyond what can be deduced directly from experience and experiment. Though it was supported by a few other physicists, and by George Berkeley in his 1721 'De Motu', Newton's black-box physics has been very wrongly either just ignored or 're-interpreted' by many including Einstein and widely taught as being a Cartesian 'dead-matter plus ether or forcefield' push why-physics. Newton's main works were his Latin 1687 Philosophiae Naturalis Principia Mathematica (or 'Mathematical Principles of Natural Philosophy' - no Gravity!!), and The Opticks published in English in 1704 and in Latin in 1706. To really study some great physics still really needs Latin.

Newton's science theory.

Though the early Newton had favoured Descartes-style mechanical push physics theory, his later major published theory work involved combining laws of force and motion in mechanics with a Gilbertian action-at-distance signal-response attraction theory, to develop laws of gravitational orbital motion around attracting 'centres of force'. This change of view may have been chiefly come from Robert Hooke around 1679, and before Newton's 'Principia', suggesting that planet orbitings might be due to an attraction force decreasing with distance compounded with inertial forces of planets momentum, though seemingly without any mention of the originator of such attraction physics and of planet inertia in space as Gilbert (though not their combination for orbits) and giving only limited support for attraction physics, and Newton had apparently considered attraction physics theory and inertia somewhat earlier in any case. Newton's library suggests that he seemingly at some point had studied Hooke's 1674 'An attempt to prove the motion of the Earth' which said the same, though Newton may not have studied Gilbert directly nor indeed Kepler's astronomy directly. (Hooke discovered his law of elasticity in 1676, but first published it only as the anagram: "ceiiinosssttuv." which he revealed two years later as the Latin 'ut tensio, sic vis' meaning "as the extension, so the force.".) By 1684 in a small manuscript he privately-distributed 'De Motu Corporum in Gyrum' (On the Motion of Bodies in an Orbit) Newton confirmed that the maths of planet motion did fit attraction physics and inertia though without significantly acknowledging Hooke or indeed Gilbert, and in his later published 'Principia' he saw gravity as governing the motions of the celestial bodies as well as of apples falling from trees. Actually the general idea that planetary bodies might be attracted by the Sun or other planetary bodies was somewhat widespread in the 1600's certainly in England, and there had chiefly come from William Gilbert though few acknowledged that fact. Gilbert, as some others, also put the idea of momuntum inertia but did not specifically consider combination of that with planetary attraction as later Hooke and Newton did but effectively they had built on Gilbert though maybe not directly.

Newton saw motions combining attraction forces and bodies momentum (mv) not kinetic energy (mv²) as some like Leibnitz wrongly supposed though that is involved in collision force motion change. But when he published, Newton publicly supported a 'blackbox physics' or 'theoryless physics', and his main work used both the 'force' terminology perhaps more readily associated with Galileo-Descartes mechanics AND the 'attraction' terminology associated often with William Gilbert action-at-distance attraction theory, and allowed that gravity might be due to signals acting across empty space to which bodies respond in line with Gilbert's physics OR might be caused by the impact force of unseen ether particles or fluid in line with Descartes' physics. But privately Newton certainly seems to have judged the evidence to favour Gilbertian action-at-distance signal physics and many of Newton's physics contemporaries believed that Newton like some other physicists did privately back Gilbert but were bullied into hiding that, so Newtonian physics cannot be really understood unless you have first studied William Gilbert's 'De Magnete' or 'On The Magnet'. And of course Gilbertian signal-response physics more firmly excludes time-reversal in nature than the Cartesian reversible-push physics commonly wrongly ascribed to Newton now.

Robert Hooke's 1674 'An attempt to prove the motion of the Earth' stated,
"I shall explain a System of the World [or an astronomy] differing in many particulars from any yet known, answering in all things to the common Rules of Mechanical Motions: This depends upon three Suppositions.
First, That all Coelestial Bodies whatsoever, have an attraction or gravitating power towards their own Centers, whereby they attract not only their own parts, and keep them from flying from them, as we may observe the Earth to do, but that they do also attract all the other Coelestial Bodies that are within the sphere of their activity; and consequently that not only the Sun and Moon have an influence upon the body and motion of the Earth, and the Earth upon them, but that ☿ also ♀, ♂, ♄, and ♃ by their attractive powers, have a considerable influence upon its motion as in the same manner the corresponding attractive power of the Earth hath a considerable influence upon every one of their motions also.
The second supposition is this, That all bodies whatsoever that are put into a direct and simple motion, will so continue to move forward in a straight line, till they are by some other effectual powers deflected and bent into a Motion, describing a Circle, Ellipsis, or some other more compounded Curve Line.
The third supposition is, That these attractive powers are so much the more powerful in operating, by how much the nearer the body wrought upon is to their own Centers.
Now what these several degrees are I have not yet experimentally verified; but it is a notion, which if fully prosecuted as it ought to be, will mightily assist the Astronomer to reduce all the Coelestial Motions to a certain rule, which I doubt will never be done true without it."

Hooke here talked 'supposition' while Gilbert earlier gave much of this as from experiment or other fact and Newton gave a maths to precisely fit many facts. And of course we do not know to what extent Hooke developed these ideas himself or maybe more likely got them from Gilbert and/or Newton or others. Certainly most of this could be found in or got from Gilbert's 'De Magnete' which Hooke had studied closely if Newton perhaps had not and which did play-down and scatter such theory ideas. But if Newton did know Gilbert then he may well have seen Hooke as wrongly presenting Gilbert's physics as his own and so wrong to credit Hooke too much ? Yet unlike Newton, Hooke did not seem to present a clear understanding of Gilbert's attraction physics as being a signal-response physics and being unsure of your full understanding of a source might seem a valid reason for not quoting that source ? Real science undoubtedly has to rest on relevant experiment, observation and mathematics, none of which Hooke himself had done on this. But Hooke wrongly insisted that his claimed supposition was the science and there is no doubt that Hooke at least helped motivate Newton, if perhaps chiefly by greatly annoying him. Newton seems to have seen Hooke, like Des Cartes, as over-inclined to 'mere theorising and hypotheses' which he certainly at times attacked as unscientific. The improved 1713 second edition of Principia like later editions included an entirely Gilbert-like Preface by Roger Cotes strongly supporting Newton's experimental physics as having disproved (Catholic) Cartesian supposition physics - and Newton added his own support for this position at the end of that second edition of Principia. Coates here also saw too many physicists of the time as 'too much prepossessed with certain prejudices', and strongly argues for Newtonian attraction physics against Cartesian push-physics but himself unlike Newton referring only to gravity and notably making no mention of magnetism. (see Principia 2nd Edn) Cotes noted that "by taking the foundation of their speculations from hypotheses", Cartesian physicists "merely put together a romance, elegant perhaps and charming, but nevertheless a romance". But strong Newton supporter Cotes who was associated with producing the second definitive edition of Newton's 'Principia' died a very young 33 with Newton complimenting "If he had lived, we might have known something". Of course Cotes' and Newton's attacks on 'mere theorising' were similar to Gilbert's as backing limiting theorising and not as entirely ruling out theorising since both Newton and Gilbert actually did some significant theorising. Newton, like William Gilbert before him, attacked 'mere theorising' as not science - and they both backed 'experiment + consistent theory or maths' as the only real science - so that the sub-title of Gilbert's 1600 "De Magnete" was basically "A new science, with many both argument and experiment proofs". As a student at Cambridge, Newton's Trinity College taught Aristotle as well as seemingly some Galileo and early Descartes and astronomy by Johannes Kepler and Thomas Streete but not William Gilbert physics, and the physics study of Cotes seems to have been similarly limited. But in Newton's lifetime Protestant europe accepted Catholic physics as well as accepting the Catholic calender that changed Newton's birthday from Christmas Day 25.12.1642 to 4.1.1643 and rather more ordinary. The success of the earlier false attacks on William Gilbert's action-at-distance physics, by Jesuits and supporters of the catholic Galileo and Descartes, greatly aided the later false attacks on Isaac Newton's related physics and helped force Newton to first compromise his physics and to then quit the physics battle.

In Principia Book 1 Section 11(or X1) final Scholium, after showing that planet orbits can be explained by some centripetal force directed towards the sun, Newton concludes that the existence of gravity as a property of bodies can be deduced from the proven existence of magnetism as a property of bodies ;

"These propositions naturally lead us to the analogy there is between centripetal forces, and the central bodies to which those forces used to be directed ; for it is reasonable to suppose that forces that are directed to bodies should depend on the nature and quantity of those bodies, as we see they do in magnetical experiments."

Magnetic force is proven to come from magnets and not from any surrounding ether or continuum as Descartes (and later also Einstein) wrongly supposed. (Einstein tried and failed to add an electromagnetic continuum to his gravity spacetime continuum.) Yet also in this scholium Newton states that he is not committing to any particular manner of operation of 'at-a-distance' forces or of 'contact' forces.

"I here use the word attraction in general for any endeavour, of what kind soever, made by bodies to approach each other ; whether (as Gilbert) that endeavour arise from the action of the bodies themselves as tending mutually to or agitating each other by spirits emitted ; or whether (as Descartes etc) it arises from the action of the aether or of the air or of any medium whatsoever whether corporeal or incorporeal any how impelling bodies placed therin towards each other. In the same sense I use the word impulse, not defining in this treatise the species or physical qualities of forces but investigating the quantities and mathematical proportions of them"

Clearly to Newton bodies moved, but experiment could not definitely decide if they were actually being pushed by others or moving themselves - there is no evidence to decide between dead matter and active matter or between 'A moves B', and 'B moves itself in response to A'. But in fact Newton did get actual attraction physics to fit the evidence and could not get the then current Descartes push-physics to fit the evidence. A better science translation of Newton's Latin 'spiritus emmisos' here maybe would not be 'spirits emitted' but 'incorporeal emissions', 'energy emissions', 'incorporeal agitating emissions' or 'energy signal emissions'? So basically Newton's 'spirits emitted' signals were one of William Gilbert's various 'effluvia' signals for magnetic, electric and gravity forces giving his action-at-distance remote-control physics which Newton basically backed. In his only other publication 'Opticks', which he published both in English and in Latin, he used the phrases 'magnetick effluvia' and 'effluvia magnetica' and only used 'spirits' for 'flammable fluids'. But in the General Scholium ending his Principia he does indicate a maybe less scientific view on 'spirits' within bodies reacting to such emissions?) And seeing gravitation as the defining property of matter, maybe matter really seemed to Isaac Newton (as to William Gilbert) to be 'that which responds to attraction signals' and so really being maybe a kind of mind - very different to the widely-taught textbook 'Newton physics' ?

Of course Newton's conclusion that his evidence strongly supported attraction physics or blackbox physics against mainstream Cartesian Atomist push-physics was (stupidly) not accepted by most physicists (who Newton noted in Principia's introduction to Book 3, had "prejudices to which they had been many years accustomed"), and it was maybe too difficult for Einstein or anyone else to address. But Newton saw his published laws of physics as correctly predicting natural events without needing to know why things happened, in the manner of 'black box' behaviour laws that relate only inputs or stimuli to outputs or responses without considering any mechanisms connecting them. Newton's blackbox physics considered hypotheses regarding currently unseens as being matters of philosophy or logic and not science, and not currently provable or disprovable by science. Newton publicly concluded that though he had disproved substantial elements of Galileo-Descartes mechanical physics, like ether vortex motion gravity and motion tides, he allowed that some future modification of a mechanical ether theory might correctly explain gravity, magnetism, electricity and light. But Newton himself really seemingly preferred to use Gilbert-style attraction theory in thinking about physics, which he also seemingly thought might more likely correctly explain gravity and some or all other forces. Proving any contact-push physics theory also clearly requires being able to definitely distinguish a zero distance from an infinitely-near-zero distance needing perfect observation or measurement which can never actually be possible. And while requiring physics and science knowledge to have limits, Newton's black-box physics failed to specify what those limits actually are or actually should be - and not all were keen on science knowledge havig limits that were themselves unknowns. The Catholic Church opposed Newton's physics more than Descartes' physics partly for him being Protestant and partly for its more strongly supporting a moving Earth. A few strong supporters of Newton's physics much overestimated its impact, as did Bernard de Fontenelle in his 1728 'The Elogium of Sir Isaac Newton' claiming "Thus attraction and vacuum banished from physics by Descartes, and in all appearance for ever, are now brought back again by Sir Isaac Newton, armed with a power entirely new, of which they were thought incapable, and only perhaps a little disguised." (see Newton contemporaries)

Newton's considerations on Descartes push-physics as against Gilbert response attraction physics is maybe best put in his Principia Book 3 Rule 3. Here he first shows how we can reason that matter has solidity and exclusive-space-occupancy, then how "we must universally allow that all bodies whatsoever are endowed with a principle of mutual gravitation." Then he concludes that the argument is stronger for the universal gravitation of all bodies than for their impenetrability. But in finding that Gilbert-like physics was somewhat more likely the true option, Newton concluded that the evidence did not exist to decide between the two theories and might well never exist, continuing with "In bodies we see only their figures and colours, we hear only the sounds, we touch only their outward surfaces, we smell only the smells, and taste the savours : but their inward substances are not to be known either by our senses or by any reflex act of our minds" - Newton could see no evidence for Descartes 'certain knowledge'. He basically concluded that the evidence did not exist to decide between taking 'mass' as the measure of the size and pushability of bodies or as the measure of bodies' ability to produce and respond to gravity signals - though he seems clearly to have privately favoured the latter.

That magnetic force emission or 'magnetick effluvia' readily penetrate solids, even dense gold, Newton saw as evidence of solids not being fully solid or fully pushable but as containing much empty space - see Opticks Book2 Part3 PropV111. (but Gilbert had taken it as evidence that magnetic effluvia are probable incorporeal or non-pushing, otherwise if gold was 90% space then magnets should push gold 10% unlike two gold balls colliding with 100% push but small pushes might be hard to detect for experiments on magnets pushing gold. But of some interest Newton's scholium words above do also allow that INCORPOREAL bodies or media might either excite responses from matter or might somehow push matter though lacking any push property in a Cartesian physics sense.

Newton's Principia claimed to explain how any kind of motion related to any kind of force, but interestingly **excluded** the motion that Gilbert had concluded was impossible to explain with any simple push physics and could only be explained by his 'magnetical' or 'attraction' signal-response physics. This is the 'compass motion' or magnetic pointing/orientation motion, additional to magnetic attraction and repulsion, and Principia omits it possibly only because Newton may have felt that it might require him to publicly accept push-physics as being disproved when he did not want to publicly do that because the great majority of his physicist peers were wrongly prejudiced in favour of that. In his Principia second edition edition Book 3 Corollary 5 "the power of a magnet falls with distance not as its square but almost as its triplicate", as Kepler also noted. It seems that the experiments this was based on concerned the ability of a magnet to turn a compass and not to attract iron but implying that both would follow the same distance law, but non-Gilbert magnetic experiment often could not be replicated. Like others then Newton in fact achieved little on magnetism or electricity, which had some tricky behaviours, and later physicists could not get electromagnetism to fit Cartesian physics theory but did not even try to get it to fit the wrongly-supposed-disproved action-at-distance physics theory so that it began its drifting into indefinite physics theory. Still some Descartes supporters certainly managed to convince themselves that Newton supported Descartes despite his published works not at all backing that. And some Cartesians like Leibnitz, in a later version of his Tentamen, even convinced themselves that Gilbert also had supported Descartes physics despite him very strongly disputing push-physics and creating his own action-at-distance physics. But Leibnitz as a response to Newton's 'Principia' in his 1689 "Essay on the causes of celestial motions" also wrongly backed Descartes' vortex theory of planet motion claimed wrongly to disprove Newton's universal gravitation. The infamous Newton-Leibnitz calculus dispute had well predated Newton's physics of 1687, but Leibnitz foolishly overall favoured Cartesian physics over the Newtonian physics that had disproved it (though he did dispute much of Descartes other philosophy and science ideas).

While many physicists have seen explaining 'at-a-distance forces' like magnetism and gravity as more problematic than explaining 'contact forces' like collision, Gilbert and Newton saw explaining both types of forces as being equally problematic. Unlike Descartes they saw trusting logical deduction or human senses to always give direct 'certain knowledge' as being unscientific - to them 'contact' could be 'small-separation' and everything really needs rigorous experimental proof, even if at that time or any time such is not possible. And mostly experimental science should be proving what is certainly more probably true, rather than what is certainly true. Proofs of substantial distances may often be more reliable than proofs of zero distances which commonly require the unproven assumption that what is not visibly large must be zero, doubtful even before microscopes. There is no good scientific proof of the actual existence of any kind of 'contact push force' though such forces may seem to exist. And a majority of scientists insisting on push-physics when the facts demonstrated by Gilbert and Newton made it very doubtful was 'science prejudice' prevailing against Gilbert-Newton sense - as it was chiefly still doing so in 2015 with Brian Cox talking nonsense about Newton on TV in England as supposed science. Today the much-used Wikipedia and Discovery Channel seem to some to often really involve the replacement of actual knowledge with popular prejudices to which many have been long accustomed ?

Newton was a professional mathematician and really an amateur physicist. While acclaimed chiefly for his physics and his mathematics, he spent much time doing experiments in materials chemistry and in optics and magnetism as well as in studying chemistry/alchemy and religion, though he published mostly physics only and he, unlike many of his time including Kepler, showed no interest in Astrology. Always a majority of scientists have set their studies to fit the prevailing views of the majority of scientists of the time but Newton strongly rejected that at least privately. And in that he agreed with the earlier great English scientist William Gilbert. Newton may have seen Astrology as chiefly concerning irrational people behaviour and so not amenable to rational study. And he may have been one of those few people who basically believe nothing, and it drives them to study many things to themselves find if they might actually contain any bits of truth or none ? But, like William Gilbert, he funded his science experiments himself from his mathematician salery - helped, also like Gilbert, by not having a wife or children to support. The also unmarried Nikola Tesla also struggled to fund his science experiments. But it seems that in his study of physics Newton cut corners bigtime and/or hid some of his physics studies, maybe thinking he might get more from experiment than from study, though that is not the biggest problem of the time for physics history. Newton was certainly very annoyed that his early groundbreaking experiments on light were misunderstood and misrepresented by many of his peers but did not let that stop him doing further experiments but did put him off further publishing for a long time. He also did useful work on sound, and produced a theory of fluids that solved problems of fluids in movement and of motion through fluids. This he applied to Descartes' supposed unseen universal fluid ether, in which many physicists came to believe, but Newton disproved substantial aspects of that and he never conceded any kind of mediating ethers nor indeed signals as proven entities though granting that action-at-a-distance needed some kind of mediation. He did in his 'Opticks' and elsewhere use both ether push explanation and attraction signal explanation to help clarify his new physics ideas, especially for physicists who supported either one of such explanations and their 'unseens'. Many at the time saw Newton as developing Gilbert's theory which supporters of Descartes' Cartesian push-physics had made very unpopular by name-calling only, but one of Newton's great originalities was in his seeing particular explanations as unnecessary to science and seeing hypotheses on unseens as being unscientific - and being the first clear proponent of a blackbox science simply predicting everything even if Newton himself did not really support it. Copernicus, Galileo and others had earlier done some black-box science, but excluded explanation only either as being more politic or as to be perhaps done later. The substantial unpublished writings that Newton left after his death showed that Newton had major concerns about which he was unwilling to publish anything in his lifetime, like his major concern with Chemistry, seemingly being fearful of peer and employer pressure prejudice while alive. Having given up with physics long before his death, he made no arrangements about publishing such after his death and did not bother to leave any will. Newton seems to have ended life caring little what the world thought of him or of his physics, and perhaps seeing the world of a very biased 1600s physics as really basicly hopelessly corrupted. After his death, Newton's library was basically dumped - and much was only chiefly accidentally rediscovered much later but with some lost forever.

Mathematics was also advanced by Newton's work on calculus, which many of his peers falsely claimed was stolen from Leibnitz though there was real evidence of the opposite. But the much bigger stealing from Newton was undoubtedly Cartesian physicists falsely claiming that his Principia physics maths supported their Cartesian physics rather than attraction physics or blackbox physics. In time most physics textbooks were teaching a Cartesian Newtonian physics wrongly as being Newton's physics as the fake-Newton still taught now. While Newton's science was presentationally mathematical and distinctly in the style of Euclid, Newton always required that experimental facts must be decisive in science and not mere logical deduction or mathematics alone. Much of Newton's time was devoted to experiments, and of course Newton's published physics mathematics was, like Gilbert's and most early mathematics, presented geometrically rather than algebraically. And while his main work Principia concentrated chiefly on gravity, it did present a complete physics of all physical forces and affects as a 'Theory Of Everything' like William Gilbert did while concentrating on magnetism. So combining Gilbert and Newton may be like producing a Unified Field Theory 'TOE' which Einstein failed at.

Newton was the chief proponent of defined mathematical behaviour laws with undefined-explanation 'black-box science', maybe chiefly because he could see no way to certainly decide or prove between the alternative Gilbert and Descartes physics explanations ('Newton's Dilemma') or between equivalent alternative explanations of light. If different theories could fit the same mathematics then maybe they were either really the same theory or were compatible image theories and descriptions that only appeared different. Newton did convince a few other scientists of his time into favouring Black Box physics that could predict everything without relying on explanations, as being the best physics possible as long as there were no proven physics theories without unseens. (Of course the classic 'black-box' hid stuff from view and so prevents explanation, but perversely a 'black-box' has now become a thing that records stuff for later viewing and so provides explanation and would be better called a 'white-box' ?) But explanation-theory retained its popularity among scientists and was even credited to Newton incorrectly by 'mainstream' Cartesians. Black-box theory was maybe fine while nature was seen as being relatively simple, but it perhaps looked less intelligible when later nature became seen as being more complex - so it could then be argued that defined explanation is then needed to help make a theory more understandable ? Or maybe some correct science theory cannot be understandable to many anyway ? Of course a science theory cannot be only a bare mathematics with no physical meaning, but it can be a mathematics whose physical meaning is not fully uniquely defined.

Newton did not well explain his two explanation-physics options, nor refer people to Descartes and Gilbert for such explanations, but most of his peers believing him to support Gilbert's physics did not want Newton saying so. Newton undoubtedly knew how badly Gilbert's earlier attraction physics theory had been treated, and correctly expected that a theory substantially based on it would likely be equally badly received especially if it referred to Gilbert. Newton's peers mostly considered Newton's published physics as going to great lengths to hide its dependance on Gilbert's physics, and Newton may then have seen that as best for physics. Of course Newton giving little explanation allowed many to misunderstand or misrepresent Newton's physics as still holds even today. It certainly appears that Newton knew of Gilbert's main experiments on magnetism, and that he himself replicated some of those experiments. But despite his published physics strongly indicating otherwise, there is little evidence that he had any substantial teaching of Gilbert or did any substantial study of Gilbert (from the library he left, it seems Newton used a weaker magnetism experiment early English textbook instead and since he generally went to lengths to study original sources for other subjects Newton was maybe warned as was Galileo that studying Gilbert was dangerous ?). From what he published it is clear that Newton did magnetism experiments that he considered important and that he supported some physics that originated with William Gilbert, yet in the large range of his papers and books left when he died reference to these are somewhat strangely absent as though he saw Gilbert as more taboo than even alchemy. Unusually at least two of Newton's acquaintances, including Robert Hooke, supported Gilbert's attraction physics theory to at least some extent (if not very openly) rather than the more popular Cartesian push physics theory - but it is not clear to what extent they influenced eachother in that respect. Hooke did put some attraction physics to Newton, but it is probable that if Newton had studied Gilbert properly then he could have developed his own physics further than he did ? (Later still Nikola Tesla did make further progress with remote-control physics, though again seemingly without the potential added benefit of a study of its basic physics as published by William Gilbert.) Newton's library shows that at some point he studied Mark Ridley's 1613 'A short treatise of magneticall bodies and motions' which basically steals much of the magnetism of Gilbert's De Magnete as his own though basically being a reasonable early English translation of that and making some small additions and briefly mentioning Gilbert's De Magnete. It uses some of Gilbert's peculiar terms like 'coition' but omits some aspects of Gilbert's effluvium signal-response physics and his static electricity studies and considerations on gravity, and gives a more God-produced astronomy based on a push magnetism as a North Magnetism joined with a South Magnetism. Newton's library also shows that he studied Hooke's friend Robert Boyle's later 1673 'Essays of the strange subtilty, determine nature, great efficacy of effluviums' on magnetism, electricity and gravity being caused by the pushings of small-particle emission effluvia. Both are dubious physics as Newton no doubt came to realise. (Though neither Hooke nor Newton publicly acknowledged Gilbert, maybe someone still has some evidence connecting them to Gilbert or Gilbertian ideas ?) Newton's library and writings suggest strongly that his main interests were mathematics, religion and chemistry/alchemy - with optics as his chief concern for his relatively minor real interest in physics ? His notes did include interest in experiments on 'changing gold into silver' (quite the opposite of actual alchemist aims) and on many other matters including a range of experiments on magnetism with no mention of Gilbert (see Newton Manuscript Note). And he may not have named his 'physics giants' on whose shoulders he claimed to have stood because he had never actually studied the physics of any of them firsthand, his notes largely referring to contemporaries ?

Newton did try publishing one short paper on a part of his optics work submitted in 1672 to the Royal Society. This first paper was a small correct non-theory technical paper on colours, colour aberration and Newton's new reflecting telescope - fully proving all that it said (and was when he supported Galileo-Descartes atomism). But amazingly the eminent physicist peer Robert Hooke immediately tried to stop the Royal Society publishing this first paper of Newton, and himself published a ridiculous factually-wrong criticism of it that was somehow widely backed. In reply to Newton rightly defending his paper in 1675 an angry Robert Hooke threatened to form his own Royal Society, yet it was widely said that Newton was unreasonable ! But the excellent light experiments of Newton were soon attacked by Robert Hooke, by Rene Descartes, and by English Jesuits in Liege and others (who generally had all failed to reproduce Newton's prism experiments which it was claimed he had unfortunately failed to sufficiently accurately explain). So science history blamed Newton.

Of course while Newton concluded that colour was a property of light itself and was not just a property of illuminated objects modifying light, which he claimed that he had 'proven definitely with a crucial experiment'. But unlike William Gilbert earlier, Newton failed to publish the exact details of his experiment and so did not help with correct replication by other scientists. But opposition by Robert Hooke, Huygens and Jesuits like Pardies was really based on a less than scientific view that hypothesis was as valid as experiment and could not be disproved by experiment. Newton at least at times also supported an anti-publication view many alchemists had favoured that some knowledge was 'not to be commu- nicated without immense damage to the world' or keep to 'high silence'. But also later Newton allowed that light itself might respond to some signals from objects or their atoms as to gravitational signals.

Then in 1684 Gottfried Leibnitz (or Leibniz) seemingly after visiting Newton and seeing some of his maths began publishing some of Newton's key mathematics as his own, but by 1690 many were claiming that Newton had stolen the maths of Leibnitz. Newton decided against publishing further papers, and though he held a higher opinion of some earlier thinkers like Euclid, he was very wary of putting his own ideas to most of his peers. With a few minor mostly anonymous exceptions and private letters to a few friends, Newton waited until he could publish his science himself complete in book form - his Principia in 1687 and his Opticks in 1704. And when they were dismissed without real disproofs by largely Descartes-supporter peers, Newton resigned his Cambridge mathematics professorship to finish with physics and he found himself a new job as head of the British Mint. He stopped his physics theory work and his physics experiments and devoted his energies to his new job instead to his death in 1727, though he did continue to basically defend his physics through the Royal Society. He had maybe decided that physics choose to proceed by lying like politics and religion both of which he also had some involvement with ? And maybe that he had been too open in publishing his physics relatively early ? In 1822 a French physicist claimed that Newton had gone mad in 1693, then making him incapable at physics and turning him to religion, but this has never been widely accepted with the evidence generally supporting a more temporary milder episode only probably caused by mercury experiments.

Newton was perceived by some as being a bit of a bully, but he was in fact clearly very afraid of his peers views of his work and so he hid or mispresented much of his work. Hence in religion Newton was a pretend Church of England believer, though he held many written but unpublished religious beliefs that were significantly opposed to Church of England beliefs then. He undoubtedly considered the two works that he published,

Principia and Opticks, the most important of his work that he wanted published while alive, but being OK with other work being published after his death even though he did not prepare anything for such.

As with Gilbert earlier, Newton's attraction physics was rubbished without real disproof as being anthropomorphic, including silly claims that it required all matter to have eyes, minds and legs - ridiculous claims that themselves involve anthropomorphic thinking. Gravity being simple can clearly need only the simplest response, and the relative nature of attraction theory really gave it more scientific power. And simple single-cell creatures like Amoeba can show various responses to light and other things though having no brain, as can a TV to a remote. But Newton's black-box theory allowing action-at-distance signal-response was soon simply ignored as though it did not exist.

To quote 'A Short Account of the History of Mathematics' (4th edition, 1908) by W. W. Rouse Ball, on Newton -
" His theory of colours and his deductions from his optical experiments were at first attacked with considerable vehemence. The correspondence which this entailed on Newton occupied nearly all his leisure in the years 1672 to 1675, and proved extremely distasteful to him. Writing on December 9, 1675, he says, `I was so persecuted with discussions arising out of my theory of light, that I blamed my own imprudence for parting with so substantial a blessing as my quiet to run after a shadow.' Again, on November 18, 1676, he observes, `I see I have made myself a slave to philosophy; but if I get rid of Mr.Linus's business, I will resolutely bid adieu to it eternally, excepting what I do for my private satisfaction, or leave to come out after me; for I see a man must either resolve to put out nothing new, or to become a slave to defend it.' "
(see www.maths.tcd.ie/pub/HistMath/People/Newton/RouseBall/RB_Newton.html)

A majority of Newton's peers were strong Galileo-Descartes push-physics supporters who would not consider alternative theories, and especially would not consider the old enemy Gilbert attraction theory. They saw Newton as an anti-Descartes Gilbert theorist and believed that Newton's blackbox position was just a fraudulent cover to disguise his backing for the hated Gilbert theory. The minority of Newton's peers who would reasonably consider alternative theory ideas, mostly took Newton at face value as supporting blackbox theory and not attraction theory - and only few of them accepted black-box theory. Nobody other than Newton gave any real consideration to attraction theory, not even to attempt disproofs of it. And Newton himself produced no disproofs of it, only disproofs of parts of Descartes mechanical physics which suffered from more rigid requirements as do many other physics theories. Newton firmly held to his blackbox-science line dividing scientific knowledge from non-scientific knowledge - with religion and explanations of gravity and other forces being areas of great interest but outside science.

Newton privately seemingly tried unsuccessfully to develop his attraction physics effluvia/emitted-spirits theory by much experimenting on novel and new materials that only chemistry could produce, additional to his published experiments in magnetism and optics which latter led to his invention of the modern reflecting telescope. His private non-catholic religious ideas were seemingly much more specific and detailed than those of catholic Descartes, but his published attraction theory emitted-spirit ideas were maybe less developed than Gilbert's published effluvia signal ideas. And early chemistry then was still being demonised by being called alchemy even by most so-called scientists who should have known better when even in 1600 Gilbert had acknowledged chemists. While the limited unpublished writings of Gilbert were eventually published after his death seemingly as he wished, though only in Latin, the many unpublished writings left by Newton strangely remain largely still unpublished. Of course the unpublished writings that Newton left when he died were not chiefly on gravity physics, but also on mathematics, religion, chemistry/alchemy and even economics.

Newton like Gilbert became acclaimed as a great scientist, while the theories of both were actually rapidly rejected without real disproof (much later Einstein did produce his 'disproof of Newton' which was eagerly accepted with nobody looking closely enough to see that the theory Einstein was actually disproving was Cartesian theory). The failure of Gilbert and Newton theory among physicists was not reflected among non-physicists, so that even today most people see their signal-attraction physics theory as correctly explaining magnetism and gravity. Newton like Gilbert did in his lifetime develop a few very strong followers in England though with strong opponents also and in a chiefly Catholic Europe mostly strong opponents. A caricature of part of Newton's physics theory became acclaimed somewhat slowly, with his real theory rejected with Gilbert's by the mob of scientific pigmy peers - and that process passing into physics history continues still now. Or maybe, being really generous, it could be said that the world was just not really ready to look at a physics that was not some simple mechanical push physics - and maybe the world is still not ready ?! Additional to fierce Catholic church opposition to early science, some opposition to early science (most often somehow directed at Newton) has been non-religious philosophical opposition as from some poets, artists and philosophers including Yeats and Goethe. Certainly it remains rare today to find an even half-reasonable scientific view of Gilbert-Newton physics outside of this website.

For comparison with other physics theories, Newton's three laws of motion were ;

1. Every body will remain at rest, or in a uniform state of motion unless acted upon by a force.

2. When a force acts upon a body, it imparts an acceleration proportional to the force and inversely proportional to the mass of the body and in the direction of the force.

3. Every action has an equal and opposite reaction OR the mutual actions of two bodies on each other are equal and opposite.

Newton's view of 'a force acting' allowed of either some kind of Descartes 'dead-matter' push action or Gilbert 'robot-matter' signal attraction action from another body. It requires the existence of 1.a force from one body AND 2.a second body acted upon by the force, with the actions of each being relative to each other. He is maybe here not clear enough that his 'force' gives RELATIVE change of motion, relative acceleration, rather than giving absolute change of motion and that all motion is DIRECTIONAL or vectorial. While push-physics requires all forces to be directly associated with an originating body, attraction physics allows some forces to exist in signals separated from an originating body though allowing that the signals themselves may be some kind of body. But for both, forces acting need BOTH an originating body AND a body acted on and forces persist (as with collision, spring and gas-pressure forces) for only as long as they are opposed. Also to Newton, two equal and opposite forces produce equal and opposite accelerations giving no motion so that force acceleration change acts primarily over time and not always also over distance or space. Newton did with substantial success mathematise Gilbert attraction physics in these respects. But the spacetime vectoral mathematics later developed by Minkowski would maybe better suit it than being wasted on some merely geometrical physics.

Current mainstream physics commonly seems to say that the gravitational force between two bodies in Newton's physics is given by the formula $F=G((m_1 m_2)/r^2)$ which may imply that their mutual attraction is proportional to the product of their masses. But a pebble doubled in mass does not fall to the ground at double the acceleration, showing only an infinitesimal increase. Though now used as an approximation for mutual gravitation in terrestrial gravitation for masses tiny compared to Earth's mass, this mainstream 'mis-equation' is only about the hypothetical one-way effect of one gravitational mass on another inertial mass and might better reflect the physics, and Newton, as $F=Gm_1(m_2/r^2)$. And, as Newton required, the mutual attraction of two bodies is the simple sum of the gravities of each. This mere force addition may seem to some to undermine mutual causation, though push action-reaction mere force addition may not seem to undermine mutual causation there. Newton did use an explicitly stated approximation for calculating the gravity of actual objects by taking the objects as zero-size point objects rather that their actual size. He concluded that experiment proved that using such center-of-gravity points commonly gives an adequate accuracy for gravity calculations, and does not give infinite gravity where two bodies touch as some still falsely claim.

There is often some misunderstanding of Newton's third law of motion - action and reaction or mutual actions being equal and opposite. Does this merely state that the inertia of a body will oppose any force applied to it ? Push reactions can seem simply explained as due to inertia plus exclusive-space-occupancy in a Galileo-Descartes type push-physics. So when A pushes on B with some force-action, B's inertia then pushes back on A with an equal and opposite force-reaction - if bodies actually can contact and push. (What determines the extent to which A and B actually accelerate here, is then the strength of the forces on each relative to the strength of their inertias.) But the equality and oppositeness of attractions or repulsions of bodies separated by some distance may seem to rule out the inertia of B causing any reaction force on A, and somehow require some different action-mutuality so that if A attracts B with some force-attraction then B must attract A with an equal and opposite force-attraction. And for a 'remote-control robot', the 'remote' body can send a signal that causes a physical action in the 'robot' body without there being any physical reaction on the 'remote' body. Newton showed that his laws of motion do apply to gravitating bodies far apart, but was maybe less clear as to exactly how they applied then. And in some chain of multiple actions and reactions, the final reaction may well wrongly appear to have no relation to the initial action.

Cartesian physics and all subsequent forms of contact-physics including Einstein's require contact action-reaction to involve zero space separation and imply zero reaction time - ie instantaneous reaction (though zero is not measurable or provable in science experiment). But all Gilbert-Newton attraction physics requires action-reaction to involve positive space separation which implies positive reaction time however small - ie non-instantaneous reaction. Of course opponents of Gilbert-Newton attraction physics have repeatedly falsely claimed the opposite holds, and still do - they nonsensically claim that 'Gilbert-Newton attraction physics requiring instantaneous reaction disproves that physics' !! Now some of Newton's astronomy physics can seem to assume or require near-instantaneous reactions to forces or near-simultaneous action and reaction - though he nowhere makes even that a specific requirement. And Newton's astronomy seemingly needing near-instantaneous gravity-signal response action may not need Einstein physics maths but that the gravity signal response involve effective signal prediction as at Information Physics. But if one (Cartesian) body collides with a second such body then they clearly collide at the same instant and the 2 bodies will seemingly exert collision forces on each other at the same instant, and give equal and opposite changes of momentum to the 2 bodies. Similarly instantaneous reaction also seems required by all push-field or push-continuum theories, which are also weaker on time directionality. A full signal-response physics clearly requires that all present physical actions are necessary responses to previous signals so that time must exist and must basically be one-directional and also that change does not just happen but must happen to some determined signal-response laws. If observation is response to signals, then signals received now will have been created earlier so that seeing the Sun 'now' is really seeing the Sun 8 minutes ago and taking nearby observations as being 'now' would generally involve little error, but taking distant observations as being 'now' could be very wrong ? And to the extent that some responses might be to some multiple set of signals, the link between some responses and individual signals can be statistical or probabilistic without any less actual determinism though allowing some apparent indeterminism. In a Gilbert-Newton action-at-distance attraction signal-response physics, actions always involve at least one observer or detector responding to at least one prior signal so that an action can be relative to at least one observer and one signal. So for any observer or detector an action is relative to the relative directionality of a received signal to that observer or detector, and its preferred frame of reference will differ both for different received signals and for different observers or detectors. And quite unlike Cartesian physics with its basially supposed one universal preferred frame of reference, an action-at-distance physics can allow greater complexity involving multiple simultaneous relativities while remaining basically the simplest physics.

Given an observer body and another body, an observer body clearly generally detects motion in the other body only relative to its own motion and generally detects its own motion only relative to the motion of the other body so that no motion can be determined as being absolute motion. From that it follows that generally neither can absolute motion energies be determined. Laws of relative motions, relative energies and relative forces alone can generally be determined. But if an observer body has some indeterminate motion then other bodies nearby may well share that same motion. And any net motion can be viewed a sum of several different component motions so that any uniform motion might be viewed as composed of (or include) cancelling opposite accelerations, and a motion uniform for a time might yet be begun and/or ended with accelerations. But a net uniform motion can seem to be a net acceleration motion, or viceversa, to an observer body that itself has some appropriate motion. A body can be at rest or in uniform motion when a force is acting on it, only if it is acted on also by some second exactly equal and opposite force. And motion energies, or kinetic energies, are subject to similar requirements.

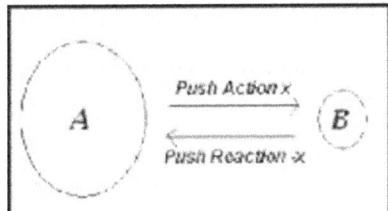 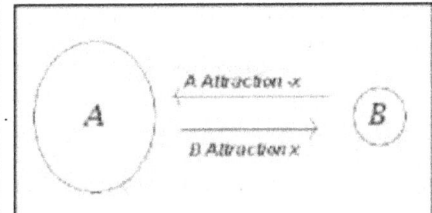

For an overview of a 'Gilbert-Newton' view of gravity and like forces see The Attraction Theory of gravity and other forces.

Motions.

The chief evidence of the operation of most physical laws of nature is found in different motions, as considered in the studies of many concerning physics such as Galileo, Gilbert, Kepler, Descartes, Newton and Einstein.

The perseverance of much natural motion like planet orbits helped convince Gilbert and Newton that space offers no resistance to, or drag on, the motion of bodies in it - and cannot affect bodies motion. But both Descartes and Einstein assumed that space can somehow push bodies and so also drag on bodies motion. The perseverance of natural planet orbits seems to some to require at least some steady force such as gravity. However, natural orbits and spins to some seemed like rest and uniform straight line motion in requiring no force to maintain them. And some even thought that uniform straight-line motion does need a force to maintain it.

Spin or rotation of a body about a central fixed point within itself, is commonly considered as for a 'perfectly solid body' or 'uni-part' body though no multi-atom body may actually be such so perhaps little is really known of actual solid body spin. Spin is physically similar to the circular motion of bodies about an external point, as of the Earth and Mars about the Sun, called orbiting or orbital revolution. Both are non-uniform motions that require forces to maintain them as well as to change them - but some forces can be persistent, like the Sun's gravity, and can be internal to a body or a system. If any multi-part object or system held together by limited forces is made to spin fast enough then its parts will fly apart. A 'perfectly solid body' is generally now taken as having parts held together by some infinite force, though short-range strong forces may actually be involved and Descartes-type physics perhaps unreasonably assume some 'uni-part' bodies needing no holding-together forces.

Some natural uniform motion velocities are probably central-attraction escape velocities and probably include atomic escape velocities of which the 'velocity of light' may well be an example. Other major natural uniform motion velocities certainly include those for wave transmission through mediums as for the 'velocity of sound'.

Another basic type of natural motion is deflection or reflection, as where the path of motion of something moving is changed when it meets another object - eg when a moving ball meets a wall or when a light ray meets a mirror. One possible explanation of some or all reflections is contact collision, of two things being unable to occupy the same space so that the parts of any motions directed to occupying the same space have their direction reversed. A second possible explanation of some or all reflections is proximity repulsion, as bodies increasingly repel each other as the distance between them falls, see Opticks Book 3 Obs X1 Query 1. But interestingly for light reflection Newton also suggested the further possible explanation of post-contact proximity attraction, where a surface strongly attracts something passing into it and pulls it back out of it. Such case of attraction mimicking repulsion might even also offer an explanation of apparent universe expansion. Of course it is maybe not clear what atomic forces would be needed for that light effect, and Newton might perhaps have done better with a simple repulsion which has attraction mathematics but with an opposite sign. And if billiard ball collisions are in fact proximity repulsions, could the extent of currently known atomic repulsion forces fully explain billiard ball collisions ? And would a perfectly elastic collision require an infinite repulsion force or just repulsion with the inverse square law ? And might post-contact proximity attraction also somehow be able to offer another possible explanation of billiard ball collision ? Certainly the maths of actual collisions and of proximity responses should show some differences that vary as closeness varies but experiments have not been designed to examine that. Newton did still see most then known light behaviours as evidence of it being a form of matter rather than just waves.

It follows from Newton's laws of motion that objects with similar velocities relative to some inertial frame of reference can attain different relative velocities only if forces do different work on them. The 'kinetic energies' of objects are measures of the work required to bring them to rest relative to some inertial frame of reference - and by definition more deceleration being required by a faster object, kinetic energies are the products of objects masses and their velocities relative to the inertial frame of reference. It follows that kinetic energies are not absolute properties of objects, but are only relative properties. But it is generally assumed that objects do have some absolute properties, which might or might not include such things as maybe 'mass' or other properties.

Objects motion can only be changed if some external force is applied to them, and for a given object a greater change in motion requires a greater force being applied. For any two different objects if a given change in motion requires different amounts of force being applied, then they are said to have proportionately different inertias. If the type of force being applied is gravitational force, then they are said to have proportionately different 'masses' or 'gravitational inertias'. But if the type of force being applied is magnetic force, then they are said to have proportionately different 'magnetic powers' or 'magnetic inertias' which will involve both their 'masses' and their 'iron percentage'. But if the type of force applied is 'contact force' or 'momentum force', then the forces and inertias involved are proportionate to the masses and gravitational inertias. Hence an objects inertia relative to gravity and momentum change is commonly called 'its inertia', despite some objects having also different forms of inertia like magnetic inertia. Like all objects non-iron objects have inertias, but they are unaffected by magnetic force with respect to which they hence have infinite inertia. So inertia is basically the responsiveness or non-responsiveness of bodies to forces or force signals. To both Gilbert and Newton, gravity and other like forces are caused by some agent or agents emitted at some high speed by objects and received by or touching other objects. And while Newton did allow that some form of Cartesian push physics might fit his mathematics as did action-at-distance physics, and Cartesians quickly claimed wrongly that their standard push physics did, no physicist or mathematician has ever proved Newton's maths actually fit any kind of Cartesian push physics unless maybe the General Relativity of Einstein be taken as such though he denied it was a push physics but he was perhaps mistaken in that. Of course Cartesian push physics and Gilbert-Newton action-at distance physics could and did share some limited common ground some of which Einstein challenged.

Motions common in larger visible objects may also be common in less easily seen microscopic objects - or may not. Hence microscopic objects do commonly show one apparently random motion called Brownian motion which may or may not have a real equivalent in larger object motion. And there is always the issue of the absoluteness and the relativity of any motion. Newton saw uniform motion as not distinguishable from a state of rest if the observer had the same uniform motion or state of rest, ie was in the same inertial frame of reference, and from that concluded that an observer could not know if his inertial frame of reference was a state of rest or some undeterminable state of uniform motion. And if gravitation is universal and necessarily non-uniform and accelerating, then maybe nowhere can there really exist any actual inertial frame of reference.

Newton's ideas overall.

Newton is best know for his work in mathematics, optics and physics, but he certainly owned books on religion, alchemy and other subjects - on which he also wrote much but published little, and many wrongly labeled him an 'alchemist' and a 'heretic Christian', and many still do so, though he did not declare publicly any real belief in either. But, as The Big Bang Theory TV show indicates, many scientists today like Science Fiction and Fantasy Gaming which they know is not real.

In editions of his Opticks from 1706, Newton discussed how microscopic forces analogous to gravity might explain some chemical phenomena and he did publish a little on simple chemistry experiments in the 1730 fourth edition of the Opticks. Newton did many chemistry experiments, but none seemed to have anything to do with the old alchemy aims of making gold or eternal youth. His main science problem almost certainly was to demonstrate exactly how gravity worked and, since magnetism and electricity show different effects with different materials as Gilbert's experiments had shown, he may have sought a substance that would impact gravity differently but found none. Newton maybe was looking for Anti-matter or Dark Matter ?!

Newton knew hot magnets did not attract and, before publishing Principia in 1680 letters to Flamstead, wondered if hot Suns might likewise not attract - clearly seeing gravitational and magnetic or electric forces as having some greater or lesser similarity as Gilbert had. Newton also wondered if emitted 'electric spirit' or signal attraction might be much stronger at atomic distances and be the most likely cause of matter cohering, and such signal attraction physics as best explaining both thinking and unthinking phenomena, in his unpublished manuscript 'The Queries' Questions 24 and 25. And elsewhere Newton wondered if such emissions might be simpler vibration energies, or favoured blackbox no-explanation attraction theory as Qu 23. And Newton did also consider a maybe perverse possible cause of gravity as the emission by bodies of a very rare material medium that got DENSER with distance from its source, so pulling bodies to the less dense area nearer to the source, as in Qy 20, Qu22 and Qu 23. A maybe less perverse explanation of gravity might be a uniform material Ether and bodies emit an Anti-Ether weakening with distance from source that 'eats' such Ether and would do so more nearer bodies and pull bodies to the less dense areas nearer to bodies. Any material Ether would need to be rare enough to give little drag but not so rare as to give little gravity pull, and while the drag would be due to rarities the gravity would be due to rarity gradients and the two not be proportionate so outer planet orbits have more drag. Of course any such simple mechanism that might explain simple gravitational force, is unlikely to be able to also explain trickier magnetic and electric forces (as Einstein's many years of failed unification work showed). Modern discoveries like gravity Black Holes and universe expansion seem to further back a Gilbert-Newton action-at-distance signal-response physics which is basically an advanced information-handling physics way ahead of its time. It seems that only a physics like Gilbert's signal-response physics can readily handle a variety of differing physical forces and maybe Standard Model 'spins' and 'colour charges'. But Newton basically saw that nature had common orbits and spins as of planets whose maths were entirely inconsistent with the maths of push-physics like that of Descartes as shown by a spinning wheel or by spinning water and so must be due to an action-at-distance physics like that of Gilbert. (See www.newtonproject.sussex.ac.uk/) The actual physics of Newton was attacked so much, especially by Cartesians, that he may have somewhat as he intended saved his physics by not totally attacking Cartesian physics - though this has allowed a sad Cartesian perversion of Newton's physics to be widely falsely taught as being 'Newton's physics'.

Todays textbook 'Newtonian physics' is basicly actually Cartesian physics stealing Newton's maths, which does not fit it well, and it is not Newton's physics for which you need to actually study Newton. So even in England 'Newtonian optics' has long been taught as a corpuscular billiard-ball Cartesian optics, though if you actually read Newton's published 'Opticks' you will see that it is actually more in line with his gravity in considering attraction forces etcetera. While accepting that matter must generally be corpuscular (or particulate), Newton did not accept that its motion could only be due to push forces but allowed of other kinds of forces and allowed of a vacuum both unlike Descartes. And while Descartes' 1637 'Dioptrics' gave his corpuscular push theory of light which was supported by the early Newton in 1672 though the push less so by his 1704 Opticks and its 1730 edition, with interesting physics Questions at the end if its 3rd Book, that he prepared before his death but was not published till after his death. See - Opticks 1730. Very unfortunately the 'Newton' widely taught worldwide is 90% Descartes and only 10% Newton, but physicists teaching this have themselves learned from textbooks teaching this fake-Newton and so firmly believe that is 'Newton' - now one more prejudice to which most physicists have long been accostumed. But in his Principia, Newton clearly indicated that he considered Rene Descartes to be his main science opponent, and not Robert Hooke who almost all 'Newton historians' have continued to falsely claim was his main science opponent. Yet Newton published strong disproofs of Descartes' physics ideas but not of Hooke's which wer clearly close to his own physics. This and other misrepresenting of Newton they support by ignoring Newton's publications and instead referring to selections of private writings despite Newton being very definite on what of his did or did not merit publishing and public consideration. Newton officially taking a 'no theories allowed' position on science might have been expected to be happily seized on by religion as favouring it, except that Newton was widely taken as really supporting Gilbert's attraction physics which catholic church Jesuits fondly imagined they had already disproved. Unlike some other scientists like Gilbert, Newton's life was not cut short, dying aged 84, before he ended to his satisfaction his science and its publishing. Having had enough of his peers wrong criticisms, he quit Cambridge and science in 1701 to fully devote himself to his British government position heading the Royal Mint which he had secured a few years earlier. Making no arrangements for further publishing, and leaving no will, confirmed Newton had given up on physics and probably on religion also. The widespread strong unreasonable opposition to Newtons science in his lifetime, lead by catholic Jesuit 'scholars' but also by fellow protestant peers, has long been and is still now also widely ignored. To Newton's time and beyond science has faced strong attack and most scientists did not strongly resist that, generally preferring instead a quieter life and following a somewhat easier path often involving only fighting eachother and rarely defending eachother. What became taught as 'Newtonian Physics' was Descartes' Cartesian physics with Newton maths falsely bolted-on though that was actually an inconsistent unworkable physics critically inferior to Newton's actual physics which significantly incorporated Gilbert's magnetical physics or attraction physics. But where Newton had repeatedly used the term 'attraction' the physics textbook writers all quietly replaced with the term 'gravitation' or 'gravity' to present it as less the attraction or magnetical physics Newton leaned towards but instead more as just a Cartesianlike gravitational push-pull mechanical physics. And while Newton did have a strong interest in problems involving both religion and alchemy, he certainly gave no significant support to religion or to alchemy and published nothing on either so that accusing him of being an alchemist is just talking nonsense !

Newton's Principia was translated into French by Émilie du Châtelet (1706-1749) and though she soundly opposed Cartesian push-physics even more than Newton she backed Newton's attraction physics maybe somewhat strangely, though like many others of the time partly including Newton, without at all considering his 'spirits emitted' action-at-distance signal-response physics or William Gilbert. She did also back Newton's optics including light being subject to proximity gravity attractions and electrical repulsions by bodies. She did of course also push her own version of 'Newton's physics' as did others but a much less Cartesian version not so far from Newton. Interestingly she also concluded that every specific action needs a specific cause, so for no action could accident or chance be a cause - a logical equivalent of Einstein's later less logical 'God does not play dice' ! See her Light and her Physics.

To read Newton's Latin original 'Principia' or an English translation see the top of our Newton's Principia section. And to read Newton's Latin and English versions of 'Opticks' see the bottom of that section.

Get our great Newtonian gravity Android App - 'Sun Pull' - in the Google Play app store to help you study or re-design the solar system better ! Or you can try it on here in our Solar System section, which also discusses what is probably chiefly needed for real actual contact with 'alien' people from other worlds. Hopefully more useful science Apps may follow?!

And you can do a good search of this website below ;

Search on this site www.new-science-theory.com, with Google.

Or do a search of the web better with DuckDuckGo -  Type web search then Enter

---

You are welcome to link to any page on this site, eg www.new-science-theory.com/isaac-newton.php

---

© new-science-theory.com, 2022 - taking care with your privacy, see New Science Theory HOME.

Sir Isaac Newton - *his major physics work 'The Principia'*

Homepage . William Gilbert . Rene Descartes . Isaac Newton . Albert Einstein .... Against Descartes...........General Image Theory

Like Descartes and Gilbert, Newton published his major works in Latin, though he did publish his Opticks first in his native English. His major work 'Philosophiae Naturalis Principia Mathematica (Mathematical Principles of Natural Philosophy)', often referred to as 'Principia', was first published in 1687 but his 1726 third edition is considered his definitive version usually translated though omitting the interesting 1713 second edition Roger Cotes preface discussed in our main Newton section here.

An English translation of 'Principia', though Newton approved none, can be read online at Principia. Or, see the Latin original Principia at Google Books Newton. Google Books also has an 1848 English translation - HERE (it starts with an added Newton Biography that should be taken with a large pinch of salt, and further adds to the end an extra 'System of the World' an earlier less-mathematical writing of Principia Book 3 which also got published separately after Newton's death maybe against his wishes and was somewhat like Gilbert's De Mundo.) - download it and read about using Google Books at the bottom of our Science History.

Newton's 'Principia' - Definitions and Axioms or Laws of Motion.

Newton's 'Principia' explains a mathematical physics allowing of two alternative explanation theories, the mechanical push-physics of Rene Descartes and the signal-response attraction-physics of William Gilbert. Newton began his Principia with two sections called 'Definitions' and 'Axioms, or Laws of Motion' which give his basic definitions and his three laws of motion. He first defines the mass of a body, the momentum of a body, the inertia (or innate force) of a body, an impressed force on a body (as accelerating it but adding nothing permanent to it), a centripetal force on a body, a centripetal force emitted by a body (diminishing with distance from it), and finally distinguishes Motive, Accelerative and Absolute centripetal force as relative to attracted bodies, to the locations of such bodies and to attracting bodies. In numbers of his definitions Newton plainly refers to magnetism and to Gilbert's theory of it (without mentioning Gilbert), especially his last three definitions with his final definition stating ;

"These quantities of forces we may, for brevity's sake, call by the names of Motive, Accelerative, and Absolute forces ; and, for distinction's sake, consider them with respect to the bodies that tend to the centre, to the places of those bodies, and to the centre of force towards which they tend ; that is to say, I refer the Motive force to the body as an endeavour and propensity of the whole (of it) towards a centre, arising from the propensities of its several parts taken together ; the Accelerative force to the place of the body, as a certain power or energy diffused from the centre to all places around to move the bodies that are in them ; and the Absolute force to the centre, as endued with some cause, without with those motive forces would not be propagated through the spaces round about ; whether that cause be some central body (such as is the loadstone in the centre of the magnetic force, or the earth in the centre of the gravitating force), or anything else that does not yet appear. For I here design only to give a mathematical notion of those forces, without considering their physical causes or seats."

and "I likewise call attractions and impulses, in the same sense, Accelerative and Motive ; and use the words attraction, impulse or propensity of any sort towards a centre, promiscuously and indifferently, one for another ; considering those forces not physically, but mathematically : wherefore, the reader is not to imagine that by these words I anywhere take upon me to define the kind or the manner of any action, the causes or the physical reason thereof, of that I attribute forces in a true and physical sense, to certain centres (which are only mathematical points) ; when at any time I happen to speak of centres as attracting, or as endued with attractive powers."

Newton takes his black-box position early, and continued it to the end, of not supporting any explanation of how forces might produce their effects and allowing that one of several explanations might be correct. This Newton also clearly did in his quite unique definition of inertia in Definition 3 ;

"The vis insita, or inate force of matter, is a power of resisting by which every body, as much as in it lies, endeavours to persevere in its present state whether it be of rest or of moving uniformly forward in a right line. This force is ever proportional to the body whose force it is ; and differs nothing from the inactivity of the mass, but in our manner of conceiving it."

It was a central conclusion of Newton that the behaviour of bodies might be conceived of in different ways by different people involving different hypotheses - and if unseens are involved then science maybe cannot prove which is correct. And to Newton inertia can be taken as the resistance of a dead body to being pushed or equally as the resistance of an active body to moving itself. To Newton the strength of mathematics in science lay in it allowing multiple explanations, specifying how things relate but not why, as by neutrally specifying physical unseens using constants or however. Of course many lesser scientists have claimed their mathematics proves some explanation, when really it cannot. Much modern physics theory now rests basically on different mathematics produced by 'mathematics experiment' or 'thought experiment' and wrongly claimed to prove different explanation theories. Newton's blackbox position disallowing mathematics from proving unique explanation theories still has some support, though many do not understand it or see it as a weakness. Of course Newton failed to note that over time as science knowledge increases so science blackboxes shrink, though they will never vanish. Increasing science knowledge encourages more theories and more claimed disprovings, while making it more difficult to fully specify a good science theory covering all. This shows especially in modern physics.

Newton's alternative-theories blackbox approach shows also in his third law of motion claimed by him to be proven by experiment in mechanics and magnetism - "To every action there is always opposed an equal reaction ; or the mutual actions of two bodies upon each other are always equal" It holds for possible push actions or for attraction actions without requiring any specified mechanisms for either unlike most other physics theories. So in his Corollary 4 which disallows gravity between themselves affecting the motion of the common centre of gravity of two or more gravitational bodies ;

"in a system of two bodies mutually acting (eg. gravitationally) upon each other, since the distance between their (gravitational) centres and the common centre of gravity of both are reciprocally as the bodies (masses), the relative motions of those bodies, whether of approaching or of receding from that centre, will be equal among themselves."

Interestingly mutuality plays a greater part in Newton's gravity theory than in that of Descartes or others, and Gilbert earlier had also made much of the mutuality of magnetic and possibly other attraction - to the extent of giving it a technical term of his own, 'coition'. Attraction Physics could almost be as well called Coition Physics. Hence in a signal physics, 'B attracts' is not really meaningful - though 'B attracts C' is. And 'Bs emit signals that attract' is not meaningful - though 'Bs emit signals that attract Cs' is. Both Gilbert and Newton realised this, but most other physicists have basically wrongly favoured 'B attracts'. Einstein and Heisenberg realised that physics had some 'observer issues' and more recently some physicists have tried to deal with the issue (perhaps unconvincingly) in adopting Relational Quantum Mechanics. But a signal-response physics along the lines of Gilbert-Newton 'attraction physics' does seem to be capable of handling this issue more meaningfully. In a signal physics, causality is 'mutual' and what causes nothing is outside science. Gilbert basically concluded that, to a real experimental science, mere dead matter if it existed would be useless - and be unobservable - and be outside science.

(A recent theory called The Final Theory claims that a non-conservation of energy problem exists for 'Newton's gravity', only by wrongly assigning to Newton a 1-way non-mutual energy 'attraction gravity' though Gilbert attraction theory requires at least 2 bodies and mutual action and Newton also held to that when he referred to attraction theory though not committing to it. Of course Descartes push gravity, as also referred to but not committed to by Newton, likewise had no energy conservation issue. But wrongly assigning theories to Newton is a common mistake that even Einstein wrongly indulged in, while replacing real experiments with his 'thought experiments'.)

This section of Principia Newton finished with an interesting Scolium showing action and reaction being equal both in mechanics 'contact' push forces and 'action-at-a-distance' forces. As part of this he shows that this not holding would allow some bodies to exhibit perpetual acceleration, so two bodies attracting each other and in contact though having different powers of attraction would show joint-acceleration with no outside force being applied which is not allowed by his first law of motion. (Newton used magnets for experimental confirmation of this in a two body situation, though magnetism is a trickier matter and he did not mention magnets differing in their power/mass ratios as magnets can but gravity cannot - Eg Magnet B of power 10 and mass 2, with magnet C of power 5 and mass 3 : 2x5 = 3x10 ?)

In his Book 1 Section 14, Newton proves that the refraction, reflection and diffraction of light can all be explained by light being attracted by and bent by gravity-like signals from materials. His light reflection involves neither particle collision surface contact reflection nor wave surface contact reflection, but below-surface attraction light bending - so light being bent by gravity is not an idea unique to Einstein as some think.

Hence writes Newton, "If two similar mediums be separated from each other by a space terminated on both sides by parallel planes, and a body in its passage through that space be attracted or impelled perpendicularly towards either of those mediums, and not agitated or hindered by any other force ; and the attraction be every where the same at equal distances from either plane, taken towards the same hand of the plane ; I say, that the sine of incidence upon either plane will be to the sine of emergence from the other plane in a given ratio."
And then, "These attractions bear a great resemblance to the reflexions and refractions of light made in a given ratio of the secants, as was discovered by Snellius ; and consequently in a given ratio of the sines, as was exhibited by Des Cartes."

Newton's 'corpuscular' light theory did not involve dead push particles, but attraction theory robot particles that responded to signals. And that was why some fiercely attacked it, while others like Hooke attacked it because it was not a wave theory. Descartes published in his 1637 'The Dioptrics' his push-physics particle optics with light being faster in dense mediums, and Fermat's 1662 theoretical claim of light being slower in dense mediums was widely dismissed as Newton, Huygens and Hooke published works on different optics from 1665 to 1704 with light being faster in dense mediums. Later experiment showed light is slower in dense mediums, but only Newton's theory of light was blamed for the early optics speed error which was claimed to be due to it being a 'corpuscular' theory and not a wave theory. Of course wave optics actually needed particles or something to be waved. In fact Newton's was an attraction signal optics and can easily be modified to allow for light being slower in denser mediums, by simply replacing attraction with repulsion which has the same mathematics with only a different sign. The real problem for Newton's optics was that his light 'fits' for interference type events seemed to have no simple explanation. But his reflection-fit/refraction-fit was akin to signal-on/signal-off or wave-peak/wave-trough or 1/0 - so taking forces or force-signals as being digital or quantal might resolve that ? Of course signal-on/signal-off may be a major property of signals but they can allow of having additional properties including directionality. It may be possible to demonstrate apparent 1+1=0 'destructive interference' of waves, of particle or signal beams or even of single particles or signals - but no simple physics theory as yet seems really able to give a proved full explanation. Of course some apparent interference effects may really just involve spatial rearrangements with nothing actually destroyed. See our Light section

Newton held strongly that some of the apparent 'seens' of science are really 'unseens' - eg when one ball is seen to hit another, the supposed contact cannot be actually seen. So even simple mechanics push theory, or impulse theory as he termed it, could rest on an unseen. This view seems supported by modern knowledge of materials atomic structure, showing that surfaces contain relatively little to make contact with, and is some evidence against contact theories of light reflection and perhaps contact push theories generally. Hence the atomic structure of glass below shows more space than atoms - and the atoms themselves now seem also to be largely empty space ;

Atomic structure of glass surface --

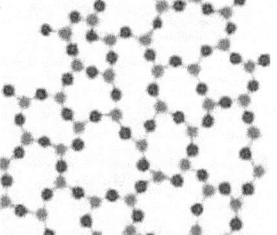

What if the only 'solid matter' is in Black Holes and the properties of Black Holes are the only real properties of matter ? Maybe this real matter only produces gravity, more in line with the real Newton ? And if 'empty matter' undermines contact-push physics theories, there may be a similar issue of 'non-full fields' also undermining field-push physics theories which require that fields totally occupy all space as well as quantum-field-push theories.

Even the early Descartes-supporting Newton saw big problems with a mechanical push physics to explain mutual action-at-distance or remote-control forces like gravity, as indicated in his 'Questiones quædam Philosophiæ' (or 'Certain Philosophical Questions') about 1663 where his musings were that any push matter cause for gravitation involved impossible contradictions - see Of Gravity and Levity. There he also assumed the existence of 'magneticall rays' causing magnetism, and on the reference frame relativity of motion mused 'we judge a thing to be moved when we see it come nigher or goe farther from some thing which our senses can perceive and so we judge not a thing to be moved in respect of the aire but of the earth or some thing'. He also concluded in his Principia Definitions Scholium that the only correct way to distinguish relative motions from absolute motions was to establish what motion-causes, or forces, were acting.

NOTE. Newton's 'action-at-distance' work or Gilbert gravity, can maybe look somewhat less certain outside gravity as in common contact situations, but electromagnetic field theory, quantum theory and relativity theory perhaps also handle such contact situations uncertainly ? And Newton's work also showed that Gilbert's theory was more easily applied to the simpler gravity phenomenon than to magnetism ! But since it is basically Gilbert theory, Newton's gravitation work should be able to be extended to also cover magnetism and electricity for a unified field theory and maybe more as Newton says eg in his Principia preface. Modern knowledge suggests for magnetism and electricity basically using for each a combination of a pair of opposite 'gravities' - and it should now be easy to modify a computer model of gravitation to do that!? The ONLY scientific attempt to seriously consider Gilbert attraction theory since Gilbert was Newton's - and he made some great progress doing so, perhaps unfinished and needing elements of modern signal theory and robot theory.

Newton's 'Principia' General Scholium.

When the first edition of Principia was published a majority of physicists supported Descartes' Cartesian physics, and wanted Newton to also support Cartesian physics though he clearly did not. Still many managed to wrongly interpret his Principia as backing Descartes though it clearly did not. In response after careful consideration, in attempted further clarification, Newton added to a second edition of his main work Principia, a final section called General Scholium including his summarised argument against the Rene Descartes ether vortices explanation of planetary motion, argument for God, his argument for black-box science rejecting hypotheses on unseens as unscientific, and a favourable view on Gilbertian inciting effluvia or 'spirits emitted' physics. These parts are ;

"The hypothesis of vortices is pressed with many difficulties. That every planet by a radius drawn to the sun may describe areas proportional to the times of description, the periodic times of the several parts of the vortices should observe the duplicate proportion of their distances from the sun; but that the periodic times of the planets may obtain the sesquiplicate proportion of their distance from the sun, the periodic times of the parts of the vortex ought to be in the sesquiplicate proportion of their distance. That the smaller vortices may maintain their lesser revolutions about Saturn, Jupiter, and other planets, and swim quietly and undisturbed in the greater vortex of the sun, the periodic times of the parts of its vortex should be equal; but the rotation of the sun and planets about their axes, which ought to correspond with the motions of their vortices, recede far from all these proportions. The motions of the comets are exceedingly regular, and are governed by the same laws with the motions of the planets, and can by no means be accounted for by the hypothesis of vortices; for comets are carried with very eccentric motions through all parts of the heavens indifferently, with a freedom that is incompatible with the notion of a vortex."

"Hitherto we have explained the phænomena of the heavens and of our sea by the power of gravity, but have not yet assigned the cause of this power. This is certain, that it must proceed from a cause that penetrates to the very centres of the sun and planets, without suffering the least diminution of its force; that operates not according to the quantity of the surfaces of the particles upon which it acts (as mechanical causes use to do), but according to the quantity, of the solid matter which they contain, and propagates its virtue on all sides to immense distances, decreasing always in the duplicate proportion of the distances. Gravitation towards the sun is made up out of the gravitations towards the several particles of which the body of the sun is composed; and in receding from the sun decreases accurately in the duplicate proportion of the distances as far as the orb of Saturn, as evidently appears from the quiescence of the aphelions of the plants; nay, and even to the remotest aphelions of the comets; if those aphelions are also quiescent. But hitherto I have not been able to discover the cause of those properties of gravity from phænomena, and I frame no hypotheses; for whatever is not deduced from the phænomena is to be called an hypothesis; and hypotheses, whether metaphysical or physical, whether of occult qualities or mechanical, have no place in experimental philosophy."

"And now we might add something concerning a certain most subtle Spirit which pervades and lies hid in all gross bodies; by the force and action of which Spirit the particles of bodies mutually attract one another at near distances, and cohere, if contiguous; and electric bodies operate to greater distances, as well repelling as attracting the neighbouring corpuscles; and light is emitted, reflected, refracted, inflected, and heats bodies; and all sensation is excited ...... But these are things that cannot be explained in few words, nor are we furnished with that sufficiency of experiments which is required to an accurate determination and demonstration of the laws by which this electric and elastic Spirit operates. THE END."

NOTES. This 'spirit' piece of Newton's General Scholium has been interpreted in several wonderfully different ways, including maybe unreasonably a Descartes material ether mediating dead matter, but it is a term used in places by William Gilbert and its mode of operation including self-acting matter, action-at-distance signal response and brain type action clearly includes the only alternative theory at the time to Descartes mechanism - the Gilbert self-acting robot signal universe. Gilbert's effluvia signals are referred to by Newton as 'spirits emitted' that 'excite bodies and sensation' in his Principia final Scholium to Book 1 Section 11. It is more loosely put in the General Scholium, but was understood as such by many scientists of the time - though Newton's publications having the often confusing habit of often not mentioning other scientists by name unless supporting some detail of his own work was unhelpful, but certainly some unpublished Newton manuscripts specifically connected his 'subtle spirit' section with electric and magnetic phenomena and Gilbert ideas or variants of them. But it may possibly have been intended by Newton to also cover a Descartes ether as well, since he considered both types of explanation and maybe others as options - though to Newton all 'outside science'. However, no matter how clearly Newton explained his blackbox physics most other physicists continued to wrongly interpret it and present it as a Cartesian physics. Indeed soon Cartesian physics with Newton's physics mathematics was widely presented as being 'Newtonian physics' - and still is today. But Newton did concede that at least some things that are at one time unobservables to science, might in the future become observables. But if everything having physical effect was observable as Einstein claimed then might different interpretations still be possible, and might people always be able to posit ever finer unobservables anyway ? Of course all bodies in attraction physics are at least in some respects observers and definable, but this is not the case for other physics theories. Yet if Newton published his physics as having 2 possible alternative explanations, Cartesian and Gilbertian, then his published disproof of Cartesian physics must logically mean his physics really being Gilbertian though avoiding publicly stating that.

Newton devoted the 10 pages of Principia book 2 section 9 to disproving Descartes' vortex theory of planetary motion, which you can read in this website's 'Newton against Descartes' section and is summarised above.

Though here defining science narrowly, as excluding hypotheses on currently unseens like causes of gravity, Newton like many scientists could greatly value ideas that he considered to be outside science - and in his case certainly hypotheses on gravity's causation. But for his science Newton stuck by his black-box theory as being the best physics possible as long as there were no proven physics theories without unseens.

Newton's 'Principia' against any ether.

Newton's final work Opticks basically supported his Principia gravity theory position - with its first proposition being that he was supporting no explanation for the working of light and again with its final Queries. He chiefly allowed as possible explanation options either a dead-particle push explanation or a robot-particle attraction theory for light, though he had earlier considered an ether wave theory. Both Principia and Opticks supported his different-causes-are-possible blackbox science conclusion. Newton's Principia book 2 section 8 was on wave theory as for sound. Newton's doubts about a wave theory of light were interestingly partly related to his strong doubts about Descartes' material ether, if there is no medium to wave then wave theory is not viable. Newton also doubted classic wave theory of light because he saw light as basically propagating in straight lines where waves propagate on every side. But mainly Newton saw waves as needing a medium, and all mediums as decelerating bodies so that any 'ether' in space should decelerate light and decelerate planets to degrees that experiments did not confirm. Hence, in book 2.7,

"since it is the opinion of some (Descartes) that there is a certain aetherial medium extremely rare and subtile, which freely pervades the pores of bodies, some resistance must needs arise ; in order to try whether the resistance, which we experience in bodies in motion, be made upon their outward superficies only, or whether their internal parts meet with any considerable resistance upon their superficies, I thought of the following experiment. I suspended ......Therefore the resistance of the empty box in its internal parts will be above 5000 times less than the resistance on its external superficies."
and,

"In the scholium attached to the sixth Section, we shewed, by experiments of pendulums, that the resistances of equal and equally swift globes moving in air, water and quicksilver, are as the densities of the fluids .... And though air, water, quicksilver, and the like fluids, by the division of their parts in infinitum, should be subtilized, and become mediums infinitely fluid, nevertheless, the resistance they would make to projected globes would be the same. For the resistance considered in the proceeding Propositions arises from the inactivity of the matter ; and the inactivity of matter is essential to bodies, and always proportional to the quantity of matter. By the division of the parts of the fluid the resistance arising from the tenacity and friction of the parts may be indeed diminished ; but the quantity of matter will not be at all diminished by this division ; and if the quantity of matter be the same its force of inactivity will be the same ; and therefore the resistance here spoken of will be the same, as always proportional to that force. To diminish this resistance, the quantity of matter in the spaces through which the bodies move must be diminished ; and therefore the celestial spaces, through which the globes of the planets and comets are perpetually passing towards all parts, with the utmost freedom, and without the least sensible diminution of their motion, must be utterly void of any corporeal fluid, excepting, perhaps, some extremely rare vapours and the rays of light."

That the real theory position of Newton on light was fully consistent with his Principia views on gravity is reflected in his opening to Opticks stating his blackbox position for light and in Opticks p376 Query 31 stating;
"It is well known that bodies act one upon another by the Attraction of Gravity, Magnetism and Electricity; and these instances show the tenor and course of nature ...."

Nearer to Einstein's time, Gilbert-Newton attraction physics was seen to have been supported by the prominent English chemist and atomic physicist Sir William Crookes (1832-1919) in one 1895 lecture stating ; "as to the nature of atoms, it seems to be capable of easiest solution by the conception that these possess - as centres of force - a persistent soul, that every atom has sensation and power of movement." This is really more Gilbert than Gilbert and combined with Newton, and was a view that he put in at least a number of lectures, so clearly Crookes strongly preferred Gilbertian physics over the Einsteinian physics that was emerging by 1905 and it may have prompted him to his experiments on spiritualism though maybe not logicly related. Sir William Crookes was maybe more openly a committed Gilbertian than Sir Isaac Newton, though again without acknowledging Gilbert. He seems to have thought that some of the atomic 'radiated matter' he studied might be some of Gilbert's 'emitted effluvia', though we still do not know if atoms can emit some more less-detectable things. But in England clearly Gilbertian action-at-distance signal-response physics long survived its falsely claimed disproof if only quietly with a select minority of English physicists that included at least Hooke, Newton and Crookes. And many physicists today claiming to know Newton's physics seem entirely unaware that in his Principia, chiefly concerning gravitational force, Newton in several key places notably falls back on magnetic force as better demonstrating his argument. And there are still some great physics ideas to be found in William Gilbert's maybe somewhat difficult 'De Magnete', or 'On The Magnet' - the nearest early science equivalent of The Golden Bough magic study of Sir James Frazer, or maybe that is this site ?

Newton stuck by his Opticks and his Principia including its General Scholium despite criticisms and false interpretations of it mainly by supporters of Descartes who resented Newton's disproofs of Descartes' theory and extensive support for Gilbert attraction physics theory. Though he did much work regarding his published disproofs of substantial aspects of Descartes mechanical push theory and little or none regarding disproving Gilbert signal attraction theory, Newton considered that his black-box theory position did not really take sides between Descartes and Gilbert explanation theories and allowed that both might be basically consistent with his own theory, and it is certainly hard to claim that Newton himself really helped to advance either theory in his trying to help advance physics.

PS. You can read the English 1730 edition of Isaac Newton's Opticks and the Latin edition at Opticks Latin- or you can learn about using Google Books at the bottom of our History of Science section.

---

otherwise, if you have any view or suggestion on the content of this site, please contact :- New Science Theory (e-mail:-vincent@new-science-theory.com) Vincent Wilmot 166 Freeman Street Grimsby Lincolnshire DN32 7AT.

You are welcome to **link** to any page on this site, eg www.new-science-theory.com/isaac-newton-principia.php

IF you like this site then Bookmark...

© new-science-theory.com, 2022 - taking care with your privacy, see New Science Theory HOME.

Sir Isaac Newton - his disproof of Descartes' science.

Homepage . William Gilbert . Rene Descartes . Isaac Newton . Albert Einstein .... Newton's Principia .... General Image Theory

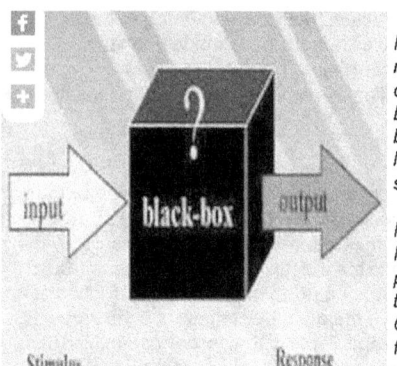

Newton's major work Principia included a substantial specific disproof of Descartes' vortex theory of planetary motion, not naming Cartes in this disproof though his was the only such vortex theory at the time. Below you can read all of Principia book 2 section 9, devoted to this disproof. Newton also argued strongly against Descartes' physics more basic requirement that space is filled with a material 'ether' substance (also required by the physics of both Aristotle and Einstein). He instead chiefly supported Gilbert's view that space must be largely really empty, but also his own view that knowing the experimental maths of nature was the limit of science.

In disproving Descartes' vortex theory of planetary motion, and some other aspects of Cartesian physics, Newton claimed to not conclude that he had completely disproved Descartes' general theory of a mechanical push universe, some modified form of which he took as one possible option beside Gilbert's signal attraction theory in his own black-box 'cause unknown' physics. He just did not prove that his good maths produced from Gilbertian attraction theory could also fit any actual valid Cartesian push physics theory - only that it might also fit some possible push physics. Newton's evidence seemed to clearly favour attraction physics. And Gilbert had also claimed to have disproved Greek-Atomist or Cartesian push-physics if maybe a bit less convincingly.

Newton's 'Principia' - Book 2.9 against Descartes' ether vortex planetary motion.

"Of the circular motion of fluids.

HYPOTHESIS.
The resistance arising from the want of lubricity in the parts of a fluid, is, caeteris paribus, proportional to the velocity with which the parts of the fluid are separated from each other.

PROPOSITION LI. THEOREM XXXIX.
If a solid cylinder infinitely long, in an uniform and infinite fluid, revolve with an uniform motion about an axis given in position, and the fluid be forced round by only this impulse of the cylinder, and every part of the fluid persevere uniformly in its motion ; I say, that the periodic times of the parts of the fluid are as their distances from, the axis of the cylinder.

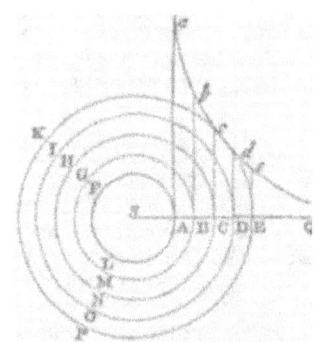

Let AFL be a cylinder turning uniformly about the axis S, and let the concentric circles BGM, CHN, DIO, EKP, etc., divide the fluid into innumerable concentric cylindric solid orbs of the same thickness. Then, because the fluid is homogeneous, the impressions which the contiguous orbs make upon each other mutually will be (by the Hypothesis) as their translations from each other, and as the contiguous superficies upon which the impressions are made. If the impression made upon any orb be greater or less on its concave than on its convex side, the stronger impression will prevail, and will either accelerate or retard the motion of the orb, according as it agrees with, or is contrary to, the motion of the same. Therefore, that every orb may persevere uniformly in its motion, the impressions made on both sides must be equal and their directions contrary. Therefore since the impressions are as the contiguous superficies, and as their translations from one another, the translations will be inversely as the superficies, that is, inversely u the distances of the superficies from the axis. But the differences of the angular motions about the axis are as those translations applied to the distances, or as the translations directly and the distances inversely; that is, joining these ratios together, as the squares of the distances inversely. Therefore if there be erected the lines Aa, Bb, Cc, Dd, Ee, etc., perpendicular to the several parts of (he infinite right line SABCDEQ, and reciprocally proportional to the squares of SA, SB, SO, SD, SE, etc., and through the extremities of those perpendiculars there be supposed to pass an hyperbolic curve, the sums of the differences, that is, the whole angular motions, will be as the correspondent sums of the lines Aa, Bb, Cc, Dd, Ee, that is (if to constitute a medium uniformly fluid the number of the orbs be increased and their breadth diminished in inflnitum), as the hyperbolic areas AaQ., BAQ, CcQ, DdQ, EeQ, etc., analogous to the sums; and the times, reciprocally proportional to the angular motions, will be also reciprocally proportional to those areas. Therefore the periodic time of any particle as D, is reciprocally as the area DdQ, that is (as appears from the known methods of quadratures of curves), directly as the distance SD. Q.E.D.

COR. 1. Hence the angular motions of the particles of the fluid are reciprocally as their distances from the axis of the cylinder, and the absolute velocities are equal.

COR. 2. *If a fluid be contained in a cylindric vessel of an infinite length, and contain another cylinder within, and both the cylinders revolve about one common axis, and the times of their revolutions be as their semi-diameters, and every part of the fluid perseveres in its motion, the periodic times of the several parts will be as the distances from the axis of the cylinders.*

COR. 3. *If there be added or taken away any common quantity of angular motion from the cylinder and fluid moving in this manner; yet because this new motion will not alter the mutual attrition of the parts of the fluid, the motion of the parts among themselves will not be changed; for the translations of the parts from one another depend upon the attrition. Any part will persevere in that motion, which, by the attrition made on both sides with contrary directions, is no more accelerated than it is retarded.*

COR. 4. *Therefore if there be taken away from this whole system of the cylinders and the fluid all the angular motion of the outward cylinder, we shall have the motion of the fluid in a quiescent cylinder.*

COR. 5. *Therefore if the fluid and outward cylinder are at rest, and the inward cylinder revolve uniformly, there will be communicated a circular motion to the fluid, which will be propagated by degrees through the whole fluid; and will go on continually increasing, till such time as the several parts of the fluid acquire the motion determined in Cor. 4.*

COR. 6. *And because the fluid endeavours to propagate its motion still farther, its impulse will carry the outmost cylinder also about with it, unless the cylinder be violently detained; and accelerate its motion till the periodic times of both cylinders become equal among themselves. But if the outward cylinder be violently detained, it will make an effort to retard the motion of the fluid; and unless the inward cylinder preserve that motion by means of some external force impressed thereon, it will make it cease by degrees.*

All these things will be found true by making the experiment in deep standing water.

PROPOSITION LII. THEOREM XL.
If a solid sphere, in an uniform and infinite fluid, revolves about an axis given in position with an uniform motion, and the fluid be forced round by only this impulse of the sphere ; and every part of the fluid perseveres- uniformly in its motion ; I say, that the periodic times of the parts of the fluid are as the squares of their distances from the centre of the sphere.

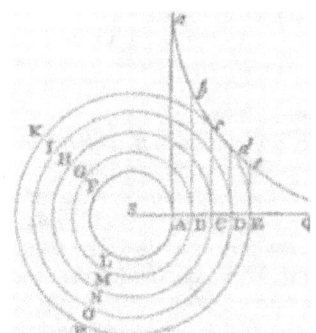

CASE 1. *Let AFL be a sphere turning uniformly about the axis S, and let the concentric circles BGM, CHN, DIO, EKP, etc., divide the fluid into innumerable concentric orbs of the same thickness. Suppose those orbs to be solid ; and, because the fluid is homogeneous, the impressions which the contiguous orbs make one upon another will be (by the supposition) as their translations from one another, and the contiguous superficies upon which the impressions are made. If the impression upon any orb be greater or less upon its concave than upon its convex side, the more forcible impression will prevail, and will either accelerate or retard the velocity of the orb, according as it is directed with a conspiring or contrary motion to that of the orb. Therefore that every orb may persevere uniformly in its motion, it is necessary that the impressions made upon both sides of the orb should be equal, and have contrary directions. Therefore since the impressions are as the contiguous superficies, and as their translations from one another, the translations will be inversely as the superficies, that is, inversely as the squares of the distances of the superficies from the centre. But the differences of the angular motions about the axis are as those translations applied to the distances, or as the translations directly and the distances inversely; that is by compounding those ratios, as the cubes of the distances inversely. Therefore if upon the several parts of the infinite right line SABCDEQ there be erected the perpendiculars Aa, Bb, Cc, Dd, Ee, etc., reciprocally proportional to the cubes of SA, SB, SO, SD. SE, etc., the sums of the differences, that is, the whole angular motions will be as the corresponding sums of the lines Aa, Bb, Cc, Dd, Ee, etc., that is (if to constitute an uniformly fluid medium the number of the orbs be increased and their thickness diminished in infinitum), as the hyperbolic areas AaQ, BbQ, CcQ, DdQ, EeQ, etc., analogous to the sums; and the periodic times being reciprocally proportional to the angular motions, will be also reciprocally proportional to those areas. Therefore the periodic time of any orb DIO is reciprocally as the area DdQ,, that is (by the known methods of quadratures), directly as the square of the distance SD. Which was first to be demonstrated.*

CASE 2. *From the centre of the sphere let there be drawn a great number of indefinite right lines, making given angles with the axis, exceeding one another by equal differences; and, by these lines revolving about the axis, conceive the orbs to be cut into innumerable annuli; then will every annulus have four annuli contiguous to it, that is, one on its inside, one on its outside, and two on each hand. Now each of these annuli cannot be impelled equally and with contrary directions by the attrition of the interior and exterior annuli, unless the motion be communicated according to the law which we demonstrated in Case 1. This appears from that demonstration. And therefore any series of annuli, taken in any right line extending itself in infinitum from the globe, will move according to the law of Case 1, except we should imagine it hindered by the attrition of the annuli on each side of it. But now in a motion, according to this law, no such is, and therefore cannot be, any obstacle to the motions persevering according to that law. If annuli at equal distances from the centre revolve either more swiftly or more slowly near the poles than near the ecliptic, they will be accelerated if slow, and retarded if swift, by their mutual attrition; and so the periodic times will continually approach to equality, according to the law of Case 1. Therefore this attrition will not at all hinder the motion from going on according to the law of Case 1, and therefore that law will take place; that is, the periodic times of the several annuli will be as the squares of their distances from the centre of the globe. Which was to be demonstrated in the second place.*

CASE 3. Let now every annulus be divided by transverse sections into innumerable particles constituting a substance absolutely and uniformly fluid; and because these sections do not at all respect the law of circular motion, but only serve to produce a fluid substance, the law of circular motion will continue the same as before. All the very small annuli will either not at all change their asperity and force of mutual attrition upon account of these sections, or else they will change the same equally. Therefore the proportion of the causes remaining the same, the proportion of the effects will remain the same also; that is, the proportion of the motions and the periodic times. Q.E.D. But now as the circular motion, and the centrifugal force thence arising, is greater at the ecliptic than at the poles, there must be some cause operating to retain the several particles in their circles; otherwise the matter that is at the ecliptic will always recede from the centre, and come round about to the poles by the outside of the vortex, and from thence return by the axis to the ecliptic with a perpetual circulation.

COR. 1. Hence the angular motions of the parts of the fluid about the axis of the globe are reciprocally as the squares of the distances from the centre of the globe, and the absolute velocities are reciprocally as the same squares applied to the distances from the axis.

COR. 2. If a globe revolve with a uniform motion about an axis of a given position in a similar and infinite quiescent fluid with an uniform motion, it will communicate a whirling motion to the fluid like that of a vortex, and that motion will by degrees be propagated onward in infinitum ; and this motion will be increased continually in every part of the fluid, till the periodical times of the several parts become as the squares of the distances from the centre of the globe.

COR. 3. Because the inward parts of the vortex are by reason of their greater velocity continually pressing upon and driving forward the external parts, and by that action are perpetually communicating motion to them, and at the same time those exterior parts communicate the same quantity of motion to those that lie still beyond them, and by this action preserve the quantity of their motion continually unchanged, it is plain that the motion is perpetually transferred from the centre to the circumference of the vortex, till it is quite swallowed up and lost in the boundless extent of that circumference. The matter between any two spherical superficies concentrical to the vortex will never be accelerated; because that matter will be always transferring the motion it receives from the matter nearer the centre to that matter which lies nearer the circumference.

COR. 4. Therefore, in order to continue a vortex in the same state of motion, some active principle is required from which the globe may receive continually the same quantity of motion which it is always communicating to the matter of the vortex. Without such a principle it will undoubtedly come to pass that the globe and the inward parts of the vortex, being always propagating their motion to the outward parts, and not receiving any new motion, will gradually move slower and slower, and at last be carried round no longer.

COR. 5. If another globe should be swimming in the same vortex at a certain distance from its centre, and in the mean time by some force revolve constantly about an axis of a given inclination, the motion of this globe will drive the fluid round after the manner of a vortex; and at first this new and small vortex will revolve with its globe about the centre of the other; and in the mean time its motion will creep on farther and farther, and by degrees be propagated in infinitum, after the manner of the first vortex. And for the same reason that the globe of the new vortex was carried about before by the motion of the other vortex, the globe of this other will be carried about by the motion of this new vortex, so that the two globes will revolve about some intermediate point, and by reason of that circular motion mutually fly from each other, unless some force restrains them. Afterward, if the constantly impressed forces, by which the globes persevere in their motions, should cease, and every thing be left to act according to the laws of mechanics, the motion of the globes will languish by degrees (for the reason assigned in Cor. 3 and 4), and the vortices at last will quite stand still.

COR. 6. If several globes in given places should constantly revolve with determined velocities about axes given in position, there would arise from them as many vortices going on in infinitum. For upon the same account that any one globe propagates its motion in infinitum, each globe apart will propagate its own motion in infinitum also; so that every part of the infinite fluid will be agitated with a motion resulting from the actions of all the globes. Therefore the vortices will not be confined by any certain limits, but by degrees run mutually into each other; and by the mutual actions of the vortices on each other, the globes will be perpetually moved from their places, as was shewn in the last Corollary; neither can they possibly keep any certain position among themselves, unless some force restrains them. But if those forces, which are constantly impressed upon the globes to continue these motions, should cease, the matter (for the reason assigned in Cor. 3 and 4) will gradually stop, and cease to move in vortices.

COR. 7. If a similar fluid be inclosed in a spherical vessel, and, by the uniform rotation of a globe in its centre, is driven round in a. vortex; and the globe and vessel revolve the same way about the same axis, and their periodical times be as the squares of the semi-diameters; the parts of the fluid will not go on in their motions without acceleration or retardation, till their periodical times are as the squares of their distances from the centre of the vortex. No constitution of a vortex can be permanent but this.

COR. 8. If the vessel, the inclosed fluid, and the globe, retain this motion, and revolve besides with a common angular motion about any given axis, because the mutual attrition of the parts of the fluid is not changed by this motion, the motions of the parts among each other will not be changed; for the translations of the parts among themselves depend upon this attrition. Any part will persevere in that motion in which its attrition on one side retards it just as much as its attrition on the other side accelerates it.

COR. 9. Therefore if the vessel be quiescent, and the motion of the globe be given, the motion of the fluid will be given. For conceive a plane to pass through the axis of the globe, and to revolve with a contrary motion ; and suppose the sum of the time of this revolution and of the revolution of the globe to be to the time of the revolution of the globe as the square of the semi-diameter of the vessel, to the square of the semi-diameter of the globe; and the periodic times of the parts of the fluid in respect of this plane will be as the squares of their distances from the centre of the globe. COR. 10. Therefore if the vessel move about the same axis with the globe, or with a given velocity about a different one, the motion of the fluid will be given. For if from the whole system we take away the angular motion of the vessel, all the motions will remain the same among themselves as before, by Cor. 8, and those motions will be given by Cor. 9.

COR. 11. If the vessel and the fluid are quiescent, and the globe revolves with an uniform motion, that motion will be propagated by degrees through the whole fluid to the vessel, and the vessel will be carried round by it, unless violently detained; and the fluid and the vessel will be continually accelerated till their periodic times become equal to the periodic times of the globe. If the vessel be either withheld by some force, or revolve with any constant and uniform motion, the medium will come by little and little to the state of motion defined in Cor. 8, 9, 10, nor will it ever persevere in any other state. But if then the forces, by which the globe and vessel revolve with certain motions, should cease, and the whole system be left to act according to the mechanical laws, the vessel and globe, by means of the intervening fluid, will act upon each other, and will continue to propagate their motions through the fluid to each other, till their periodic times become equal among themselves, and the whole system revolves together like one solid body.

SCHOLIUM.
In all these reasonings I suppose the fluid to consist of matter of uniform density and fluidity ; I mean, that the fluid is such, that a globe placed any where therein may propagate with the same motion of its own, at distances from itself continually equal, similar and equal motions in the fluid in the same interval of time. The matter by its circular motion endeavours to recede from the axis of the vortex, and therefore presses all the matter that lies beyond. This pressure makes the attrition greater, and the separation of the parts more difficult; and by consequence diminishes the fluidity of the matter. Again; if the parts of the fluid are in any one place denser or larger than in the others, the fluidity will be less in that place, because there are fewer superficies where the parts can be separated from each other. In these cases I suppose the defect of the fluidity to be supplied by the smoothness or softness of the parts, or some other condition ; otherwise the matter where it is less fluid will cohere more, and be more sluggish, and therefore will receive the motion more slowly, and propagate it farther than agrees with the ratio above assigned. If the vessel be not spherical, the particles will move in lines not circular, but answering to the figure of the vessel; and the periodic times will be nearly as the squares of the mean distances from the centre. In the parts between the centre and the circumference the motions will be slower where the spaces are wide, and swifter where narrow; but yet the particles will not tend to the circumference at all the more for their greater swiftness; for they then describe arcs of less curvity, and the conatus of receding from the centre is as much diminished by the diminution of this curvature as it is augmented by the increase of the velocity. As they go out of narrow into wide spaces, they recede a little farther from the centre, but in doing so are retarded ; and when they come out of wide into narrow spaces, they are again accelerated; and so each particle is retarded and accelerated by turns for ever. These things will come to pass in a rigid vessel; for the state of vortices in an infinite fluid is known by Cor. 6 of this Proposition.

I have endeavoured in this Proposition to investigate the properties of vortices, that I might find whether the celestial phenomena can be explained by them; for the phenomenon is this, that the periodic times of the planets revolving about Jupiter are in the sesquiplicate ratio of their distances from Jupiter's centre; and the same rule obtains also among the planets that revolve about the sun. And these rules obtain also with the greatest accuracy, as far as has been yet discovered by astronomical observation. Therefore if those planets are carried round in vortices revolving about Jupiter and the sun, the vortices must revolve according to that law. But here we found the periodic times of the parts of the vortex to be in the duplicate ratio of the distances from the centre of motion; and this ratio cannot be diminished and reduced to the sesquiplicate, unless either the matter of the vortex be more fluid the farther it is from the centre, or the resistance arising from the want of lubricity in the parts of the fluid should, as the velocity with which the parts of the fluid are separated goes on increasing, be augmented with it in a greater ratio than that in which the velocity increases. But neither of these suppositions seem reasonable. The more gross and less fluid parts will tend to the circumference, unless they are heavy towards the centre. And though, for the sake of demonstration, I proposed, at the beginning of this Section, an Hypothesis that the resistance is proportional to the velocity, nevertheless, it is in truth probable that the resistance is in a less ratio than that of the velocity ; which granted, the periodic times of the parts of the vortex will be in a greater than the duplicate ratio of the distances from its centre. If, as some think, the vortices move more swiftly near the centre, then slower to a certain limit, then again swifter near the circumference, certainly neither the sesquiplicate, nor any other certain and determinate ratio, can obtain in them. Let philosophers then see how that phenomenon of the sesquiplicate ratio can be accounted for by vortices.

PROPOSITION LIII. THEOREM XLI.
Bodies carried about in a vortex, and returning in the same orb, are of the same density with the vortex, and are moved according to the same law with the parts of the vortex, as to velocity and direction of motion.
For if any small part of the vortex, whose particles or physical points preserve a given situation among each other, be supposed to be congealed, this particle will move according to the same law as before, since no change is made either in its density, vis insita, or figure. And again; if a congealed or solid part of the vortex be of the same density with the rest of the vortex, and be resolved into a fluid, this will move according to the same law as before, except in so far as its particles, now become fluid, may be moved among themselves. Neglect, therefore, the motion of the particles among themselves as not at all concerning the progressive motion of the whole, and the motion of the whole will be the same as before. But this motion will be the same with the motion of other parts of the vortex at equal distances from the centre; because the solid, now resolved into a fluid, is become perfectly like to the other parts of the vortex. Therefore a solid, if it be of the same density with the matter of the vortex, will move with the same motion as the parts thereof, being relatively at rest in the matter that surrounds it. If it be more dense, it will endeavour more than before to recede from the centre; and therefore overcoming that force of the vortex, by which, being, as it were, kept in equilibrio, it was retained in its orbit, it will recede from the centre, and in its revolution describe a spiral, returning no longer into the same orbit. And, by the same argument, if it be more rare, it will approach to the centre. Therefore it can never continually go round in the same orbit, unless it be of the same density with the fluid. But we have shewn in that case that it would revolve according to the same law with those parts of the fluid that are at the same or equal distances from the centre of the vortex.

COR. 1. Therefore a solid revolving in a vortex, and continually going round in the same orbit, is relatively quiescent in the fluid that carries it.

COR. 2. And if the vortex be of an uniform density, the same body may revolve at any distance from the centre of the vortex.

SCHOLIUM.
Hence it is manifest that the planets are not carried round in corporeal vortices; for, according to the Copernican hypothesis, the planets going round the sun revolve in ellipses, having the sun in their common focus; and by radii drawn to the sun describe areas proportional to the times. But now the parts of a vortex can never revolve with such a motion.

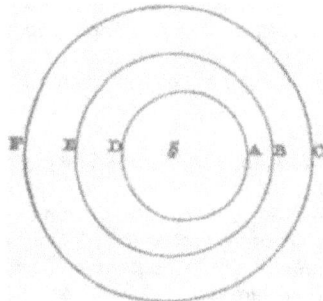

Let AD, BE, CF, represent three orbits described about the sun S, of which let the utmost circle CF be concentric to the Sun ; and let the aphelia of the two innermost be A, B ; and their perihelia D, E. Therefore a body revolving in the orb CF, describing, by a radius drawn to the sun, areas proportional to the times, will move with an uniform motion. And, according to the laws of astronomy, the body revolving in the orb BE will move slower in its aphelion B, and swifter in its perihelion E; whereas, according to the laws of mechanics, the matter of the vortex ought to move more swiftly in the narrow space between A and C than in the wide space between D and P; that is, more swiftly in the aphelion than in the perihelion. Now these two conclusions contradict each other. So at the beginning of the sign of Virgo, where the aphelion of Mars is at present, the distance between the orbits of Mars and Venus is to the distance between the same orbits, at the beginning of the sign of Pisces, as about 3 to 2; and therefore the matter of the vortex between those orbits ought to be swifter at the beginning of Pisces than at the beginning of Virgo in the ratio of 3 to 2 ; for the narrower the space is through which the same quantity of matter passes in the same time of one revolution, the greater will be the velocity with which it passes through it. Therefore if the earth being relatively at rest in this celestial matter should be carried round by it, and revolve together with it about the sun, the velocity of the earth at the beginning of Pisces would be to its velocity at the beginning of Virgo in a sesquialteral ratio. Therefore the sun's apparent diurnal motion at the beginning of Virgo ought to be above 70 minutes, and at the beginning of Pisces less than 48 minutes; whereas, on the contrary, that apparent motion of the sun is really greater at the beginning of Pisces than at the beginning of Virgo, as experience testifies; and therefore the earth is swifter at the beginning of Virgo than at the beginning of Pisces; so that the hypothesis of vortices is utterly irreconcileable with astronomical phenomena, and rather serves to perplex than explain the heavenly motions.

How these motions are performed in free spaces without vortices, may be understood by the first Book; and I shall now more fully treat of it in the following Book."

PS. It should be noted that also the mathematics of the motion of a spinning solid disc (in line perhaps with the ideas of Aristotle or early Kepler), do not match the mathematics of the actual motion of the planets around the sun. So neither the mathematics of the motion of a spinning solid disc nor of the motion of a spinning fluid, match the mathematics of the actual motion of the planets around the sun. This seemingly proves that no mechanical push-physics can explain gravity (or magnetism) or be an acceptable physics. Only the mathematics of Gilbert-Newton signal-response attraction physics match the mathematics of the actual motion of the planets around the sun and can explain gravity. Newton showed that no simple push-physics can explain planet motion, though possibly some as yet undemonstrated several different forms of pushings in combination might somehow. And so fake textbook 'Newtonian physics' that tries to present magnetism and gravity as just different forms of Descartes' one push-force, rather than them being signal-response forces, cannot be shown to work ! But still Newton, like some other physicists of the time, was bullied out of committing to Gilbertian action-at-distance signal-response physics for a no-physics-is-provable blackbox physics. Of course some later physicists like Einstein did try to develop a more suited mathematics using different theories, but even where their mathematics looks good their theories maybe remain doubtful. And the modern discovery of gravity Black Holes and universe expansion seems to really back Gilbert-Newton attraction physics further and not any kind of mechanical push-physics, with a signal-response physics being basically an advanced information-handling physics.

Though never publishing it, Newton seems to have considered that he had also disproved Descartes' theory of terrestrial gravity as he conjectured in his unpublished notes 'Certain Philosophical Questions'. For terrestrial gravity to be due to some matter pushing bodies towards the Earth, as per Descartes, must require perfect penetration which contradicts pushing - and matter causing gravity by pushing must also push itself and that cause further contradictory effects.
See http://www.newtonproject.sussex.ac.uk/view/texts/normalized/THEM00092 at 97r
(Note that reception plus re-emission could appear to be penetration, but gravitational attraction by pushing has further problems)
At 113r-v there Newton also conjectured that collision motion must be due to a force like gravity, and said that a thing that penetrates all matter he terms a 'spirit' - though William Gilbert earlier had preferred the term 'non-corporeal body'.
[Try colliding two magnets North-to-North in a tube at different velocities, and observe their collision and rebound ?] [In similar manner to Newton's disproof of Cartesian planet orbit vortex theory, it should be possible to disprove the general Cartesian small-particle-push theory of matter-penetrating magnetism and gravity. The experiments could involve a series of metal meshes of differing hole size allowing 1%, 50%, 99% air penetration and measuring the actual push forces produced for each to give a penetration/push-force equation to find if the Cartesian theory is or is not practicable physics ?]

As well as disproving several aspects of Descartes mechanical theory like his planet motion vortex theory, Newton also disproved Galileo's mechanical theory of Earth tides in general favour of the earlier Gilbertian theory that tides were caused chiefly by the attraction of the Moon. Leibnitz wrongly supported Descartes push-vortex planetary motion against 'Newton's' Gilbertian action-at-distance planetary motion. But like many then wrongly assumed action-at-distance physics as Newton's idea, though it came from William Gilbert who both Robert Hooke and Kepler had certainly studied and who had both influenced Newton. Often Newton avoided ascribing poorer Descartes theories to Descartes when disproving them, maybe as a kindness towards Descartes that was not returned by his opponents who could only create lies about Newton 'having a bad personality'. Newton was maybe also showing some kindness towards Descartes and others in not naming those he considered science giants on whose shoulders he stood ? He certainly seems to have considered that most if not all of his physics peers of that time were very second rate scientists, though refraining from saying so. But Newton's not naming those he considered science giants was no doubt also part of his determined efforts to avoid himself being associated with the then demonised William Gilbert whose physics he privately favoured ?

That gravitational force is produced by objects only proportional to their inertia or mass, seems proven by Galileo's on-Earth experiments, by Newton's proof that in-space planet motions seem consistent with that and more recently also by near-Earth space measurements of variations in Earth's gravity by NASA's orbiting GRACE project. (Newton did demonstrate that gravitational attraction could maintain solar system orbits for a very long time, though he did not examine all possible solar system gravity issues - for more on this see our Solar System Problems.)

And that gravity decreases with distance from a producing object was demonstrated by numbers of physicists including Cavendish in 1798 (see Vision Learning) and was also recently confirmed for short distances by a University of Washington project as in Physical Review Focus at http://focus.aps.org/story/v7/st8

Galileo showing that all objects tend to fall to the surface of the earth with the same acceleration, is evidence that response to gravity seems proportional to inertia or mass.

Of course Einstein later claimed that Newtonian gravitation does not always hold accurately, with some claimed evidence of that.

otherwise, if you have any view or suggestion on the content of this site, please contact :-
New Science Theory (e-mail:-vincent@new-science-theory.com)
Vincent Wilmot 166 Freeman Street Grimsby Lincolnshire DN32 7AT.

You are welcome to **link** to any page on this site, eg www.new-science-theory.com/isaac-newton-principia.php

IF you like this site then Bookmark

OR maybe make a small donation ;

PayPal Donate

(it will help with site development, and just possibly with some experiments long planned but never afforded.)

© **new-science-theory.com**, 2022 - *taking care with your privacy, see* New Science Theory HOME.

Albert Einstein - *spacetime relativity theory*

Homepage . William Gilbert . Rene Descartes . Isaac Newton .... Einstein's continuum . Blackbox Einstein .... General Image Theory
- Site Search at bottom v -

Albert Einstein (1879-1955) developed a theory of the universe based on a 'spacetime continuum', somewhat like Descartes' earlier dead-matter mechanical push universe with its 'space ether'. Gravity was an integral part of Einstein's spacetime-continuum ether, and light and other electromagnetic signals somehow propagated through it at a constant speed, the speed of light, with time variable. This relativity theory chiefly came from the relativity of light signals carrying information to human observers and its apparent mathematics, though his theory included no actual electromagnetic actions despite him trying in vain for many years to find a way to include such. His relativity theory was basicly a gravity centered one-force physics though Einstein concluded that his gravitational push-continuum should have a second electromagnetic push-continuum work with it, but could not prove how and Cartesian physics had a similar problem with multiple forces but Descartes supposed a fanciful solution with selective-pushings although pushing by his own space-exclusion definition should not be selective. And correctly resolving this multiple-forces issue could just maybe also resolve Newton's failure to prove the correct option between a Cartesian push-physics and a Gilbertian action-at-distance physics. But Einstein's theory with its good maths never really included a fully defined mechanism for gravitational or other actions despite needing it, though he never conceded that it was a poor physics and continued flogging his dead horse. Einstein did his science sit-and-think like Descartes and sold his thinking as 'thought-experiment' which as being based on mathematics he claimed to be logical and necessarily-right or certain if maybe a fiction-physics as 'thought' is really the opposite of experiment-doing. Einstein claimed his theory was both consistent with Newton physics and disproved Newton physics, if on a quite wrong textbook view of his physics as the fake-Newton Cartesian corpuscular push-physics that Newton himself had disproved and not the Gilbertian action-at-distance signal-response physics that Newton actually favoured if not himself proved.

Einstein's science theory.

When asked "Did you stand on the shoulders of Newton?", Einstein replied, "No, I stood on the shoulders of Maxwell" - basically letting slip that he, like most modern physicists, had really only studied a small fraction of physics of special interest to him. But of course James Clerk Maxwell had developed electromagnetic physics, unlike Newton's gravitational physics which in fact more closely related to Einstein's relativity physics. And really with his 'thought experiments' Einstein stood more on the shoulders of the philosopher Descartes as being more a philosopher of science than a scientist and encouraged a great increase in 'philosopher science' presenting as 'theoretical science' ? While Einstein presented his physics as an explanation physics, it perhaps really was a mathematics-only no-explanation physics like Newton's or at best an explanation physics whose mechanism is unspecified so that users can take it as a push-physics or action-at-distance or other physics ? Einstein did believe that physics should have actual mechanism explanation as Newton did though he failed to provide any, and unlike Newton did not indicate that alternative explanations might share the same mathematics. If Newton did not publicly commit to any mechanism fearing adverse response from peers, Einstein seems to have failed to commit to any mechanism more from lack of proof for any and both apparently chose to hide these facts. He may have developed a better mathematics but not a better science ? He himself seemingly saw that his spacetime continuum could not affect any motion in it by any push mechanism, but he failed to challenge most of the supporters of his theory who have wrongly taken it to have such a mechanism. Einstein's actual physics contained several major internal contradictions that he himself somehow accepted, and some major omissions that he tried and failed to deal with. But his supporters mostly did not really understand his physics and took it as a Cartesian push-physics where matter balls actually pushed a rubber-like continuum. This fake-Einstein physics is still somehow widely supported by a good number of physicists today, though Einstein himself considered that option disproved basically by Newton. So as proven partly by Gilbert and fully by Newton, 1.) all solid, liquid or gas mediums that have push properties show demonstrable drag on the motion of bodies in them not shown in the actual motion of bodies in empty space and 2.) rotations of any such medium cannot move bodies in them matching the actual motions of planetary bodies, nor medium tensions. And if spacetime continuum motion involves no push mechanism, the only non-magic mechanism possible would seem to be some form of Gilbert-Newton signal-response attraction physics mechanism. Einstein's physics was a somewhat anti-Cartesian physics (in at least partly rejecting push-action) though he thought his was an anti-Newton physics since he like most physicists of the time wrongly saw Newton's physics as being fully Cartesian. In reality Einstein's physics is rather more like Newton's than it seems, excepting chiefly that Einstein did take a unique view of time, though for space Newton favoured Gilbert's view of it being actually nothing rather than Einstein's Cartesian-ether-type continuum and he never offered any real disproof of action-at-distance physics theory while he happily used the new radio/tv remote technologies. However the maths of Einstein's physics theory did seem a real advance over Newton's in better predicting more in astronomy. But while still backed mainly by astronomers, some of the major claimed predictions of Einstein's physics like two-way time-travel remain unproven mere theorisings and have perhaps to date really helped only science fiction especially now that multiple-universes have been added to spacetime though still no experiment has really confirmed the existence of either.

Up to Newton's time, and indeed for a good time beyond, physicists and astronomers were almost all agreed that the physical universe followed basically simple laws of behaviour, and that their observations and experiments showed that - though explanation of it was not so fully agreed. But by Einstein's time technology and experiments had become more sophisticated and seemed to be showing that the physical universe followed more complex laws of behaviour, perhaps even defying logic. Little effort was put into trying to develop Newtonian physics, and instead new tricky physics theories were developed - mostly by returning to the early-Kepler method of trying to produce physics theories from mathematics only.

Forcefield theory was already taking a view, more in line with Gilbert and Newton, that force or energy could have forms other than just Descartes mass-in-motion. And Descartes mobile-indiscriminate-push-matter ether was to H.A.Lorentz a rather different 'force-ether' present everywhere and basically immobile-discriminate-push-energy with light being an ether wave, and Einstein at first took that as proven and deduced that a direct consequence of the stationary ether was that the absolute velocity of light with respect to the ether is a constant, independent of the motion of the source of light or the observer. Lorentz took the ether as being the ONLY valid non-accelerating 'inertial frame of reference' for light.

The Special Theory of Relativity

By 1905 Einstein had concluded that the immobile Lorentz ether was disproved by the Michelson-Morley experiment and that light was not an ether wave, and that any observer frame of reference in which Newton's law of inertia holds for that observer (for some period of time) is an 'inertial frame of reference'. And all observer frames of reference (and only such frames) at rest or moving with constant velocity with respect to a given inertial frame of reference are also inertial frames of reference. An observer could determine that it shared some inertial frame of reference with things that it saw as following Newton's law of inertia.

Thus far is simple Newton, but Einstein concluded that for velocity it requires that, for a common frame of reference if one observer moving at some unknown absolute velocity **v** fires a projectile at a known relative velocity u_1, and if a second observer absolutely at rest sees the projectile relative velocity as some unknown velocity **u** then keeping the speed of light as a constant **c** ;

$$u_1 = (u-v)/1-(uv/c^2)$$

The unknowns involved prevent this equation from ever actually being directly proved, but this is claimed to have been indirectly proved. The nearest experimental equivalent will normally involve two observers having unknown absolute velocities but a known relative velocity to eachother and known relative projectile velocities. But this 'relativity maths' had been previously developed by FitzGerald and Lorentz from quite different theory.

Einstein in 1905 asserted that all the laws of physics take the same form in any 'inertial frame', including them having the same constant velocity of light relating to time determination with time being a relative variable. This was basically a new alternative to the theory of Irish physicist George FitzGerald, supported by Lorentz, which was then current and had space or distance being a relative variable in motion involving the 'FitzGerald Contraction' rather than Einstein's 'Time Dilation' with time as a relative variable in motion. (earlier physics had taken motion as distance over time with both distance and time fixed.) Einstein's Special Theory of Relativity universe also involved a somewhat new kind of 'force-ether' or 'field' that he called a 'space-time continuum' which worked by 'some unspecified non-pushing mechanism akin to pushing'. Evidence for absolutes being variable, whether space or time, is rather weaker than the extensive firm evidence of relatives being variable as per Gilbert, Doppler and many others. As absolute or relative Motion is Distance moved in a Time, absolute or relative distance or time change is commonly easily viewed as relative motion change. And for a William Gilbert action-at-distance signal-response universe almost everything is an observer and responds to force-signals including variable relative signals of which there is substantial evidence and indeed variability is inherent to relatives but not to absolutes. So the maths of Einstein may well hold for an appropriate action-at-distance physics.

Hence the 1887 Michelson-Morley experiments (on our moving Earth, to demonstrate its motion through an immobile light-ether) showed light apparently having no velocity variation in a vacuum when such was expected in the Descartes-style ether that Lorentz assumed in vacuums. Many concluded that this disproved all Descartes' push-forces physics, and many wrongly thought that also meant all Newton physics - but it proved only that a vacuum does not affect the passage of light much, or at least its wave velocity, - or simply that measuring any velocity within a moving system cannot reflect the systems velocity ? Physical detectors/observers could be even themselves be automatically adjusting reported signal velocities for relative velocity. Though it has long been taken that all objects on the Earth share the Earth's velocity, this experiment did weaken the then current modified-Descartes' Lorentz ether theory - but not maybe Descartes own ether vortex theory claiming the ether pushed the Earth along and so they would have the same velocity ? Fitzgerald saw the likely explanation as being motion length contraction as actually real, but some as being just apparent due to using light for length measurement and requiring light to have a constant velocity. Somehow Einstein and his peers claimed that this experiment crucially proved his theory (eg. Einstein's 1912 manuscript on the special theory of relativity pp.18.) - though maybe just proving light reflection conserved velocity and that light which can travel through space is not any ether medium-wave as any relative medium motion changes the apparent velocity of any wave of that medium, and some other experiments were also perhaps more justifiably claimed to be proofs supporting at least Einstein's maths.

Michelson-Morley interferometer speed of light experiment ;

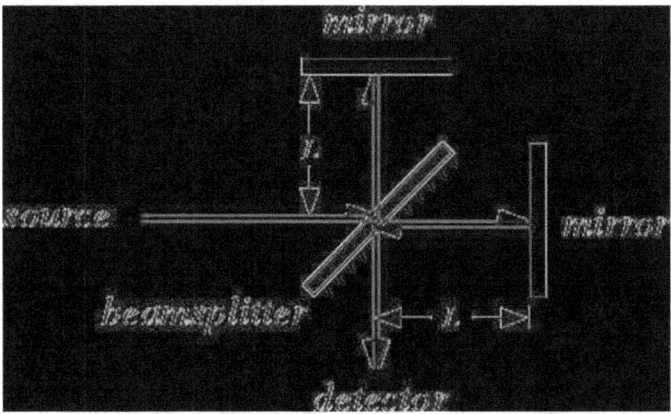

While for Newton all velocity physical effects were to relative velocity, for Einstein some velocity physical effects were to absolute velocity. But accelerating particles 'absolutely' to near the speed of light does not seem to have given any of the predicted Einstein effects. And perversely Einstein took his no-magnetism-or-electricity theory's constant speed of light from Maxwell's magnetism-and-electricity-only theory's equations ! But 1895 to 1905 was a period with very public experiments by Marconi on remote-communication and by Tesla on remote-control and remote-power, seemingly clearly backing action-at-distance physics as by William Gilbert and apparently favoured by Newton. Yet physics theorists of the time including Einstein somehow just ignored all such experiments and continued the ridiculing of action-at-distance remote-action physics quite wrongly. Hence writing to Max Born in 1948 Einstein stated "I cannot seriously believe in quantum mechanics because the theory cannot be reconciled with the idea that physics should represent a reality in space and time, free from spooky actions at a distance." Yet there is no evidence that Einstein ever actually studied the real action-at-distance physics of William Gilbert, nor indeed even the other classic spooky action-at-a-distance physics of Francis Bacon. And undoubtedly this also holds for most if not all more recent physicists.

Newton and Einstein both produced substantial works on light being particulate or corpuscular quanta, rather than waves in any ether, but Newton moved to his 'either push-particles or robot-particles might hold' black-box theory position while Einstein took a 'particle-wave duality' position on developing his own continuum ether theory with a Dualist non-consistent theory of light both being particle and being wave (or more accurately perhaps Particle and medium-less Energy-Packets for which wave maths held, with at least the latter not clearly defined). Einstein basically took experiment as both proving and disproving light being a wave. Newton and Einstein both gave gravity a substantial part in their physics but Einstein failed to integrate other forces and left magnetism and electricity to an isolated electromagnetic forcefield theory seemingly involving some other ether or continuum. And competing non-ether physics ideas continued in emerging quantum theories.

Until he developed his spacetime continuum theory of relativity, Einstein had like Newton been a bit of a black-box mathematical laws physicist though with leanings towards Descartes mechanical universe explanation, but his physics from then relates much to spacetime continuum localisations and curvatures. These Einstein ideas were to some extent along the lines of force field theory that had been developed for electromagnetism, and to which he also increasingly committed, and was basically a new energy-ether version of Descartes' matter-ether all-dead-matter push-physics with only humans doing any signal-response 'thinking' or 'observing'. But for anything, including Einstein's spacetime continuum, to have variable curvature it must be a pushy thing and a pushing thing - which Einstein insisted his spacetime continuum could not be in his physics ? So his physics did not really give any actual explanation for gravity or how it works. And with no real push mechanism for his energy spacetime continuum, Einstein's physics maybe favoured some type of 'field' physics that leaned more to some undefined 'information-field' or 'signal-field' than to any kind of a 'push-field' that he still seemingly inclined to. Yet not all who support his physics seem to support a push position, and Einstein himself was maybe less definitely committed to any physics mechanism than Gilbert and Descartes were to theirs. So his general relativity physics theory was perhaps really never fully defined. Maybe he thought that Newton allowing alternative mechanism options made it OK science to commit to no mechanism for a physics, though their seems no evidence Einstein ever really studied Newton.

Einstein's famous equation $E=mc^2$ defined his postulated inter-convertibility of 'mass' and 'energy' as two forms of matter, with c being the speed of light in a vacuum having to be invariant to any non-accelerating observer even if moving towards or away from the light. Einstein's matter was generally seen as involving Descartes 'dead-mass' and somehow also a 'dead-energy' as a form of that, though it perhaps better suits a Gilbert-Newton 'energetic-matter' as allowing energy both being located in bodies and outside but activating bodies more than just being the motion of bodies as Descartes held. For Einstein's physics, actual $E=mc^2$ conversions between mass and energy perhaps really held only for photon emitting and absorbing, though the equation $E=mc^2$ might be claimed to also fit Descartes kinetic energy - or at least maybe a 'potential energy' for a body if it is accelerated or decelerated. (and for eg graviton emitting and absorbing, the equation might perhaps be $E=mg^2$ if gravitons have a differing base emission velocity g ?) Descartes kinetic energy basically did remain but now as one form of energy only, and with normal changes in it being claimed to give bodies changes in 'potential mass' or 'relativistic mass'.

Most physicists from Gilbert and Galileo onwards had taken the measure of the amount of matter or 'mass' of any object as being its resistance to motion change or its 'inertia', though often this matter property was not precisely defined. This was generally considered independent of an object's velocity or temperature and might today be termed 'rest mass'. Einstein concluded that matter motion energy (and maybe other energy) was a property of matter that is separate from but convertible into the 'rest mass' property of matter, so any object should also have a 'relativistic mass' that increases with the object's absolute velocity or temperature. There then may be issues about whether the different effects that objects can show (which eg might or might not include gravity production and/or inertia) are due to their 'rest mass' or their 'relativistic mass'. And some physicists seem to take it as two kinds of matter being convertible into each other, each with some largely unspecified sets of properties.

Newton's force-gravity physics seems to require that gravity is produced by and affects only bodies with mass. Einstein's space-gravity physics seems to require that gravity is produced by all bodies with mass or energy, and affects all bodies with mass or energy. There seems no real proof that either is fully right in this respect, and maybe either or both might better be modified, as accurate measurement of zero masses or energies is at least almost as tricky as accurate measurement of near-zero gravities.

Maybe energy generally is not gravitationally equivalent to mass, ie a faster billiard ball does not gravitationally attract more than a slower billiard ball and a hot billiard ball does not gravitationally attract more than a cold billiard ball ? Energy in the form of photons may even be a special case and be gravitationally equivalent to mass, ie photons may gravitationally attract and be gravitationally attracted ? But if that is the case, then should we expect that a beam of white light passing close by a massive body would be split into its rainbow colours as Newton showed happens with a prism ? Or might variation in photon energy involve a non-gravitational component akin to matter motion energy ? Or photons might perhaps just be smaller faster neutrinos, unique less in their speed than simply in them being one of the basic building-blocks of which other 'elementary particles' are composed ?

Taking observers and light, or more broadly 'electromagnetic radiations', as maybe more fundamental than time and space, maybe came close to adopting a Gilbert-style signal theory but Einstein went elsewhere with his spacetime relativity. And his theory has perhaps produced some confusion of the properties of matter and the properties of energy, especially for matter-related energy like gravity. Einstein perhaps began the modern physics ascribing of properties to things without proof of such properties being consistent with other properties they had been proved to have. And often failing to precisely define what 'mass' and 'energy' exactly are in their theories.

Heisenberg and others claimed that there were limits beyond which no observer could get exact knowledge of nature, so that scientific predictions could at most be predictions of probabilities and essentially Newtonian blackbox science. But science rests on multiple observation and not just on individual observations. Einstein supported full-prediction laws of nature science and held that a valid theory's necessary 'unseens' like his spacetime continuum would actually be observable if only indirectly. But force fields and spacetime continua perhaps fit uneasily with eachother and uneasily with the many discrete quantum effects that nature seems to actually exhibit and have led to much work on developing a quantum mechanics physics generally including human observer uncertainty though not always also dropping all fields or continua.

The General Theory of Relativity

Einstein soon added gravity to his theory in his 1915 General Theory of Relativity now involving a space-time-gravity continuum. He postulated that masses (and maybe also energies ?) somehow locally curve his spacetime continuum (maybe by pushing it ?) and that the continuum curves somehow accelerate masses in it (maybe by pushing them ?)(and maybe also energies ?). Einstein concluded that gravity works 'somehow' though his general relativity theory still seems to definitely rule-out any kind of push-physics or Descartes-style gravity mechanism while not ruling-in any other kind of mechanism such as a Gilbertian signal or 'emitted spirits' gravity mechanism. This addition of gravity to his spacetime continuum basically came from Einstein's assumed Equivalence Principle saying that acceleration was equivalent to gravity, a perhaps arbitrary limitation of Newton's force definition claim that force was whatever produced acceleration and applied to all gravity, magnetism, collision or touch forces and any other forces. And gravity acceleration is not the same as touch acceleration. Touch-accelerate a platform that a man is on, at eg 100g, and parts of his body in contact with the platform accelerate 100g but other parts have inertia so the body flattens and you get a squashed-dead man. But gravity-accelerate a man 100g and all parts of his body accelerate 100g and the man is OK ! Einstein's claim that his 'one-force' gravity is not a physical force like magnetism is really a ridiculous claim. Many similar experiments have been done on both gravity and magnetism showing that they are basically similar physical forces with their similarities clearly noted by both William Gilbert and Isaac Newton. Einstein really failed to explain their similarities or their differences.

And of course if, as Einstein's theory required, Gravitational Attraction is equivalent to Acceleration, then how is that consistent with eg Electrical

attraction or repulsion not also being equivalent to Acceleration ? The strengths of such natural forces as electric charge and magnetism seem to have no relation to the mass of the objects producing them, while the strengths of such natural forces as both gravity and collision do seem to exactly reflect the mass of the objects producing them. Not all physics theories seem able to account for, or reconcile, these facts easily - if at all. There is also some evidence that gravity has some relation to other forces that hold atoms together, with increasing gravity maybe reducing 'spontaneous' radioactivity. Hence Einstein's limited 'Gravity Equivalence Principle' like his 'Time Dilation' seems a doubtful extra assumption for a physics.
See eg The Equivalence Principle from http://arxiv.org/ftp/arxiv/papers/0908/0908.3885.pdf

However Einstein could now claim some consistency with Newton on gravity maths at least, though not with Newton blackbox or all-forces physics, and Newton would no doubt have strongly opposed Einstein's theory as requiring 'unscientific hypotheses about unseens'. Einstein's maths does seem to better fit the well known precession of the planet Mercury, but that remains a 1-off coincidence until it is compared with Newton's maths for many bodies - as for the good number of moons of Jupiter that have orbit gravities stronger than Mercury ? Planets maybe look better than Suns at holding bodies in closer orbits, perhaps due to Suns emitting ignored push-force radiations that most affect closer bodies ? Einstein devoted years to trying to modify his theory to handle all forces for a non-arbitrary 'Unified Field Theory', but he could not manage this and neither has anyone yet from General Relativity. Einstein held to the basics of his continuum physics theory and, though agreeing that substantial evidence of quantal phenomena in nature did make his continuum theory doubtful, he thought that there must be one right theory and he did not consider quantum physics a better physics theory option. Of course there is motion in the universe so gravity is something variable, and to Einstein that makes spacetime variable and measurement variable - giving a much trickier science than assuming space and time to be fixed.

His theory did have some absolute rigid requirements chiefly of his continuum and of the particular velocity termed the speed of light. And if we removed Einstein's continuum relativity explanation from Einstein theory then we would have a no-explaining black-box Einstein theory maybe more complex and so less easily understandable than Newton's as well as covering much less. Of course, though some may be happy with the general idea of black-box science, many will complain that 'they do not really explain anything' - which supporters will say is fine if they correctly predict everything, but the absence of an explanation can maybe also make them harder to understand. Some modern physicists support theories that involve extra dimensions as explanation, though to many this does not itself explain anything and such theories might be better presented as black-box ? There are certainly plenty of proven cases of maths needing extra variables for reasons other than dimensionality that such physics 'explanation' seems to ignore. The maths of Einstein's theory certainly seems to predict better than the maths of Newton's theory in some limited areas, but that in itself is perhaps no proof of Einstein's postulated explanation - and as an explanation it maybe smacks of a mathematician's attempt at a logically simple universe basically like the Harmonies and Geometries theories of the early Kepler ? Albert Einstein's relativity universe explanation even he considered to be at least incomplete. Support for it largely come from astronomers and maybe helped take physics back to an early-Kepler mathematical-imagination physics scenario that still prevails. Einstein physics now is perhaps chiefly supported by astronomers and 'cosmologists'. Einstein repeatedly claimed that Newton's ideas supported his own, though it may merely have been that Einstein managed to construct his maths to match Newton's maths under some conditions. For planetary body motion, to the extent that he defined his mechanism for gravity it seems a basically similar push mechanism to Cartesian mechanisms which Newton proved are not compatible with his planetary motion or with actual planetary motion. For Einstein's theory to be viable it seems to need a gravity mechanism different to its claimed neo-Cartesian mechanism.

Einstein also showed no understanding of the real nature of the attraction theory that Newton used as one possible explanation theory (eg. thinking that it required faster-than-light action and nothing being emitted by bodies) and Einstein largely ignored Newton black-box theory and the wider cover of Newton physics. He seems not to have substantially studied Newton, and still less Gilbert. Yet Einstein confidently claimed that he had disproved the basics of all Newton theory - and Einstein assumed that Newton had disproved all prior physics. But any real reading of Newton and Gilbert physics contradicts most of these claims.

To read of Einstein's limited understanding of Newton's physics, and of his concern on quantal experiments disproving his own continuum field physics, read Einstein at www.pbs.org/wgbh/nova/physics/einstein-on-newton.html. (in this Einstein was also perhaps at the very least very tasteless in unnecessarily bringing up yet again some of the early unsubstantiated mud-slinging of Newton as being a claimed maths thief and liar)

Einstein developed his basic relativity thory ideas while working as a patent clerk being an amateur physicist. His general relativity physics involved a necessary acceptance of contradiction in physics theory, though many physicists somehow came to support it only as a non-contradiction theory. He took the acceptance of wave-particle duality in light physics as the general acceptance of contradiction in a physics theory. And not just allowing of contrary interpretations and contrary mathematics, which Newton had allowed as a blackbox philosophical option, but allowing of contradiction in actual experiments and in actual nature. Einstein said that nobody fully understood his theory, seemingly meaning that everybody misunderstood it. But, understood or misunderstood, key physicists proceeded to misrepresent Einstein's physics as had happened with previous physics - and that continues.

In relation to gravity, the behaviour of Einstein's spacetime continuum is commonly taken as not far from the behaviour of a Descartes' push ether and so involving some of its problems that were well addressed by Newton. However Einstein required that his continuum could not actually push objects but somehow 'helped direct their motions' (maybe better with a signals-giving-responses mechanism ?) It is not possible to directly detect his claimed 'spacetime continuum' or its claimed curvatures. It is another physics 'unseen' like the ethers of Descartes push physics and the signal effluvia or spirits of Gilbert-Newton attraction physics, so that claimed 'indirect evidence' for one of them can perhaps equally be taken as being indirect evidence for any of them. And if bodies do somehow tend to move along lines of equal gravity or equal space curvature (and not in a straight line), then should gravity and space curvature around the Sun be spherical and cause planet orbits to be circular and not elliptical as they actually are ? But really Einstein's gravity mechanism was poorly specified and so was, and remains, misunderstood or misrepresented by most physicists.

In 1931 and 1952 a modern edition of Newton's Opticks was published with a Preface by Einstein in which he specifically also claimed that Newton's (blackbox) optics theory was a forerunner of the 'Wave/Particle Duality' light theory that he supported (which is maybe better termed Energy-Packet/Particle Duality as it generally involves no medium that can wave). But this involves a silly interpretation of Newton's actual optics, which were fully based on Newton's blackbox theory which did not allow contradiction within a theory and only allowed of multiple theory explanations as philosophically possible though unprovable IF they were consistent with the same maths - and Duality theory involves multiple parts of one physics theory with different contradictory maths. So this claim of Einstein of Newton-compatibility was plain ridiculous. If blackbox theory acceptance of alternatives leads to anything along those lines, it must be to an 'image theory' of science theories like that proposed elsewhere on this website. (PS. Newton's optics explained known light phenomena mathematically as well as did wave theory - the only 'problem' being that his 'light fits' when light passed close to atoms was not understood - and was presumably a microscopic quantal effect. Anyway, the fact that the initial write-up of a theory does not explain every phenomenon in the universe is no proof that it cannot be amplified to do that and certainly is not a sufficient disproof of any science theory)

In reality Einstein only disproved some of Descartes push-physics, adding to Newton's disproof of some of it, though trying to retain some Cartesian fundamentals like its definition of 'mass' or 'matter'. And the Descartes view of matter as dead stuff whose chief property is fixed space occupancy requiring contact pushing, fits uneasily with Einstein's space variability. General Relativity opposes and maybe disproves some of the essentials of Cartesian physics but not really the essentials of Gilbert-Newton physics. Relativity physics was actually a basic part of Gilbert signal attraction physics in its 'mutuality' which Newton physics also incorporated but Einstein ignored. William Gilbert 'mutuality' and 'coition' physics was basically relativistic and did not rest on fixed co-ordinate requirements or the like as Einstein supposed. And Newton's theory took such as only a matter of convenience and not a theory requirement either. All 'attraction' forces on a body and resulting motions are in both magnitude and direction relative to another body. It is also generally not proven that Einstein-supported 'field physics' maths cannot also be derived from Einstein-opposed 'signal attraction physics'. And it is generally not proven that Einstein-supported 'relativity' maths cannot also be derived from Einstein-opposed 'attraction mutuality physics'. These may well involve image compatibilities.

Though signal observer relativity was no doubt rightly central to Einstein's physics, his was a physics which itself had only matter, energy and continua and no observers or signals within it so that his observers and signals are weakly defined. Most attempts to incorporate observation and measurement into physics are maybe too narrowly human-oriented or 'anthropomorphic'. In line with both Relational Quantum Mechanics and Gilbert-Newton Attraction Physics, it can reasonably be posited that no physical event can happen without some information or signal being observed and responded to. Then the key requirements of the physical universe would seem to be not particles and/or waves or humans; but information emitters, information responders and response time ? Gilbert response physics and its Newton derived attraction physics did include observers and signals within the physics and Einstein might better have worked from that to have observers and signals better defined. Relativity basically took light and other 'electromagnetic signals' as emissions from bodies that, a Gilbert robot-matter Attraction theory supporter might complain, do little substantial in the universe except happen to inform human observers. Observers and signals are really bodies outside Einstein's actual physics and not essential bodies in it, unlike Gilbert signal theory or attraction theory as used by Newton though not fully publicly committed to by him. Einstein physics is really less a relativity physics than Gilberts. And almost all that we now know about the universe has come from electromagnetic or other signals, and perhaps nothing has as yet been learnt from any mechanical or force ether or spacetime continuum indicators. And interestingly in modern signal theory, the difference between digital and analogue signals is basically the difference between particles and waves.

Classical experimental physics theory certainly had holes so that Einstein could push his fictional-experiment 'relativity' physics theory. While many now claim that real physics started with Einstein, there is a maybe stronger argument that Einstein ended any chance of real physics theory and confirmed Catholic Inquisition 'just ideas' science-fiction physics theory based on his 'thought-experiment' or 'fictional-experiment' method. Of course there is some small chance that any science based only on thought or fiction may be right, but generally science based on actual experiments has a bigger chance of being right. 'Thought-experiments' can easily give results that conflict with the results of real experiments. Science theory was no longer to be based on substantial prior experimental fact, instead conjectural theories would hope to later find one or two selected facts to fit them and call that 'proof'. Einstein physics was challenged chiefly by quantum mechanics and its standard model(s), which has involved substantial real experiment on particles but has maybe struggled on its theory side.

To some at least, support for Einstein's relativity theory is support for its mathematics only - in line with Newton's blackbox theory position that science is only about predictive mathematical description of natural phenomena, and that explanations are unnecessary philosophy. From that position, Einstein's mathematics might allow of several different explanations of the physical universe - different image theories. Perhaps, in adopting duality physics and contradiction in physics, Einstein thought that he was merely expanding on Newton's black-box compatible alternative-theories physics. But there was a rigorous logic to Newton blackbox physics, and none to duality physics or contradiction physics. Einstein really rejected real science for a sci-fi magic version, and has been followed in that by too many who should know better.

For comparison with other physics theories, Einstein's three laws of motion would be ;

1. Every body will remain at rest, or in a uniform state of motion unless acted upon by a force or a spacetime curvature.

2. When a force or a spacetime curvature acts upon a body, it imparts an acceleration proportional to the force or the spacetime curvature and inversely proportional to the mass of the body and in the direction of the force or the spacetime curvature.

3. Every action has an equal and opposite reaction.

Atomic physics - what is an atom ?

Most early physicists assumed that gravity and the like were atom behaviours, and that atoms must be basically simple and improved knowledge of atoms would clarify the laws of physics for gravity etcetera. But atomic physics study has shown atoms to actually be complex, more in line with William Gilbert atom behaviour theory than with Descartes simple billiard-ball atoms. Early experiments seemed to show atoms as basically electromagnetic with electrons orbiting protons, but soon were taken as involving more parts and more behaviours. Atoms can absorb and emit light and other EM radiation, and can absorb and emit different particles, with some atomic events seeming simple immediate events and some involving cumulative excitement delay. And the behaviour of atoms including the photoelectric effect and spontaneous radiation seem to show that generally hitting an atom with a large particle causes a large immediate clear atomic radiation effect while some small things may have a delayed cumulative effect that can be hard to link to cause. Most atomic experiment has been on 'hitting' atoms with big stuff, though no actual contacts have ever been observed. And this is perhaps more 'abnormal' atom behaviour - and far from clarifying gravity and electromagnetic forces, atomic physics has had to assume that at least two new additional very different and really unproven atomic forces also exist.

While gravity and electromagnetic forces can be demonstrated to have basically infinite range with strength decreasing with distance, the supposed Strong and Weak atomic forces are claimed to somehow have a limited small range - and a Nobel prize was issued to David Gross et al for the claimed discovery that the short-range strong atomic force INCREASES with distance. The claim of a multitude of forces at work within an atom is problematic for Einstein physics, and atomic physicists generally adopt some version of quantum theory often with forces said to be based on particle exchange emission rather than on fields or space continua. Their 'particles' include as yet undetected gravitons for gravity, and others for electric charge and magnetism, as yet with little evidence. If atoms physically appear mini-solar-systems, their behaviour and forces seem more complex rather than simpler ! Modern atomic science atoms are looking too complex for Newton or Einstein theory but are maybe looking better suited to being Gilbert signal emission robot behaviour atoms ? There has been much debate in physics recently on whether the Graviton exists, though no debate on whether a Graviton might be a momentum-push particle, an energy quantum or perhaps a particle or energy quantum signal ?

And our 'elementary particles' like electrons may yet be found to themselves be complex systems. The mathematics of elementary particles and of photons allows of a humble 5+ MeV electron possibly being a complex composite system of eg 5,000,000+ 1 eV photons and/or other components ?

Light

Gilbert claimed that the physical universe works 'like light', while Descartes' optics had light as a push in his material ether medium, and both Newton and Einstein produced works on light as particulate radiation (or 'corpuscular' or 'quantal') without committing fully to them - and both considered light as subject to gravity. Yet others produced theories on light as waves in a medium, and support has at times swung between different theories of light. Light certainly shows some complex radiation, transmission and absorption behaviours not all of which seem easily explained by one theory ? Hence it basically travels in straight lines while waves spread all around, and a denser medium makes normal pressure waves travel faster but makes light travel slower. Several formulations of wave-particle duality theory have not given anything agreeable, and some experiments claiming to follow light paths may involve light absorption and re-emission or combine responses to light with responses to some Gilbert signal emitted by light photons.

Einstein relativity theory has to assign only to light the unique absolute property of velocity invariance even relative to moving observers. The normal almost-constant speed of light for stationary observers is reasonably understandable and may be simply the escape velocity from some very short-range atomic particle attraction force, as $c = \sqrt{2Fr}$, where F is force and r is atomic particle radius. (but that would seemingly require light to have mass and be attracted by gravity and maybe another force or forces) Or if light emission is by a repulsive force as of the electron using repulsion acceleration force signals emitted at velocity c, then if the light emission reaches velocity c it would then cease to receive the repulsive acceleration signals and would so emit at the repulsion force signal velocity c. Both attraction escape velocity and repulsion force signal velocity explanations of c would raise the issue of exactly what forces they could relate to. But it is certainly not proved that stuff like water and glass are not largely vacuum. And if a 'non-vacuum' is simply a vacuum with a few bodies in it, then light slowed in a 'non-vacuum' is probably light slowed in a vacuum and so is probable evidence against a key part of Einstein's theory ? Of course for spinning bodies an escape velocity should have a range of c +/- spin velocity. And Einstein velocity invariance is an absolute property that is not a normal property of waves that wave a medium and is not a normal property of particles either. Yet Gilbert signal attraction physics has a simple natural relative property, signalness, that can apply to light and maybe some other things only when they are acting as signals in eliciting signal responses from another body. And signalness is a natural relative property, and looks much stronger than the unnatural absolute light velocity-invariance property needed by Einstein relativity. (If Einstein's observer was blind and relied only on sound signals then his relativity physics would collapse.)

Einstein's theory seems to be supported by the fact that particle accelerators to date generally cease to accelerate particles that have reached speeds close to the speed of light. But this is confined to only electromagnetic acceleration of charged particles, which could be explained in a signal response physics by a response time. Einstein non-response theory always assumes a zero response time which looks maybe unlikely ? Signal saturation and other established signal theory effects could also possibly be involved. Numbers of astronomical observations and of physics lab experiments seem to have shown some massive particles moving at velocities very close to or even exceeding the speed of light. This evidence generally concerns neutrinos and appears to be some real evidence against Einstein's physics. Of course Newton insisted that no fixed velocity, even the velocity of light, can really be distinguished from rest - and so like Gilbert based his science on acceleration rather than on velocity with $F=ma$ rather than $F=mv^2$ suggesting Einstein's $E=mc^2$ may be shakier ground. Einstein's claim for c as a velocity limit for all motion also seems confined to rectilinear motion and maybe does not cover spin motion ?

His amazing c is linked to his view of <u>time</u>, as being merely a property of his gravity-curved space ether or continuum and as not being independent of space and gravity as most previous physics held. If experiments indicate that two events seem always linked and seem always to happen 'at the same time', then it seems that one of the events is the cause of the other but also that such experiments cannot prove which event is the causal event. If anything indicates that one of the events is causal, then that event must be taken to precede its effect by some 'response time' even if too small to detect. And if causal events need not involve motion, then this 'time' need not basically relate to motion or to space as commonly assumed but rather to causation generally. If, as Gilbert showed, a magnet can induce magnetisation in a nearby piece of iron with no apparent motions involved by some forces involving the working of some causes and effects, then that seems not to involve any changes of motion or of occupancy of space, yet causes may be deduced to be working and to be preceding effects so that 'time' may be deduced here with NO observed motions or space changes being involved. Of course such deduction may need to be backed by other evidence of such forces from motion experiments, but that need not confine 'time' to only motion ?

Light interacts with atomic particles and most is known about its interaction with electrons which look much like simple particle collision type interactions, though little is actually known about simple particle collisions if they exist at all. Logically perhaps light looks like a class of particles normally bound to electrons by some attraction force the escape velocity from which is c. While light is said by some to be waves of a range of frequencies, it can act more like uncharged particles of a range of masses though perhaps not responding to gravity like normal matter particles. And a quantal or non-continuous emission can be of distinct single things OR of distinct sets of things like firing 3 missiles or a bunch of shot. Neutrons were at first claimed to be light wave photons, but they act quite differently and interact with atom nucleons more than with atom electrons.

Two interesting types of light-electron interactions are those called the Photoelectric Effect and the Compton Effect :-

The Photoelectric Effect involves different atoms emitting electrons in response to absorbing incoming photons.
1. A quantal response threshold normally applies, no electrons being emitted if the energy/frequency of each photon received is too low. (so normally one 4ev photon gets a response but two 2ev electrons gets no response - however some much lower-level of response is also produced in the latter case, seemingly whenever two lower energy photons are received simultaneously as Sipila et al 2007 at www.iop.org/EJ/article/1367-2630/9/10/368/njp7_10_368.html)
2. If electrons are emitted, the energy of each emitted electron is normally proportional to the energy/frequency of each above-threshold photon received.
3. If electrons are emitted, the number emitted is normally proportional to the number of above-threshold photons received.
This 'Photoelectric Effect' is better called the Photoelectron Effect, as the electric charge seems to not be involved at all in its mechanism. And this got Einstein his Nobel prize though it seems maybe more in line with Descartes push-physics in requiring light photons to provide all the energy or push to power the resulting effect. It is not light acting as a signal of small energy triggering some bigger energy response, but like Descartes having light punch and push nerves. But the Photoelectric Effect is a material emitting electrons in response to absorbing light of at least some Threshold Frequency - and higher light frequencies give emitted electrons higher energy. Discovered in 1887 by Heinrich Hertz, in 1900 Max Planck suggested that light was quantal in it then in 1905 Einstein published his small theory paper interpreting the photoelectric effect as involving quantal light and was awarded the 1921 Nobel Prize in Physics "for his discovery of the law of the photoelectric effect" though this small work of his really achieved little or nothing. And a logical quantal light theory interpretation of the photoelectric effect might seem to be a particle-set response as requiring receipt of some set of particles within some time period. Einstein's was a small theoretical work of interpretation of somebody's photo-electric experiment which may not be the best possible interpretation. Light theory currently requiring duality is certainly unsatisfactory.

The Compton Effect seems to involve light photons hitting electrons and losing some energy/frequency as though being a photon mass momentum collision. But here when a higher energy photon interacts with an atom, it can give up some of its energy and create a new additional photon of some lesser amount of energy (and lesser frequency). The energy of the new additional photon does not depend on the energy of the initiating photon, so that the Compton Effect looks more like a signal-response effect and unlike the Photoelectric Effect. It seems to require some electrons in a material being less constrained than atomic electrons usually are. This can be affected by the energy state of the electron, and has a much bigger multi-photon response than does the Photoelectric Effect.

Many take the Photoelectric and Compton Effects as proving that light is quantal, which may be true though generally responses being quantal does not as Einstein concluded require signals being quantal, though it certainly shows that atoms can respond to some subsidiary properties of things. Photoelectric Effect and Compton Effect responses show a directionality range similar to ball collision or to some spherical force repulsion. (and if a central attraction force has some emission Escape Velocity such as c, then a central repulsion force should have some equivalent absorption Entry Velocity such as c ? This might suggest emitted photons having some property differing from absorbed photons, but this does hold for emitted electrons and absorbed electrons.) It has been shown that atoms can gain momentum by absorbing a photon having one energy and re-emitting a photon having some different energy, the energies concerned being quantised, see https://physicsworld.com/cws/article/news/2009/dec/09/quantum-trampoline-measures-gravity

A quantal signal theory of light that could alone explain both wave and particle responses might perhaps be a Particle Set theory of light, where light is emitted as a set of say 3 particles and 1 particle set is 1 photon. Some physical effects could then be to its set properties and some physical effects be to its particle properties. Set properties could include equivalents to wavelength but not have the same relationship to velocity that simple waves have. Einstein got his 1921/22 Nobel Prize not for discovering the photoelectric effect but for his mathematical explaination of the photoelectric effect theorising that light is particulate or quantal as Newton thought, but like others he went with the limiting assumption that quantal effects proved single-particle action and excluded particle-set action. And he did not have his gravity or spacetime continuum quantal and he opposed quantum mechanics physics theory.

But a wave theory of light could perhaps also alone explain both wave and quantal responses using all-or-none response mechanisms as a mechanical clock can convert continuous spring pressure to digital ratchet motion. 'Duality' theory, as in taking light or matter as both being a wave and being not a wave, of course involves blatant logical contradiction and as such should not be unconditionally acceptable in science. Even if nature actually behaves in apparently contradictory ways, good science seems to require that there must be some non-contradictory explanation behind it. So at most it may be reasonable science to say that light seems to show both wave and non-wave behaviours and the explanation for it is not known and not to support contradictory 'duality theory'. Or for multiple theories to be logically acceptable they must fit the conditions set by General Image Theory science, as in Doppler Effect mathematics applying equally both to a wave and to any series of regular emissions as it truely does in fact.

- a particle-set 'photon'.

- a wave.

Indeed the logical particle-theory interpretation of the photoelectric effect having a 'wavelength' requirement seems to be a particle-set interpretation. Light theory currently requiring duality is certainly unsatisfactory and suggests the need for new experiments, perhaps not just on light itself but on a range of particle beams and on a range of pressure waves in a variety of scenarios to clarify the actual properties of them. Pressure waves have perhaps been fairly thoroughly studied, but particle rays much less so - especially uncharged particle rays like neutrons that seem to react little with matter and so are almost unseens and hard to detect refraction, diffraction etc in. It may be that in similar circumstances both behave similarly or not to some extent, hence diffraction at material edges seems a wave property but might also be a particle ray response to quantal signals from material edges or something else. On other differing interpretations of 'double-slit' light diffraction experiments, see - quantum light theory.

If waves are motions of matter that repeat regularly and can be described with wave mathematics, then all events that repeat regularly can be described with 'wave mathematics'. And maybe any regular quantal signal or any regular quantal observation can be described with wave mathematics ? Or if matter existence involves regular repeat events then maybe matter can be described with wave mathematics without matter being waves ? So proof that a non-wave something can be described with wave mathematics is maybe no proof that the something is a wave, and even less is it proof that the non-wave something is both a wave and a non-wave. (Or if time is itself something and is quantal such that it repeats regularly, then maybe time can be described with wave mathematics. And then matter in time can be described with wave mathematics, without time or matter being waves ? Proof that a non-wave something can be described with wave mathematics is no proof that the something is a wave, and even less is it proof that the non-wave something is both a wave and a non-wave.)

In some circumstances a laser spot on a wall can be observed to move along the wall faster than the speed of light. While here nothing actual is moving faster than light, this is an observable illusion of something appearing to move faster than the speed of light. But strangely astronomy and particle physics seem to never report observing this type of illusion, and that maybe raises an issue as to the reliability of some astronomy and some particle physics ? And astronomical 'evidence' of 'gravity bending light' or 'spacetime bending light' seem to not fully match laboratory experiment, and with our limited knowledge of the actual physics of extraterrestrial regions could be mere refraction or diffraction type events. And some areas of space with strong gravity fields may also have strong magnetic or other fields as well confounding some claimed effects Also, pulsating quasars show redshifts in line with the redshifts of other astronomical bodies, but their pulse timings do not seem to be related to their redshifts as relativity theory should imply - see www.newscientist.com/article/mg20627554.200-time-waits-for-no-quasar--even-though-it-should.html Attempts to 'explain' this quasar problem have been weak and include positing invisible astronomical bodies. Many key physics issues now seem not logic or maths issues, but experiment issue and all possible experiments have not yet been done. And interpretation of experimental results involving light acting as a signal certainly needs to consider signal theory interpretation, as in our Light as a signal section.

Isaac Newton demonstrated, and many experiments since have confirmed, that objects respond to gravity from other bodies as though gravity signals travel at a speed much greater than the speed of light and the same seems to hold also for electric and magnetic forces. Einstein gravity maths may explain this but several other explanations have been suggested. One possible explanation for this observed effect suggested by William Gilbert attraction theory could be that response to gravity, electric and magnetic signals may involve a signal anticipation mechanism (akin to eg anticipator thermostats). So a simple mechanism for this (tending to cancel at least some of the normal delay effect of a signal taking time to travel) could be response requiring a set of multiple signals and its directionality being to the last of the set ? See Information Physics.

Some kind of signal response mechanism seems really needed in the perhaps dubious Shifting Gravity Theory proposed by Daniel Emilio at http://home.earthlink.net/~danielemilio/a_shifting_theory_of_gravity.html. That basically needs particle gravity response to be basically a William Gilbert robot-response, but many signal-response mechanisms can have mechanical equivalents such as using valve, escapement and other mechanisms.

In most field and ether theories including Einstein's, forces are basically tied to their sources as is the Sun's gravity and can only be modified by modifying the source (ie. the Sun). But in a Gilbert style signal theory when graviton signals are emitted by the Sun (like light) they are separated from it and may allow of signal modification as by gravity-shields or gravity-magnifiers - though none such have yet been discovered. And signal theory can offer other effects as signal thresholds, signal saturation, response maxima and reaction time are normal phenomena in any signal theory, but their equivalents in other forms of physics theory when present can often appear perhaps more arbitrary ? Of course Einstein died before remotes and computers became common and his and other 'modern physics' have failed to incorporate the main modern technology ideas that were anticipated in Gilbert's physics and its part-development by Newton.

England in 2013 saw the somewhat unusually Einstein-supporting particle physicist populist professor Brian Cox claim in a speech at the British Science Festival that time machines are possible, "though only for travel forward into the future". If Einstein's theory depends on time-travel then, perhaps conveniently for it, no time-traveller could ever come back to prove they had actually time-travelled !! Of course it is not clear that Einstein's General Relativity requires time to actually vary rather than just to appear to vary by it being relative like velocity. Of course now more than 100 years after Einstein's 'spacetime curving' theory, technology that can 'curve spacetime' has still not yet been developed which can only be evidence that Einstein's physics is inadequate (and maybe that as yet all physics theories developed to date have been inadequate ?). If physical forces like magnetism and gravity involve responses to signals, then responses must follow signals and time must exist and must be one-directional ? Einstein like Descartes wrongly saw theory, and in his case specifically mathematical theory rather than just logical theory, as being more significant to science than experiment. as he claimed in eg his 1933 Oxford University lecture 'On the method of theoretical physics'. This shows Einstein having a badly mistaken understanding of the basic nature of mathematics, additional to a badly mistaken understanding of the fundamentals of science that may really rest on a view of the universe as being a creation of a mysteriously-logical God more than of the universe being actually logical ? Hence he did claim as a supposedly significant science argument that 'God does not play dice'. Einstein's science theory does contain numbers of significant problems but his biggest mistake was undoubtedly like fellow theoretician Descartes in rating science theory above the study of the actual universe. While being apparently a liberal Jew, Einstein basically followed the chief requirement of the Catholic Inquisition trial of Galileo that he "must present his science as being only thought". Einstein also inclined to a Pope argument that what he was claiming was too complex for ordinary non-genius people to understand, so most people should just trust him and believe what he claimed by faith. But many who want people to believe some nonsense make such a claim. It is true that a universe whose basic workings are simple, can have some multitude of workings at the same time so the totality may appear complex and will give some complex mathematics. Dwelling on some such derivative complexities while ignoring the basic simplicities is very bad science and is shown by too many modern physics theorists, who should really be looking at viewpoints more like that of Isaac Newton and of William Gilbert. Such science is basically understandable by the majority of people without needing all possible trickier mathematics to be understood. But still 2017 saw NASA backing doubtful theory claims that solar system orbiting in not fully explained by Einstein's or Newton's equations, but requires an as yet unseen mystery 'Ninth Planet' a big planet far beyond Neptune's orbit.

A new 2016 seemingly anti-Einstein anti-spacetime-continuum finding has been reported concerning the Sun's magnetic field. Apparently like its gravitational field, the magnetic field of the Sun seems near-spherical and so unaffected by the Sun's movement through space though Einsteinians had predicted a comet-tail shape. Of course Einstein's Relativity theory in fact failed to cover electromagnetism at all, but does seem to imply a strongly non-spherical gravitational field which also seems not backed by the evidence ? See - On the Sun's magnetic field. Newton predicted no effect from the Sun moving through space because all solar system bodies share that motion so that there is no relative motion involved and he saw only relative motion as giving physical effects as he showed for gravity between multiple bodies. Newton like Gilbert saw space as containing no kind of ether and so no ether drag on anything that could give physical effects such as claimed often by Einsteinians though Einstein himself had claimed that his continuum ether involved no drags or pushings. But clearly many 'Einsteinians' support a Cartesian version of Einstein's physics. Some physicists do talk of pushing of masses somehow caused by 'energy-momemtum' though it is not clear if that has any actual meaning ?

The chief requirement of Einstein's relativity physics was that masses locally curve its spacetime-continuum, though it specified no mechanism for this. Newton had posited two alternative possible mechanisms for his gravity physics, of which he seemed to favour an action-at-distance signal-response mechanism over some possible push-physics mechanism. Einstein really specified no mechanism but did seem to fully exclude a push-physics mechanism which maybe leaves possible some action-at-distance signal-response mechanism if his spacetime continuum can somehow respond to signals and can curve itself proportionately in response to masses 'gravity signals' ? Einstein perhaps ended up with his physics in a somewhat similar position to Newton's ? But having no specified mechanism for the theory's main requirement, there is perhaps really no specified way to prove or to disprove the theory. Certainly proving or disproving some minor bits of its maths cannot really prove or disprove Einstein's relativity. Maybe confirming its spacetime continuum by achieving the time-travel that it implies might be an actual proof of it, but now more than 100 years after its 1905 publishing we still have no time-travel. Indeed the theory has produced no technology at all and maybe should not really qualify as a scientific theory ?

Einstein's fame really grew after the USA's 1945 atomic bomb use, widely credited to Einstein undoubtedly largely wrongly though he had somewhat helped promote atomic bomb development and use but was not himself involved in either. (Einstein probably thought of atomic bomb use against Germany not Japan. But if Russia substantially defeated Germany, Russia also seemed likely to defeat Japan and America may have used atomic bombs against Japan to prevent them surrendering to Russia. When Japan did surrender to America their emperor got leniency that he probably would not have got from the Russians ? Maybe the Japanese emperor even favoured the atomic bomb use as allowing his leniency ? And atomic bombs really followed the discovery of radioactivity, of which nuclear fission is one kind.) It seems that late in life Einstein also basically abandoned physics, as he turned down his doctor's recommended deathbed surgery saying "I have done my share; it is time to go. I will do it elegantly."

But despite modern quantum physics development like string, loop and other quantal theories that seem supported mostly by 'particle physicists' and only some of which use field and particle-wave duality ideas, it can perhaps be said that nobody has yet successfully published a real disproof of Einstein's physics theory ? But the same can perhaps be said also of Gilbert-Newton attraction physics theory ? In current physics, the first statement by C.A.Mead in his introduction to his 2000 'Collective Electrodynamics' is that "the last 7 decades of the 20th century will be characterised in history as the dark ages of theoretical physics" - and perhaps it has not ended yet. In the rest of his work Mead claims to prove that the universe consists only of electromagnetic waves and fields with no medium - his maths look good and others have backed such waves, but waves in nothing and fields of nothing as not nothing ? For other relevant views of physics theory now see our String Theory, and for Black Hole, Dark Matter, Universe Expansion and other claimed phenomena see our Gravity section.

Einstein, unlike Newton, Descartes and Gilbert, published none of his science in Latin - sticking largely to his native German. English translations to date seem largely to be on his relativity theories dealing with trickier phenomena. If we ever find a good explanation of his relativity theory for ordinary phenomena, as to how gravity works for planets, moons and comets and how collision energy transfer between bodies works with his $E=mc^2$ (how that works for emission and absorption of electromagnetic waves [or photons] seems obvious), then we will add it here.

As the closest we can find for now, you can read good English translations of Einstein's interesting 1920 lecture on Ether and the Theory of Relativity and his 1910 non-relativity lecture on Electricity and Magnetism at our Einsteins Ether.
And through Google Books you can read an English version of Einstein's 1916 Relativity.

Or for the best source of Einstein papers see http://alberteinstein.info/
Or to read another physicist Many-Minds Relativity view of Einstein's relativity see http://claesjohnsonmathscience.wordpress.com/article/many-minds-relativity-yvfu3xg7d7wt-5/.

You can do a good search of this website below ;

Search on this site www.new-science-theory.com, with .

Or do a search of the web better with DuckDuckGo - Type web search then Enter

If you have any view or suggestion on the content of this site, please contact :- New Science Theory
Vincent Wilmot 166 Freeman Street Grimsby Lincolnshire DN32 7AT.

OR if you like this site you could maybe make a donation ;

It will help with site development, and just possibly with some key basic physics experiments long planned but never afforded.
[PS. and you may perhaps help make history for science ?]
(The fictional time-travel and multi-universe type ideas of modern physics theory have long totally discouraged certain lines of physics experiment despite there being strong reasons to believe them to be very promising if not essential lines of experiment. Some such lines of experiment considered here identified as early as the 1960s seem still to have had no work done on them and there is maybe not much more time here for this. Science funding both government and private unfortunately now all goes to basically safe standard mainstream science, and no money at all goes to any really innovative risky science though that might pay a thousand times greater.)

You are welcome to link to any page on this site, eg www.new-science-theory.com/albert-einstein.php

© new-science-theory.com, 2021 - taking care with your privacy, see New Science Theory HOME.

Albert Einstein - ether and electromagnetic fields

Homepage . William Gilbert . Rene Descartes . Isaac Newton . Albert Einstein .... Blackbox Einstein .... General Image Theory

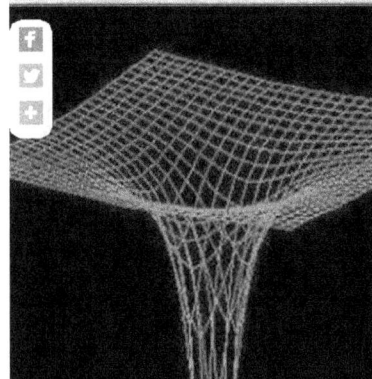

Below is a good English translation of Einstein's 1920 lecture on Ether and the Theory of Relativity, explaining his spacetime continuum as basically a forcefield form of Descartes mass ether, like Johannes Kepler's push forcefield, and showing his entirely conventional gross misunderstanding of supposed 'Newton's ether theory' which was really Cartesian physics theory that Newton gave little or no support to.

You can also read online or download an English translation of Einstein's 1910 lecture on electricity and magnetism. Einstein presents it basically as a theoretical construct with no attempt to explain what an electromagnetic force field might actually be, though later committing more to force fields while still giving a blackbox explanation (try a web search on "how force fields work" !).
Or read online or download free Einstein's 'Lecture Notes', PDF 4.77mb to load.

Ether and the Theory of Relativity

Albert Einstein, an address delivered on May 5th, 1920, in the University of Leyden.

"How does it come about that alongside of the idea of ponderable matter, which is derived by abstraction from everyday life, the physicists set the idea of the existence of another kind of matter, the ether ? The explanation is probably to be sought in those phenomena which have given rise to the theory of action at a distance, and in the properties of light which have led to the undulatory theory. Let us devote a little while to the consideration of these two subjects.

Outside of physics we know nothing of action at a distance. When we try to connect cause and effect in the experiences which natural objects afford us, it seems at first as if there were no other mutual actions than those of immediate contact, e.g. the communication of motion by impact, push and pull, heating or inducing combustion by means of a flame, etc. It is true that even in everyday experience weight, which is in a sense action at a distance, plays a very important part. But since in daily experience the weight of bodies meets us as something constant, something not linked to any cause which is variable in time or place, we do not in everyday life speculate as to the cause of gravity, and therefore do not become conscious of its character as action at a distance. It was Newton's theory of gravitation that first assigned a cause for gravity by interpreting it as action at a distance, proceeding from masses. Newton's theory is probably the greatest stride ever made in the effort towards the causal nexus of natural phenomena. And yet this theory evoked a lively sense of discomfort among Newton's contemporaries, because it seemed to be in conflict with the principle springing from the rest of experience, that there can be reciprocal action only through contact, and not through immediate action at a distance.

It is only with reluctance that man's desire for knowledge endures a dualism of this kind. How was unity to be preserved in his comprehension of the forces of nature ? Either by trying to look upon contact forces as being themselves distant forces which admittedly are observable only at a very small distance and this was the road which Newton's followers, who were entirely under the spell of his doctrine, mostly preferred to take; or by assuming that the Newtonian action at a distance is only apparently immediate action at a distance, but in truth is conveyed by a medium permeating space, whether by movements or by elastic deformation of this medium. Thus the endeavour toward a unified view of the nature of forces leads to the hypothesis of an ether. This hypothesis, to be sure, did not at first bring with it any advance in the theory of gravitation or in physics generally, so that it became customary to treat Newton's law of force as an axiom not further reducible. But the ether hypothesis was bound always to play some part in physical science, even if at first only a latent part.

When in the first half of the nineteenth century the far-reaching similarity was revealed which subsists between the properties of light and those of elastic waves in ponderable bodies, the ether hypothesis found fresh support. It appeared beyond question that light must be interpreted as a vibratory process in an elastic, inert medium filling up universal space. It also seemed to be a necessary consequence of the fact that light is capable of polarisation that this medium, the ether, must be of the nature of a solid body, because transverse waves are not possible in a fluid, but only in a solid. Thus the physicists were bound to arrive at the theory of the "quasi-rigid" luminiferous ether, the parts of which can carry out no movements relatively to one another except the small movements of deformation which correspond to light-waves.

This theory also called the theory of the stationary luminiferous ether moreover found a strong support in an experiment which is also of fundamental importance in the special theory of relativity, the experiment of Fizeau, from which one was obliged to infer that the luminiferous ether does not take part in the movements of bodies. The phenomenon of aberration also favoured the theory of the quasi-rigid ether.

The development of the theory of electricity along the path opened up by Maxwell and Lorentz gave the development of our ideas concerning the ether quite a peculiar and unexpected turn. For Maxwell himself the ether indeed still had properties which were purely mechanical, although of a much more complicated kind than the mechanical properties of tangible solid bodies. But neither Maxwell nor his followers succeeded in elaborating a mechanical model for the ether which might furnish a satisfactory mechanical interpretation of Maxwell's laws of the electro-magnetic field. The laws were clear and simple, the mechanical interpretations clumsy and contradictory. Almost imperceptibly the theoretical physicists adapted themselves to a situation which, from the standpoint of their mechanical programme, was very depressing. They were particularly influenced by the electro-dynamical investigations of Heinrich Hertz. For whereas they previously had required of a conclusive theory that it should content itself with the fundamental concepts which belong exclusively to mechanics (e.g. densities, velocities, deformations, stresses) they gradually accustomed themselves to admitting electric and magnetic force as fundamental concepts side by side with those of mechanics, without requiring a mechanical interpretation for them. Thus the purely mechanical view of nature was gradually abandoned. But this change led to a fundamental dualism which in the long-run was insupportable. A way of escape was now sought in the reverse direction, by reducing the principles of mechanics to those of electricity, and this especially as confidence in the strict validity of the equations of Newton's mechanics was shaken by the experiments with b-rays and rapid cathode rays.

This dualism still confronts us in unextenuated form in the theory of Hertz, where matter appears not only as the bearer of velocities, kinetic energy, and mechanical pressures, but also as the bearer of electromagnetic fields. Since such fields also occur in vacuo i.e. in free ether the ether also appears as bearer of electromagnetic fields. The ether appears indistinguishable in its functions from ordinary matter. Within matter it takes part in the motion of matter and in empty space it has everywhere a velocity; so that the ether has a definitely assigned velocity throughout the whole of space. There is no fundamental difference between Hertz's ether and ponderable matter (which in part subsists in the ether).

The Hertz theory suffered not only from the defect of ascribing to matter and ether, on the one hand mechanical states, and on the other hand electrical states, which do not stand in any conceivable relation to each other; it was also at variance with the result of Fizeau's important experiment on the velocity of the propagation of light in moving fluids, and with other established experimental results.

Such was the state of things when H. A. Lorentz entered upon the scene. He brought theory into harmony with experience by means of a wonderful simplification of theoretical principles. He achieved this, the most important advance in the theory of electricity since Maxwell, by taking from ether its mechanical, and from matter its electromagnetic qualities. As in empty space, so too in the interior of material bodies, the ether, and not matter viewed atomistically, was exclusively the seat of electromagnetic fields. According to Lorentz the elementary particles of matter alone are capable of carrying out movements; their electromagnetic activity is entirely confined to the carrying of electric charges. Thus Lorentz succeeded in reducing all electromagnetic happenings to Maxwell's equations for free space.

As to the mechanical nature of the Lorentzian ether, it may be said of it, in a somewhat playful spirit, that immobility is the only mechanical property of which it has not been deprived by H. A. Lorentz. It may be added that the whole change in the conception of the ether which the special theory of relativity brought about, consisted in taking away from the ether its last mechanical quality, namely, its immobility. How this is to be understood will forthwith be expounded.

The space-time theory and the kinematics of the special theory of relativity were modelled on the Maxwell-Lorentz theory of the electromagnetic field. This theory therefore satisfies the conditions of the special theory of relativity, but when viewed from the latter it acquires a novel aspect. For if K be a system of co-ordinates relatively to which the Lorentzian ether is at rest, the Maxwell-Lorentz equations are valid primarily with reference to K. But by the special theory of relativity the same equations without any change of meaning also hold in relation to any new system of co-ordinates K' which is moving in uniform translation relatively to K. Now comes the anxious question: Why must I in the theory distinguish the K system above all K' systems, which are physically equivalent to it in all respects, by assuming that the ether is at rest relatively to the K system ? For the theoretician such an asymmetry in the theoretical structure, with no corresponding asymmetry in the system of experience, is intolerable. If we assume the ether to be at rest relatively to K, but in motion relatively to K', the physical equivalence of K and K' seems to me from the logical standpoint, not indeed downright incorrect, but nevertheless inacceptable.

The next position which it was possible to take up in face of this state of things appeared to be the following. The ether does not exist at all. The electromagnetic fields are not states of a medium, and are not bound down to any bearer, but they are independent realities which are not reducible to anything else, exactly like the atoms of ponderable matter. This conception suggests itself the more readily as, according to Lorentz's theory, electromagnetic radiation, like ponderable matter, brings impulse and energy with it, and as, according to the special theory of relativity, both matter and radiation are but special forms of distributed energy, ponderable mass losing its isolation and appearing as a special form of energy.

<u>More careful reflection teaches us, however, that the special theory of relativity does not compel us to deny ether</u>. We may assume the existence of an ether, only we must give up ascribing a definite state of motion to it, i.e. we must by abstraction take from it the last mechanical characteristic which Lorentz had still left it. We shall see later that this point of view, the conceivability of which shall at once endeavour to make more intelligible by a somewhat halting comparison, is justified by the results of the general theory of relativity.

Think of waves on the surface of water. Here we can describe two entirely different things. Either we may observe how the undulatory surface forming the boundary between water and air alters in the course of time; or else with the help of small floats, for instance we can observe how the position of the separate particles of water alters in the course of time. If the existence of such floats for tracking the motion of the particles of a fluid were a fundamental impossibility in physics if, in fact, nothing else whatever were observable than the shape of the space occupied by the water as it varies in time, we should have no ground for the assumption that water consists of movable particles. But all the same we could characterise it as a medium.

We have something like this in the electromagnetic field. For we may picture the field to ourselves as consisting of lines of force. If we wish to interpret these lines of force to ourselves as something material in the ordinary sense, we are tempted to interpret the dynamic processes as motions of these lines of force, such that each separate line of force is tracked through the course of time. It is well known, however, that this way of regarding the electromagnetic field leads to contradictions.

Generalising we must say this: There may be supposed to be extended physical objects to which the idea of motion cannot be applied. They may not be thought of as consisting of particles which allow themselves to be separately tracked through time. In Minkowski's idiom this is expressed as follows: Not every extended conformation in the four-dimensional world can be regarded as composed of worldthreads. The special theory of relativity forbids us to assume the ether to consist of particles observable through time, but the hypothesis of ether in itself is not in conflict with the special theory of relativity. Only we must be on our guard against ascribing a state of motion to the ether.

Certainly, from the standpoint of the special theory of relativity, the ether hypothesis appears at first to be an empty hypothesis. In the equations of the electromagnetic field there occur, in addition to the densities of the electric charge, only the intensities of the field. The career of electromagnetic processes in vacuo appears to be completely determined by these equations, uninfluenced by other physical quantities. The electromagnetic fields appear as ultimate, irreducible realities, and at first it seems superfluous to postulate a homogeneous, isotropic ether-medium, and to envisage electromagnetic fields as states of this medium.

But on the other hand there is a weighty argument to be adduced in favour of the ether hypothesis. To deny the ether is ultimately to assume that empty space has no physical qualities whatever. The fundamental facts of mechanics do not harmonize with this view. For the mechanical behaviour of a corporeal system hovering freely in empty space depends not only on relative positions (distances) and relative velocities, but also on its state of rotation, which physically may be taken as a characteristic not appertaining to the system in itself. In order to be able to look upon the rotation of the system, at least formally, as something real, Newton objectivises space. Since he classes his absolute space together with real things, for him rotation relative to an absolute space is also something real. Newton might no less well have called his absolute space "Ether"; what is essential is merely that besides observable objects, another thing, which is not perceptible, must be looked upon as real, to enable acceleration or rotation to be looked upon as something real.

It is true that Mach tried to avoid having to accept as real something which is not observable by endeavouring to substitute in mechanics a mean acceleration with reference to the totality of the masses in the universe in place of an acceleration with reference to absolute space. But inertial resistance opposed to relative acceleration of distant masses presupposes action at a distance; and as the modern physicist does not believe that he may accept this action at a distance, he comes back once more, if he follows Mach, to the ether, which has to serve as medium for the effects of inertia. But this conception of the ether to which we are led by Mach's way of thinking differs essentially from the ether as conceived by Newton, by Fresnel, and by Lorentz. Mach's ether not only conditions the behaviour of inert masses, but is also conditioned in its state by them.

Mach's idea finds its full development in the ether of the general theory of relativity. According to this theory the metrical qualities of the continuum of space-time differ in the environment of different points of space-time, and are partly conditioned by the matter existing outside of the territory under consideration. This space-time variability of the reciprocal relations of the standards of space and time, or, perhaps, the recognition of the fact that "empty space" in its physical relation is neither homogeneous nor isotropic, compelling us to describe its state by ten functions (the gravitation potentials g), has, I think, finally disposed of the view that space is physically empty. But therewith the conception of the ether has again acquired an intelligible content, although this content differs widely from that of the ether of the mechanical undulatory theory of light. The ether of the general theory of relativity is a medium which is itself devoid of all mechanical and kinematical qualities, but helps to determine mechanical (and electromagnetic) events.

What is fundamentally new in the ether of the general theory of relativity as opposed to the ether of Lorentz consists in this, that the state of the former is at every place determined by connections with the matter and the state of the ether in neighbouring places, which are amenable to law in the form of differential equations,; whereas the state of the Lorentzian ether in the absence of electromagnetic fields is conditioned by nothing outside itself, and is everywhere the same. The ether of the general theory of relativity is transmuted conceptually into the ether of Lorentz if we substitute constants for the functions of space which describe the former, disregarding the causes which condition its state. Thus we may also say, I think, that the ether of the general theory of relativity is the outcome of the Lorentzian ether, through relativation.

As to the part which the new ether is to play in the physics of the future we are not yet clear. We know that it determines the metrical relations in the space-time continuum, e.g. the configurative possibilities of solid bodies as well as the gravitational fields; but we do not know whether it has an essential share in the structure of the electrical elementary particles constituting matter. Nor do we know whether it is only in the proximity of ponderable masses that its structure differs essentially from that of the Lorentzian ether; whether the geometry of spaces of cosmic extent is approximately Euclidean. But we can assert by reason of the relativistic equations of gravitation that there must be a departure from Euclidean relations, with spaces of cosmic order of magnitude, if there exists a positive mean density, no matter how small, of the matter in the universe. In this case the universe must of necessity be spatially unbounded and of finite magnitude, its magnitude being determined by the value of that mean density.

If we consider the gravitational field and the electromagnetic field from the standpoint of the ether hypothesis, we find a remarkable difference between the two. There can be no space nor any part of space without gravitational potentials; for these confer upon space its metrical qualities, without which it cannot be imagined at all. The existence of the gravitational field is inseparably bound up with the existence of space. On the other hand a part of space may very well be imagined without an electromagnetic field; thus in contrast with the gravitational field, the electromagnetic field seems to be only secondarily linked to the ether, the formal nature of the electromagnetic field being as yet in no way determined by that of gravitational ether. From the present state of theory it looks as if the electromagnetic field, as opposed to the gravitational field, rests upon an entirely new formal motif, as though nature might just as well have endowed the gravitational ether with fields of quite another type, for example, with fields of a scalar potential, instead of fields of the electromagnetic type.

Since according to our present conceptions the elementary particles of matter are also, in their essence, nothing else than condensations of the electromagnetic field, our present view of the universe presents two realities which are completely separated from each other conceptually, although connected causally, namely, gravitational ether and electromagnetic field, or as they might also be called space and matter.

Of course it would be a great advance if we could succeed in comprehending the gravitational field and the electromagnetic field together as one unified conformation. Then for the first time the epoch of theoretical physics founded by Faraday and Maxwell would reach a satisfactory conclusion. The contrast between ether and matter would fade away, and, through the general theory of relativity, the whole of physics would become a complete system of thought, like geometry, kinematics, and the theory of gravitation. An exceedingly ingenious attempt in this direction has been made by the mathematician H. Weyl,; but I do not believe that his theory will hold its ground in relation to reality. Further, in contemplating the immediate future of theoretical physics we ought not unconditionally to reject the possibility that the facts comprised in the quantum theory may set bounds to the field theory beyond which it cannot pass.

Recapitulating, we may say that according to the general theory of relativity space is endowed with physical qualities; in this sense, therefore, there exists an ether. According to the general theory of relativity space without ether is unthinkable; for in such space there not only would be no propagation of light, but also no possibility of existence for standards of space and time (measuring-rods and clocks), nor therefore any space-time intervals in the physical sense. But this ether may not be thought of as endowed with the quality characteristic of ponderable media, as consisting of parts which may be tracked through time. The idea of motion may not be applied to it."

Einstein's above rejection or 'disproof' of action-at-distance or remote-control in physics is basically an entirely unreasonable rejection of the majority of science experiments on gravity, magnetism and electricity from Galileo and Gilbert onwards. His theory backs push-contact control for everything in the face of massive evidence supporting remote-control in physics. You clearly do NOT need to push-contact your TV to change channels, for almost everything now remote works - using signal-contact involving no pushings. The TV itself acts in response to remote-emitted signals or effluvia or spirits emitted.

Einstein's ether does seem so insubstantial that Newton's disproof of Descartes ether might not apply. But so insubstantial clearly also as to perhaps make it much harder to impute activity to it rather than to matter. And maybe so insubstantial as to limit it to consisting of only information and suggest a signal theory ? So basically Einstein's theory needs his spacetime-continuum ether to be 'both pushing and not pushing'. But insofar as they are defined, both Descartes' ether and Einstein's continuum perhaps have properties that are 'almost nothing' properties - like almost-zero inelasticity and almost-zero mechanical push. Of course science has always had experimental problems with distinguishing an almost-zero from an actual zero, reflecting difficulty with comparing theories resting on such difference. Is 'empty space' actually nothing or is it some particular almost-nothing ? And what of physics theories resting on an almost-nothing which require that its effects be far more than nothing ? Only if space is nothing does it have no problems of its creation or destruction, or of it being finite or infinite, or of it producing drag or other effects on bodies moving in it etcetera. Nobody has shown any space being destroyed or created, any finite space ends or any space motion drag, nor has an Einsteinian 'space wormhole' been found or made, so the weight of scientific evidence still seems to favour space being nothing and weighs against physics theories that cannot handle that.

Einstein's physics as he stated above requires that remote-control cannot exist - and we all now know that is totally wrong and that a signal can produce a distant response by a reception 'contact' that need involve no push. But he attempts to take Descartes' position of equating contact with push, though without the space-occupancy logic that Descartes had supporting that. And Einstein's spacetime continuum implies two-way time and the possibility of two-way time travel, but more than a hundred years after his theory nobody has got a marble to even one-way time-travel so there is no really substantial evidence to support this. Einstein also claimed to support James Clerk Maxwell's electromagnetism which at least half-backed Michael Faraday's claim that his own 'Lines of Force' vacuum-tension-pull physics theory was better than some undefined 'action-at-distance' physics theory though his theory is of quite limited use and in some conflict with gravity as Einstein later found seemingly needing vacuum-tension-pull that produces no pull-drag. (See Maxwell) Einstein spent his last 40 years trying to add electromagnetism to his gravity relativity theory without success, maybe not wanting to let down the many people who had come to believe in his conjectured spacetime continuum. And he like other more recent physicists published his theories in an ad hoc manner, with no write-up directly comparable to Newton's Principia to readily show where they are compatible or incompatible to identify their proof issues. Trying to compare any more recent physics like String Theory, QM or even neo-Einstein theories like those of Dewey Larson (claiming to use Einstein's maths) or Randell Mills is almost impossible. Near his end in 1955 Einstein gave up and he declined doctor recommendations of surgery choosing to instead die.

You are welcome to **link** to any page on this site, eg www.new-science-theory.com/albert-einstein.php

IF you like this site then

OR maybe make a small donation ;

(it will help with site development, and just possibly with some experiments long planned but never afforded.)

© new-science-theory.com, 2022 - taking care with your privacy, see New Science Theory HOME.

Albert Einstein - blackbox force fields and light

Homepage . William Gilbert . Rene Descartes . Isaac Newton . Albert Einstein .... Einstein's continuum .... General Image Theory

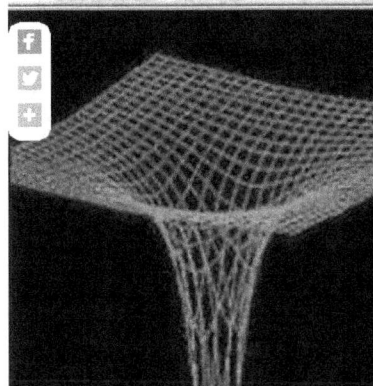

Below are two good English translation extracts from Einstein's 1912 lecture manuscript on his Special Theory of Relativity, giving some of his then black-box thinking on 'multiple continuum ethers' to explain how electromagnetic forcefields might work with a spacetime gravity continuum but not necessarily to be taken as correct explanation. Einstein here also seems to require that his spacetime gravity continuum must work by some form of unexplainable magic, and gives one of his various definitions of the principle of the constancy of the velocity of light, not specifically including constancy relative to moving observers.

-- Pictured here is the dubious rubber-sheet analogy of a bit of Einstein's spacetime continuum near a single gravity source. How could it really work, and how could the theory really work at all for common multiple gravity source situations like a solar system ?

A bit of black-box Einstein.

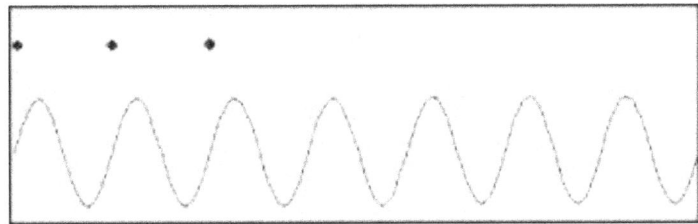

Albert Einstein, 1912 Manuscript on the Special Theory of Relativity.

A piece of Einstein's earlier black-box thinking on how electromagnetic fields work [with his energetic-matter still being pushed around like dead-matter ?]. (Manuscript page 6.)

"Lorentz conceives of electricity as being bound to corpuscles of molecular dimensions (electrons in the broader sense), a conception whose validity is hardly doubted today. But complications are thereby created for the theory, in that one is dealing here with field quantities that vary rapidly with location and that are to be replaced, then, by suitable mean values.

One can avoid these complications without doing any essential damage if one proceeds in the following way. According to the picture that Lorentz's conception gives, we have to conceive an electrically polarizable body in the following way. In every unit volume of a body in an electrically neutral state there are present at least two approximately evenly distributed kinds of electrons of zero total charge. But these are not freely movable; instead, they are linked to matter by elastic forces (in the simplest case). An electric field displaces the positive and negative electrons from their equilibrium position by means of oppositely directed forces. In this process, the electromagnetic field varies extremely rapidly with location.

We avoid this by conceiving of the positive as well as the negative electrons of the same kind as being combined into continua. In the simplest case, we have to picture an inertia-free electrical continuum of positive density and, in the case of an electrically unexcited body, one of equally great negative density, linked elastically to the matter. If we also wish to represent the conductivity of the body, we introduce, in addition, two further electrically opposite density-continua that can move relative to the body by overcoming a kind of friction.

There is nothing strange in the introduction of several continua at the same location if one realizes that this is only an idealization aimed at avoiding mathematical complications."

One Einstein definition of the principle of the constancy of the velocity of light, not specifically including constancy relative to moving observers. (Manuscript page 16.)

"Hence, in accordance with Lorentz's theory we can proclaim the following principle, which we call "the principle of the constancy of the velocity of light":

There exists a coordinate system with respect to which every light ray propagates in vacuum with the velocity c.

This principle contains a far-reaching assertion. It asserts that the propagation velocity of light depends neither on the state of motion of the light source nor on the states of motion of the bodies surrounding the propagation space."

Einstein's time measurement assuming zero reaction times to light signals, first put in his 1905 paper 'On the Electrodynamics of Moving Bodies' introducing his relativity theory ;

" ....evaluating the time of events by stationing an observer with a clock at the origin of the coordinates, who assigns to an event to be evaluated the corresponding position of the hands of the clock <u>when a light signal from that event reaches him</u> through empty space."

Bodies tell Einstein's continuum how to curve, and the continuum tells bodies how to move.

In Einstein's general relativity bodies impose curves on his time-space-gravity continuum, and the continuum imposes motion on bodies. Although push-physics analogies are often used to 'explain' this, the theory does not specifically involve any push-physics mechanism and indeed does not specify any clear mechanism for this. Gravitational forces of any kind are completely abolished as controlling the motion of planets or other bodies, and somehow space-curves do this - logically by pushings but seemingly without having any push properties since the continuum is non-material ?

As you can read in our 'Einstein's Continuum' section, Einstein concluded that "The ether of the general theory of relativity is a medium which is itself devoid of all mechanical and kinematical qualities, but helps to determine mechanical (and maybe electromagnetic) events." This seems to leave his continuum(s) as more information entities like Gilbert's signal effluvia. But if this leaves the improved maths of Einstein's theory with an unrealistic explanation, then his theory must be basically taken (as Newton wanted his theory taken) as a blackbox theory with real explanation unknown.

Of course it is a general failing of modern physics theories, supposed to replace part or all of Newton's theory, that they are not written as his Principia or even as its chief parts for comparability - but are instead written in ad hoc manner often in brief articles chiefly with reference only to black holes, wormholes or other exotic claimed phenomena.

Otherwise, if you have any view or suggestion on the content of this site, please contact :- New Science Theory
Vincent Wilmot 166 Freeman Street Grimsby Lincolnshire DN32 7AT.

You are welcome to link to any page on this site, eg www.new-science-theory.com/albert-einstein.php

IF you like this site then Bookmark

OR maybe make a small donation ;

PayPal Donate

(it will help with site development, and just possibly with some experiments long planned but never afforded.)

© new-science-theory.com, 2022 - taking care with your privacy, see New Science Theory HOME.

Information physics, and natural signal response theory

Homepage . William Gilbert . Rene Descartes . Isaac Newton . Albert Einstein .... Gravity .... General Image Theory

A Natural Signal Information Physics ?
There have been a number of attempts at developing an Information Physics, though maybe none yet have been developed far. But from the classical action-at-distance signal-response physics of William Gilbert, forerunner of Newton's attraction physics, it is possible to construct a general natural-signal information physics that takes all objects in the universe as being one or more of three general classes of information objects, namely;
1. Objects that are different kinds of Natural Signals emitted by some source objects and naturally carrying some information regarding their source and their journey from their source.
2. Objects that are Signal Emitters or sources of some such Natural Signals.
3. Objects that are Signal Responders and respond to some information from some Natural Signals such as by motion directed towards or away from those signals apparent source as seems the case with eg gravity and magnetism. And it may be that objects that emit some one kind of Natural Signal, may also respond to that kind of signal as seems the case with gravity and magnetism for example and maybe some other physical forces. So it may be of some interest to now consider such a general signal information physics concerning natural signals as communicating information, starting from a consideration of the basics of signal-response theory.

The basics of signal-response theory

The basics of signal-response theory can be taken as being simply that any body can be a signal, relative to some observer body that can respond to it. And that any body can be an observer, relative to some signal body to which it can respond. So in basic signal-response theory a response can take any form and can be to any property or properties of a signal that can be taken as carrying data or information, so that signal responses may or may not reflect the basic nature of the signal itself but might just eg reflect any aspect of the signal which might involve anything that had previously happened to the signal. Responses to signals are to the signal, and only indirectly or approximately are they responses to signal sources subject to appropriate signal noise concerns. Though it may not normally happen in nature any signal can in principle be made to give any response, including producing any other signal as a response, so that it can be possible for example to see sounds or to hear colours. And any signal needs to be interpreted to show what it signifies, but any signal can be interpreted differently by different interpreters. And a signal emitter causing a response in a signal receiver, may or may not also involve the signal receiver being able to cause a similar reverse response in the signal emitter.

The basics of a signal-response physics

A signal-response physics like William Gilbert's, partly incorporated into Newton's physics, is necessarily an information physics, which no kind of push-physics can really be. But natural physical actions must have some general basic causal relations involving some Occam simplicity though nature can have some complexities as due to several simplicities adding to a relative complexity. Hence natural physical signal responses would need to accord with physical observation and experiment.

Unlike any Cartesian-style push-physics where a cause supplies exactly the energy for an effect, signal-response theory generally does not assume any one specific mathematics, eg signal strength or signal frequency need not always decrease as the square of the distance from its source and a response strength might be fixed over some range of signal strengths or might vary with more than one aspect of apparent signal strength or frequency and apparent signal directionality even of multiply refracted signals or signals from relocated sources. Of course particular pieces or forms of natural signal-response are likely to normally actually follow some specific mathematics, as many natural signals may normally avoid any significant transit modification. So in nature gravity and magnetism do show that different bodies may respond in regular ways to different natural signals, but this need not follow how modern technology can link almost any response however complex to almost any signal however simple - as also can the human mind. And in many cases of natural physical action no direct signal detection may be possible and so can only be inferred from observed response reactions. And it is clearly the case that signal-response phenomena especially can involve small-causes-big 'Butterfly Effects'.

Some or all existent objects certainly emit natural signals that indicate their colour, magnetism, mass, motion or other properties of the signal emitting object. Emitted signals indicate or reflect properties of their emitters, so that responses of other objects to received signals are responses to properties of signal emitters and are caused by the signal emission to the extent that the signal does not get modified in transmission or by signal reception. In signal theory signals are basically anything to which some signal detector can produce some response, so that signal-response physical actions must involve two separate related phenomena being signal and response. If light is taken as being a signal, then different physical systems might be expected to show some different responses to light. This appears to be the case with at least some light-related phenomena like reflection and refraction. It has even been shown that punching holes in thin plates can increase OR decrease the amount of light that appears to penetrate a plate, see Physics World light And while in nature there are many cases of bodies affected by light as a source of heat or energy, there are many cases in nature and in technology of bodies responding to light as a signal. Of course Newton concluded that colour was a property of light itself and was not just a property of illuminated objects modifying light, which he claimed 'to have proven definitely with a crucial experiment'. But unlike William Gilbert earlier, Newton failed to publish the exact details of his experiment and so did not help with correct replication by other scientists. And also later Newton allowed that light itself might respond to some signals from objects or their atoms as to gravitational signals.

The nature of light itself

In any physics that does not take light as being a signal, light impacting different physical systems may be taken as being different behaviours of light itself. This can lead to taking reflection, refraction, diffraction, photoelectric emission, Compton emission etcetera as being light behaviours. And some of these apparent light behaviours can be taken as evidence for light itself being an ether wave, a quantal particle, or either or both. But in a physics that takes light as being a signal, light impacting different physical systems can be taken as evoking different detector responses. This leads to taking reflection, refraction, diffraction, photoelectric emission, Compton emission etcetera as being responses to light signals. And some of these being responses can be taken as giving no evidence for light itself being of any specific nature if the nature of responses is not fully determined by the basic nature of the signals involved.

Natural responses to signals

In nature responses are not fully determined by the nature of signals, but reflects only some one property or few properties of a signal. Hence some detectors can give digital quantal responses to some natural continuous signals, or give analog continuous response to natural digital signals. See eg Digital to Analog Converter or Analog to Digital Converter - though these sources may not be the best. And the different magnetic responses (as attraction, orientation and magnetization) to the same signal can operate at very different ranges, so that apparent 'signal range' can clearly be less a property of the signal than an indicator of response sensitivities.

Modern 'signal processing' is predominantly electronic and often involves systems using designed program calculation methods in producing signal responses of any designable form irrespective of the signal involved, but other physical systems can respond in various ways to different signals using only basic physical responses. And it is perhaps that kind of non-designed signal response that is of more fundamental relevance to physics. Hence mechanical clocks can respond to an analog spring pressure with ratchet-gear digital responses. (And even computational physics can be basically simple resting on 0/1 or On/Off states, so that eg atoms for some phenomena involving one signal may have two states allowing two different responses. Some recently have even proposed a physics on that basis like the New Kind of Science of Stephen Wolfram.) And in his 'Opticks' Newton considered light reflection and refraction as possibly light itself responding to signals from atoms, see Light.

Response to signals might often be proportional to signal strength or intensity and might involve signal strength thresholds. For example, it might be that some moth shows no response to light below some low-threshold light intensity, but shows an attraction response to light intensities above that low-threshold intensity and then shows a repulsive response to light intensities above some high-threshold intensity. And somewhat similar threshold signal-intensity effects might also apply to signal-response in forces like gravity or the strong nuclear force, see our string theory and gravity sections.

And the 'butterfly effect' loved by time-travel fiction theorists, which is a real problem as for computer modelling, rests on a basic of signal-response or remote-control information physics that a small low-energy low-information signal can cause a big high-energy high-information response. See Butterfly Effect. A similar class of issue to the Archemedes 'Law of the Lever' issue, and to a tiny germ killing an elephant or to a small button-press making a flying remote-control drone land on the ground. Or indeed, as with rats and plague, some cause and its effect may be mediated by something smaller.

Information in physics

Signal theory generally locates information, either intentional information or unintentional natural information, in signals - but a basically non-information physics tends to trying to locate information either in physical bodies themselves or in ill-defined 'observers'. So a range of issues can arise such as ;
1. physical bodies either do or do not carry some natural information before an observer observes ?
2. signals either do or do not carry some natural information before they are detected ?
3. physical bodies either do or do not carry some natural information before they emit signals ?
4. signals either do or do not carry natural information reflecting the full nature of their source ?
These and other related issues have not always been properly addressed by physics theories to date. A signal physics may seem better able to handle this, though no doubt non-signal image theories of a signal theory would be compatible also.

All forces as signals

It is of course possible that all physical forces may work by natural signal response, as proposed first by William Gilbert for magnetism, electricity and gravity. And the apparent-contact push force could be a short-range proximity-signal force action-at-short-distance involving no actual contact or push. This would allow of a natural signal information physics 'theory of everything'.

Responses to signals can involve issues like signal thresholds, response times, signal noise, excitation states, conditional response and signal summation. Depending on the particular signal response parameters involved, signal response systems may also be capable of looping or of hanging. And for some signal response systems a numbers of factors may vary the probability of some signal giving some response. Avoiding the use of signal theory, current physics struggles poorly to explain much.

And while most modern physics theory may have no natural place for time, natural signal information theory physics in fundamentally involving response to signals does fundamentally involve time as a consequence. What basically distinguishes a response event from a signal event is simply time, with signals being causes and responses being subsequent effects. If an attraction response cannot precede an attraction signal, then the universe is not time reversible and has one-direction time inbuilt. Many other physics theories by default predict a time reversibility that is quite contrary to many confirmed experiments and observation.

Newton's gravity mathematics works well for the orbitings of planets and moons, though for bodies to be attracted to the actual location of other gravitational bodies by signal rays directly emitted by them would generally seem to require either fixed relative locations or signals having infinite velocity or instantaneous response. So basically Newton demonstrated, and many experiments since have confirmed, that objects respond to gravity from other bodies as though gravity signals travel at a speed much greater than the speed of light. Hence moving-body gravity does not seem to show speed-of-light propagation aberration delays. Objects give a gravity response to a moving body that is not directed to where the body was when light left that body but to a position ahead of that. The same seems to hold also for electric and magnetic forces. One possible explanation of this fact may be that gravity, electric and magnetic signals actually propagate at a speed greater than the speed of light. But several other explanations have been suggested. One possible explanation for this observed effect suggested by William Gilbert attraction theory could be that response to gravity, electric and magnetic signals may involve a signal anticipation mechanism (akin to eg anticipator thermostats). So a simple mechanism for this (tending to cancel at least some of the normal delay effect of a signal taking time to travel) could be response requiring a set of multiple signals and its directionality being to the last of the set ? - as below with a response needing a set of three gravitons ;

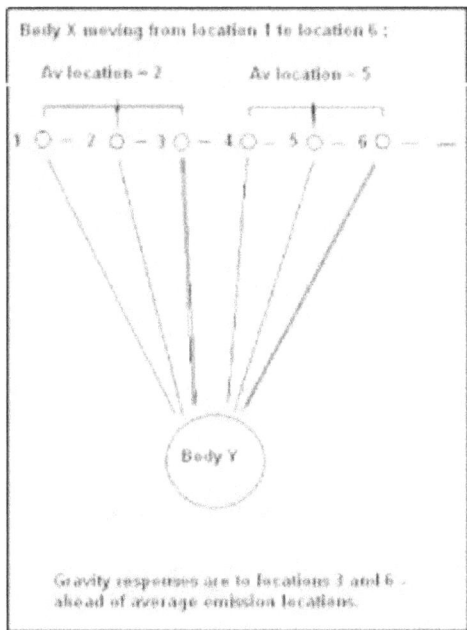

Basically it is possible for some signal response to include an effective signal prediction without it needing any actual signal prediction, and the above anticipation mechanism could work both at a macroscopic averaged level and at the microscopic quantal level. And there could also be response differences to approaching/receding signal source motions besides response to static-source signals. Of course alternative anticipatory signal response mechanisms are conceivable, but anticipatory signal response mechanisms would involve specific testable predictions for astronomy and physics. Hence the above mechanism could show different response effects at high or low gravitation intensities, and could also of course involve the effect varying with the direction and angle of the motion trajectory. The direction of signal reception could be set to directly force the direction of response. Hence if gravity response has minimum and maximum response times then more than 3 signals received in the minimum time might give a 3-signal response or some other second gravity response law, and less than 3 signals received in the maximum response time might give a 3-signal response or some other third gravity response law so that gravity might at different gravity strengths involve three gravity laws rather than one and give a gravity maths nearer Einstein's than Newton's though in a signalised Gilbert-Newton attraction physics. Clearly experiments in gravity extremes could resolve this, and similar effects might also apply to other physical forces.

Of course generally natural physical emissions including perhaps light and gravity signals should be emitted in some direction with some velocity in that direction, but generally with additionally also a velocity component that reflects any velocity present in the emitter. Testing for such will present problems, especially if one velocity is usually much smaller than the other. There are related consequences for the emitter on an 'action and reaction are equal and opposite' basis, and other consequences if an emission does not involve such 'velocity-carrying'. And motion velocity or acceleration of bodies may confer properties on them that motion direction does not, though measurement is generally direction specific and direction dependent.

The 3-signals signal anticipation mechanism given above gives apparent faster-than-light response and could perhaps also explain both averaged macroscopic-body orbits and quantised microscopic-body orbits. At macroscopic distances emitted signals will tend to being larger numbers of signals averaged, but at microscopic distances emitted signals will tend to be infrequent individual signals. If receiver response to a signal-emitting object orbiting around it requires the receiver sending a directional response signal, that signal will be received by the orbiting emitter (and not miss the orbiting emitter) only if it is orbiting at some specific appropriate velocity so that possible orbits would be confined to some specific quantal values and so give a new quantal atomic orbits explanation. And such quantal attraction motions would add to an object's non-quantal continuous inertial motion. Hence it is possible to build a gravitating robot or a missile/asteroid detector with such anticipatory 'faster-than-light' response. And, if fully programmable, such can be programmed for extra anticipation - though that need not always work well as the Moon's motion may be well predictable but not all other motion is.

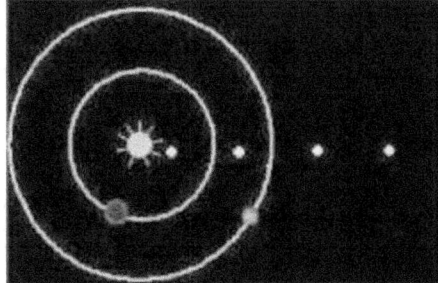

(The above illustrates only the point made, not any actual scenario, and the shape of quantal orbits might actually have to be more ellipsoid polygonal.)

A basic signal theory view of Newton 2-body gravitation might reasonably involve a background signal flux and 2 body fluxes something like below. And though a difference in background gravitation will have do direct force impact on the relative motion of 2 bodies, it could have an indirect impact if it changes the extent of gravity anticipation by the two bodies.

'Noise' in a signal will often reduce signal information, but there are exceptional situations when adding more 'noise' to a signal can somewhat perversely preserve more information. This may basically apply where attempts to remove some 'noise' from a signal, removes too much information, see Youtube example.

Einstein's gravity mathematics also looks an improvement on Newton's at distances much larger than solar-system distances, but both lean towards implying a contracting universe while distance light red-shift evidence seems to indicate an expanding universe involving some 'Dark Energy'. However a Natural Signal Information physics can allow a distance red-shift of gravity signals which would reduce their frequency and that cut distance gravity strength which could help cause universe expansion without needing any 'Dark Energy' and so improve on Einstein.

And current gravity theory for Black Holes seems to involve a gravity excess that implies some 'Dark Matter'. However a Natural Signal Information physics can allow near Black Holes a local gravity signal blue-shift which would increase their frequency and that increase local gravity strength without needing any 'Dark Matter' and so improve on current Black Hole theory also. For physics as cause-effect science an Information Physics can be naturally based on signal-response action as attested by magnetism and other physical phenomena, though of course Information for some may seem to have a relation to Observation, Measurement, Probability or maybe to Computation or maybe even to Thinking or Intelligence or indeed to God.

PS. 'Natural Signals' or 'Natural Physical Signals' in a signal physics of gravity, magnetism etc, are basically the 'Indexes' of Charles Sanders Peirce 1939-1914 (see Peirce.) Though as he noted, in some cases an observer may incorrectly mistake an artificial intentional sign for a natural signal, as from eg the human use of natural signals to carry added intentional information. And a natural signal physics can reconcile the apparent conflicts between classical physics and probability quantum mechanics physics with appropriate noise concerns. This was demonstrated well by Peter Warwick Morgan of Yale in a 2021 lecture, which you can see at - Signal Physics. For other interesting physics on Youtube see - William Gilbert. And there are some further interesting bits of this Natural Signal Information physics theory in other sections of this website, as in our Light section.

But such a new information physics certainly needs appropriate new experiments, and it is looking like nobody will realistically fund that.

You are welcome to **link** to any page on this site, eg www.new-science-theory.com/information-physics.php

OR if you like this site then you could maybe make a donation ;

It will help with site development, and hopefully with some key new physics experiments long planned but never afforded that may do some real good.
[PS. and you may perhaps help make history for science ?]
(The fictional time-travel and multi-universe type ideas of modern physics theory have long totally discouraged certain lines of physics experiment despite there being strong reasons to believe them to be very promising if not essential lines of experiment. Some such lines of experiment considered here identified as early as the 1960s seem still to have had no work done on them and there is maybe not much more time here for this. Science funding both government and private unfortunately now all goes to basically safe standard mainstream science, and no money at all goes to any really innovative risky science though that might pay a thousand times greater.)

Otherwise, if you have any view or suggestion on the content of this site, please contact :- New Science Theory
Vincent Wilmot 166 Freeman Street Grimsby Lincolnshire DN32 7AT.

© new-science-theory.com, 2022 - taking care with your privacy, see New Science Theory HOME.

Gravity, Black Holes, Dark Matter, Expanding Universe

Homepage . William Gilbert . Rene Descartes . Isaac Newton . Albert Einstein .... Light .... General Image Theory

That 'gravitational force' is produced by objects only proportional to their inertia or mass, seems proven by Galileo's on-Earth experiments, by Newton's proof that planet motions seem consistent with that, and it being demonstrated with laboratory masses by Cavendish in 1798 (see Vision Learning Gravity). And that 'gravitational force' decreases with the square of the distance from a producing object seems proved by Newton's gravitational planet motions and by the 1798 Cavendish experiments.

Of course there are claims that this does not hold accurately always, mostly based on astronomical evidence of apparent amounts of gravity and apparent amounts of matter in space seemingly having both localised and universe-wide discrepancies. Some try to explain such apparent gravity discrepancies by assuming the existence of local Black Holes and universe-wide Dark Matter, but with maybe little if any direct evidence. Modern physicists make major 'counter-intuitive' or ridiculous claims about gravity, but the most interesting ideas regarding gravity were in fact by William Gilbert and then better mathematized and formulated by Isaac Newton to an extent.

Gravity and its causation

Applied external forces generally, including 'pushes', seem to accelerate bodies in inverse proportion to the mass of the body. That bodies responses to gravity seem likewise inversely proportional to their inertia or mass, is consistent with Newton - and with Galileo demonstrating that all objects fall to the surface of the earth with the same acceleration, independent of the density or inertia or mass of the falling object. That applies to all kinds of force, but to date gravity is the only natural force with good evidence of also being produced in proportion to a source objects mass (but generally bigger magnets also give stronger magnetism, and a larger number of electrons give stronger electric attraction).

Of course these facts are maybe not full proof that gravity is actually an external force pulling bodies, only that gravity works like there is an external attraction 'force'. Gravity is 'universal' or indiscriminate, in affecting all material mass bodies, unlike eg magnetic force but like 'contact push force'.

It is easy enough to build mobile robots that each emit signals proportional to their mass and each accelerate themselves towards another in proportion to the strength of received signals and in inverse proportion to their own mass. Though such gravity-robots can be built and could be very useful for gravity research, it seems that no physicist has actually tried building them to date. Now anyone can download free a practical Ebook manual "Build A Remote-Controlled Robot" by David Shircliff from Hotfile.com at http://hotfile.com/dl/94090930/a5870ef/Build_A_Remote-Controlled_Robot.rar.html Such 'gravity robots' would be dynamic gravity model mimic equivalents of William Gilbert's dynamic magnetic Terrella models of the magnetic Earth which he used for many of his interesting magnetism experiments. Of course such 'gravity robots' could only mimic gravity with response programs as Newton's laws or others, while Gilbert Terrella models involved actual magnetism. Gravity would have to be a much stronger force than it is to have useful 'Gravity Terrella' models - small black holes might do well if they can be produced and controlled !

That gravity production and response can be mimicked by robots emitting signals and responding to such signals, is of course no real evidence on whether gravity is actually produced and actually works in such a manner or not. But that can certainly be made to work and help explicate gravity and perhaps other physical forces while not needing billions of funding, though a realistic equivalent push-physics gravity model maybe looks less attainable.

Isaac Newton produced the first good gravity theory, apparently based on William Gilbert's 'attraction physics', and then Albert Einstein produced a spacetime continuum gravity theory related to Johannes Kepler's field-push physics. But some have claimed that gravity has issues that may challenge both gravity theories.

Too much gravity.

Dark Matter ?

Gravity can work at different levels at the same time - as in attracting an apple to the ground and in holding the Moon in orbit around Earth. It can produce actual or potential accelerations of various bodies, with varied possible effects at very large or small distances.

But the motions of galaxies appear to some to require much more gravity than the visible components of such galaxies should produce. This has led some to conclude that there must exist some Dark Matter producing extra gravity, perhaps based on uncharged and maybe slow-moving neutrino-style particles (of small or big mass) that interact little with normal matter and so will give little evidence of their existing. Of course dark matter can go with attraction theory physics which has no requirement that matter emit light, and indeed has no general requirement for light - unlike Einstein's relativity theory. But dark matter could also maybe go with some other physics, and dark matter might also require some as yet undiscovered dark forces ?

Instead of dark matter for 'missing gravity', Modified Newtonian Dynamics or MOND gravity theory basically involves gravity response maintaining a minimum value even when gravity signals drop - so giving more gravity than expected as at galactic-plus distances. This is of course compatible with an attraction physics signal-theory in which response to signals might be expected to have some minimum level. For recent evidence supporting MOND gravity see Galactic Gravity. Of course some signal response systems can involve all-or-none digital-type response, and/or signal-threshold response, response delay times or other signal response effects. So identifying all applicable parameters for a signal-response gravity covering all circumstances may not be easy and may need much more accurate and complete data than is now available.

Black Holes ?

Some areas of the universe that appear to produce more gravity than their visible components ought, are thought by some to contain super-compressed matter as 'Black Holes', whose gravity is claimed to be extremely strong and to be able to even prevent the emission of light or other radiation from itself. Of course this requires that light can respond to gravity as objects with mass do. But in General Relativity gravity is basically only space curvature, so to confine emitted light would seem to only need the spacetime continuum being 'closed down' locally - however in that case the gravity within a black hole should also be confined and have no effect beyond it ?! A basic conflict with the evidence if General Relativity was true, so black holes certainly cannot be evidence for that theory. And black holes can go with attraction theory physics which can allow of light being attracted by some force, with Newton giving possible explanations of the reflection and refraction of light by local attraction acting on light. Black holes could also maybe fit with some other theories.

Of course for anything, including Einstein's spacetime continuum, to have variable curvature it must be a pushy thing and a pushing thing - which Einstein insisted his spacetime continuum could not be in his physics ? So his physics does not really give any actual explanation for gravity or how it works.

A simpler possible explanation of both dark matter and black holes might be that light emission and lower matter density is associated with charged particles. Uncharged particles and gatherings of them may be more common in the universe generally than apparent on earth. A gathering of neutrons or neutrinos could be both darker or blacker and denser, without involving any strange theory. Neutrons or neutrinos look more like simple 'Descartes-atoms', and indeed it may be that charged particles and more light emitting matter based on them are actually the somewhat rarer phenomenon.

Too little gravity.

Gravity acting as an attraction force, it perhaps should cause the universe to be contracting. And there does seem to be evidence of at least some gravity contraction in that most galaxies seem to have a greater concentration of matter nearer their centres. However there is claimed evidence for an expanding universe, basically resting only on observation showing that light over longer than galactic distances appears to lose more energy - though there could well be other more likely explanations for such observation. The evidence for the universe expanding, is largely apparent Hubble light redshifts being greater for more distant galaxies. The Doppler relative-velocity Effect (apparent change in frequency and wavelength of a wave for relative movement between it or its source and an observer) may be applicable to starlight. So physicists now commonly assume that the received amplitude of starlight must be a linear measure of the distance of its source, and star redshifts being related to starlight amplitudes is taken as being a measure of universe expansion. But the Tired Light Theory of Fritz Zwicky posits that star redshifts are a measure of the energy depletion of the amplitude of light from travel over great distances and indicating a non-expanding or even contracting universe. And over such large galactic distances very small reductions in the speed of light over very small gravity gradients may give another possible mechanism as Hubble favoured in his 'Tired Light' Theory?

Those claiming that the universe is actually expanding, generally offer variations around two types of explanation for such expansion ;

A. The universe began with an explosion and momentum maintains its expansion at some fixed velocity.
B. There is a stronger repulsive force produced by 'Dark Energy' working against gravity expanding the universe at some fixed acceleration. Apparent Hubble light redshifts being greater for more distant galaxies, if due to the Doppler relative-velocity Effect being applicable to light, seemingly supports a B type expansion rather than an A type expansion. However B needs a suitable repulsive force and some posit Dark Energy for that, but there is little supporting evidence. Of course there could seem to be other possibilities, one being a gravity expansion involving gravity from outside the currently visible universe as from an external shell of matter or from 'invisible dimensions'. However universe expansion, and Dark Energy, is disputed by some astronomers though supported by many. See [No Dark Energy](#).
C. The universe began with an explosion and momentum maintains its expansion at a fixed velocity but with centralised gravity decelerating bodies nearer the centre more strongly than bodies further from the centre. This should give Doppler red shifts that are stronger towards the universe centre than towards the universe edge radially, but with some blue shifting tangentially.
D. The universe is gravitationally contracting, with centralised gravity accelerating bodies nearer the centre more strongly than bodies further from the centre. This should give Doppler red shifts radially towards and away from the universe centre but with some blue shifting tangentially.

The explosion Big Bang explanation, A, alone should give no Doppler redshifts - while B, C and D explanations should give differing redshifts more radially than tangentially to different extents. Einstein time-dilation gravitation-redshifting predicts some redshifting from higher-gravity locations and blue-shifting from lower-gravity locations. Many current astronomers support a general space-expanding explanation, and some even a FitzGerald matter-shrinking explanation, giving Doppler-equivalent redshifts. What explanation, or combination of explanations, of apparent universe-expansion is more likely depends on having exact numbers for redshifts, distances, velocities and masses - and current astronomy numbers are maybe not very exact, but if the universe is expanding that does not itself seem to favour any of the general physics theories particularly.

There is debate about the claimed rate of expansion of the universe, with some claiming that it is around 9% faster than some others are claiming. There is also claimed to be evidence that redshift universe expansion may be somewhat weaker - or dark matter repulsion be weaker, or gravity be stronger, or whatever - at longer distance and at later time. (see eg [Afshordi, Geshnizjani and Khoury](#)) But a signal-response gravity physics might perhaps rather predict some closer-distance stronger-signal above-threshold attractive response proportional to signal strength, becoming at some greater-distance weaker-signal a below-threshold repulsive response still proportional to signal strength ? The strong nuclear force might also more logically work that way, as involving some closer-distance stronger-signal above-threshold responses being repulsive proportional to signal strength, but some greater-distance weaker-signal below-threshold responses becoming attractive but still proportional to signal strength, rather than the 'counter-intuitive' or ridiculous strong nuclear force of [David Gross](#) ?

Classical relative motion involves the Addition of Velocities Effect which basically says that for any two bodies moving towards each other, at velocities v1 and v2, their relative velocity is v1 + v2 with opposite motion being a -v. The Doppler Effect simply applies this to periodic emission motions, such as are commonly found in waves in mediums. For waves, their frequency is their velocity times the inverse of their wavelength, or is their period per second, as F=v/L. Periodic particle beams, eg of particles emitted each 5 seconds as their period, can have equivalent measures including frequency, velocity and period length. Hence a positive velocity of a signal detector relative to any periodic signal, adds to the relative signal velocity and so increases the signals apparent frequency and decreases its apparent wavelength or period length as F=(v1+v2)/L, and no accelerations or acceleration forces are required for such Addition of Velocity or Doppler effects. (an increased apparent frequency can be called a blue-shift and its opposite a red-shift).

Einstein claimed that light uniquely does **not** show classical relative velocity effects, but does show both acceleration effects and gravitation effects from his claimed Acceleration-Gravity Equivalence Principle. Light passing a massive body will be deflected towards the body as its speed is reduced more in regions of greater gravity or under greater acceleration - and it will hence also suffer some reduction in frequency (red-shift)...predicted effect values are greater than classical motion effect values alone, but in itself that still allows that the classical effect may hold but with some extra factor also applying.

Spherical Gravity ?

Newton showed that the strength of gravitational attraction seems to decrease in proportion to the square of the distance from a source object, and one explanation of that might be something emitted spherically from the source and diluting with distance with zero attenuation, as would the surface area of expanding spheres around it. The surface area of spheres is proportional to the square of their radius.

Of course at present the only gravity detectors we have are other gravity sources responding to gravity, which perhaps cannot distinguish gravity being actually directed spherically from gravity being directed to other gravity sources ? The fact that bodies like galaxies and solar systems seem generally to be flat discs, rather than being spherical, may cast some doubt on gravity being actually spherical and require another explanation for Newton's inverse square law ? Newton's inverse square law for gravity is of the form $G(d) = Go/(d.d)$, in line with unattenuating spherically diluting signals, but may not precisely hold for all distance scales. Non-spherical attenuating part-diluting signals should mean an equation form $G(Xd) = (Go/(\pi.((d.TanX).(d.TanX)))) - 10alog(d)$, which could possibly match Newton's law over some range of equation values and might have wider application also ?

At the atomic level, spherical non-discrete forces may seem to fit more with field, wave or space continuum ideas while non-spherical discrete forces may fit more with body-body digital signal ideas. The fact that electrons seem confined to very specific atomic orbits maybe better fits a non-spherical non-continuum force holding them, and if one force is non-spherical and digital then maybe all such forces are also.

Many scientists and mathematicians have considered the sphere to be the most 'perfectly ordered' of shapes, but in nature the spherical is often in fact the most disordered. If something basically has some specific linear emission directionality, then lots of things having random linear emission directionalities will average an approximately spherical emission directionality. A spherical directionality can be effectively no directionality or random directionality. The Sun seeming to emit both light and gravity spherically does not prevent either such emission at the atomic level from perhaps being directionally linear emissions. And the claim that particles and medium-waves differ in the former propagating linearly and the latter spherically may hold only at some general approximation levels reflecting the extent to which mediums traversed do or do not disorganise their transit.

If gravity basically involves straight-line body-to-body signals then part of signal dilution with distance could be due to relative body-body motion and might also include some movement anticipation with apparent faster-than-light response as considered near the bottom of our main section on Einstein. Of course if that holds between two elementary particles, a large isolated body having vast numbers of such particles could be expected to leak some gravity signals spherically and that leakage might reasonably approximate to Newton's inverse square law with a little extra attenuation. The gravity between two large bodies would be leakage gravity plus some body-body gravity that might about balance any extra attenuation. Distinguishing and quantifying the various factors in such gravity would not be simple.

If bodies emit gravitons only in response to gravitons received, and if the probability of a body emitting a graviton in response to a graviton received is proportional to the mass of the body, then two isolated bodies at relative rest should maintain some graviton emission intensities directed at each other proportional to their masses. And if there are also additional background random gravitons of some intensity then, in response to that, the two bodies should also maintain some additional spherical graviton emissions with intensities proportional to their masses ?

The orbits of artificial Earth satellites seem to support Earth's gravity being spherical, and the directionality of Earth's gravity signal emission being independent of the directionality of its gravity signal reception from other bodies since it does not seem to be significantly stronger facing the Moon or Sun ? Of course Earth's tides do not require big pulls from the Moon and Sun, for the Moon being about 1×10^{-7} g. Where gravities are strong is interesting but where gravities are very weak may be very interesting but difficult to detect and measure. Of course gravitational bodies can move, have tides, collide, contract, expand or explode and show other change producing gravity perturbations or waves that may be very hard to detect if distant. And different gravity theories may also predict differences in gravity that may be very hard to detect.

A somewhat improved version of Descartes old particle-push gravity theory was propounded first by Nicolas Fatio in 1690 and then maybe independently by Georges-Louis Le Sage in 1748 and can be termed the particle push Shadow-Gravity theory. Supposedly proven and in line with general Cartesian physics, it claimed that bodies shielding each other to an extent from 'universal gravity particles' would be attracted to each other in accordance with Newton's laws of gravitation. It requires space everywhere having lots of some randomly moving fast particles (or maybe waves) of unknown origin, but it has been claimed that they would create excessive drag and heat that is not observed and involves other problems. Most physicists rejected this theory with Newton rejecting the Fatio version and Maxwell the Le Sage version. Of some small interest is the fact that Le Sage's father in the 1720's to 1740's seems to have supported Gilbert-Newton attraction physics against Fatio-Le Sage Cartesian gravity physics, publishing in 1743 "Truth is not always probable. In physics, the principle of impulse is most probable; but that of attraction is established fact." He was aware of the Cartesian physics preference for impulse (contact forces) as the means of conveying every causal effect, and the apparent difficulty of explaining gravity that way.

Newton raised the drag issue for gravity mechanisms that involve push - and the issue holds for any particle, quantum, field, ether, or continuum mechanism that works by push - since push should produce drag and/or heat and there is strong evidence that space produces very little drag or heat for planets or other bodies.

Of course some kind of push gravity may still be possible with the right mechanism, which might need most of the push to somehow convert to eg spin energy instead of drag or heat ? Maybe even some field-push or continuum-push theory not yet fully specified ? Of course that would seem to need a response mechanism of some kind and so might still favour an attraction gravity or signal-response gravity that seems more able to avoid the problem perhaps ?

2010 in England sees an interesting publicised addition by the Royal Society for the first time to the internet of one physics related manuscript relating to gravity and Isaac Newton, but maybe adding to long-running lies rather than to the truth ? A 'friend' of Newton in his 1752 'Memoirs of Sir Isaac Newton's Life', regarding an around-1666 event, seems to translate Newton's idea of gravitational attraction as referring to 'a drawing power' - which might be a pull ? In William Stukeley's words, "as when formerly the notion of gravitation came into his mind. Why does that apple always descend perpendicularly to the ground ... assuredly the reason is that the Earth draws it. There must be a drawing power in matter. If matter thus draws matter ; it must be in proportion of its quantity. Therefore the apple draws the Earth, as well as the Earth draws the apple." - from the Royal Society manuscript at http://royalsociety.org/library/turning-the-pages/

This maybe does not help clarify whether Newton actually first thought of gravitation as being an attraction or as being a pull or in line with his later published position as being possibly either. Newton's own words include no gravitation 'drawing', only "attraction(signal response) OR impulse(push-pull)". Rather perhaps Stukeley exemplifies how Newton's actual gravitation theory ideas were misrepresented while his physics mathematics were misappropriated for a Descartes mechanical physics when they perhaps better fitted a William Gilbert effluvia-signal-processing attraction information physics.

Strong sources of magnetic force can be moved and otherwise controlled by a scientist, unlike strong sources of gravitational force. And everything responds similarly to gravity, but only some things respond to magnetism and some magnetic effects are said to work at greater distance than others, and some to work slower. If magnetic signals go to the same distance and are the same speed but responses and response times differ then that would seem to prove that a signal-response effect is indeed involved and William Gilbert concluded that he had proven that for magnetic force. But the nature of gravitational force does not allow such direct experimental proof for how gravity works excepting that the mathematics for gravity are basically consistant with that for magnetism so that it must also be a similar signal-response force. Both Gilbert and Newton seem to have believed this though Newton did not fully commit to it publicly. Gravity may be big but magnetism is maybe really key to physics.

For an overview of 'Gilbert-Newton' gravity see The Attraction Theory of gravity and other forces or Attraction Physics
(en Français - La théorie d'Attraction de gravité et d'autres forces),
(auf Deutsch - Die Attraktivität Theorie von schwerkraft und andere kräfte).

Two significant general gravity issues

Two possibly significant general issues have been raised relating to gravity, and they may well be inter-related issues ;

1. Does a mass with more energy generate a greater gravitational attraction than the same mass with less energy, ie does energy like mass also generate gravity.
2. Does a gravitationally accelerating body actually show any net gain or net loss of mass and/or energy from its gravitational acceleration ?

While there is evidence that any such effects must be small, in line with gravity being a weak force, there seems to be no further real experimental evidence to date on such effects ?

Newton certainly proved that gravitational attraction seems to normally transmit at some very fast speed, and seems to normally work in straight lines - to at least some good approximation for most common circumstances. There remain issues about the exactness and the universality of both these aspects of gravity, with some claims for a gravity speed-of-light fixed velocity and for gravity bending like light if not having some other light-like properties. To date there seems to be no evidence that gravity or magnetism reflects or refracts like light does. Gravity affecting many objects to some extent that may be very weak, it may never be possible to prove that something is entirely unaffected by gravity or is entirely 'massless'.

James Clerk Maxwell's 1867 'demon' thought experiment linked thermodynamics and information which maybe really needs William Gilbert's action-at-distance signal-response physics to explain it though most physicists have strangely settled for simple push-physics or non-physics information 'explanations' ?

2010 sees a 'holographic information physics' being 'logically' developed from string physics by Dutch physicist Erik Verlinde and others. Variations in entropy (or temperature-like) directional information gradients with matter location exist, and they somehow give directional pressures or forces acting as gravity. This physics seems to require that statistical entropy information has some actual existence and is more fundamental than matter, energy, space or time - which may be impossible to actually prove. And some supporters of M-theory basically posit that the universe is an information hologram.

But since the Verlinde physics 'forces' seem to lack a mechanical push mechanism, such types of information physics seem to require matter to be able to detect, and be able to itself respond to, directional information. But gravity would then require matter to respond not to single information bits but to statistical gradients of many information bits, yet this Verlinde 'information physics' includes no information processing mechanism. And while an information processing physics may be possible, a more discrete information processing physics (with statistical entropy information a maybe less used derivative) may look more likely and may more readily fit an attraction physics. And of course Verlinde physics maybe lacks rigorous definitions of information, energy, mass and other key elements in the theory, such that it is hard to determine if the theory is logically consistent. It seems a weak attempt at applying an ill-defined physics jargon to what looks a possibly good mathematics. (see http://arxiv.org/PS_cache/arxiv/pdf/1001/1001.0785v1.pdf and http://arxiv.org/abs/1001.5445v2)

Massless Mini-gravity ?

Massive bodies that respond to gravity generally also themselves produce gravity, both being in proportion to their mass. Might there also be some massless bodies or energies that respond to gravity but themselves produce no gravity ? And might that include light ? Or might there also be some massless bodies or energies that do not respond to gravity but themselves do produce gravity ? And might that be 'Dark Matter' ? The universe does not show many significant concentrations of energy that are not associated with any mass. And gravity being as weak a force as it clearly is makes any mini-gravity effects like these practically undetectable with current technology for now at least.

Unusual Gravity claims

Gravity has been claimed to also have weaker effects producing motion in a direction other than attraction's normal directionality. Hence gravity has been claimed to have a 'Geodetic' or 'de Sitter' effect such that the gravity of a fixed body will produce precession in a body orbiting it. Gravity has also been claimed to have a 'gravitomagnetic' or 'Lense-Thirring' effect such that the gravity of a rotating body will produce rotation in a fixed body near it. These claimed gravity effects may seem doubtful, since comparable proved properties of magnetism affect chiefly body alignment and not body precession or body rotation. Magnetic bodies suspended over our magnetic Earth do not seem to show regular rotation or to show precession, and only show a small 'rotation' to reaching some fixed alignment position ? Of course gravity could maybe show some different effects, but the evidence is maybe not strong.

The Strong Force that applies to some sub-atomic particle has been claimed to actually be gravity, though stronger by a factor of around 10^{38} over a very short distance range only. It has been claimed that then Einstein's General Relativity equation $k = Gs.(8\pi/c^4)$ where Gs is the 'strong gravity' value of the G of normal gravity, can predict the masses of strong-force Hadron composite sub-atomic particles - but not of non-strong-force Lepton elementary particles. For these claims of gravity taking two forms, no explanation seems to have been posited yet, but if gravity is a response to signals then there being two types of responses might well be more readily explainable [see http://arxiv.org/ftp/astro-ph/papers/0701/0701006.pdf].

Attempts to explain gravity as a small difference between electric charge repulsion forces and simple electric charge attraction forces, face the problem of similar charges seeming to distribute similarly (eg negative charge particles orbiting outside positive charge particles) with such attraction being between dissimilar charges. Hence negative charges being on average 1% closer to other negative charges and positive charges being on average 1% further from other positive charges, leaving opposite charges on average the same distance apart, need not affect net attractions (or net repulsions). The same holds for any regular dipole distribution for simple electric charge forces. Experiment does not seem to support any universal electric charge distribution in matter that could give a universal gravity effect that way, without the addition of perhaps debatable secondary field effect assumptions.

Of course both Newton and Einstein did much on gravity, though gravity is basically the simplest of forces in being just attraction. They both made no real attempt to explain the much trickier force of magnetism as shown by William Gilbert's published experiments. Indeed although some more modern physics theories do appear to explain some part of magnetism, chiefly its attraction and repulsion effects, none explain all of the various magnetism effects as well as Gilbert's own action-at-distance physics theory. Physicists today of all theory inclinations could really do with a close study of Gilbert's 1600 'De Magnete' especially perhaps in its most recent translation. Attempts to replace Einstein's gravity theory with a quantum gravity theory that works have been various and all unsuccessful to date. Maybe a better quantum gravity theory would be a quantised signal theory gravity, more in line with a William Gilbert style action-at-distance physics ? Clearly if gravity is one push thing, then evidence for it being quantized is simply evidence for gravity being quantized. But if gravity is a dual thing involving both signals and responses then an apparent quantization of gravity could be just the quantization of responses to gravity, whose signals themselves might still be non-quantal ? Or the two could actually be differently quantal ?

Distant stars all around Earth show an about even distribution of redshifts claimed to be due to universe expansion, though that would seem to require that Earth is located at the center of the universe which seems highly improbable. There is increasing evidence for redshift 'quantization' that suggests redshifts may not be due to the universe expanding (and incidentally that there may not be 'missing mass' dark matter). There is also some evidence that redshifts may be slowly reducing with time, possibly due to the speed of light slowly decreasing quantally with time (maybe due to gravity or other energy fields slowly increasing quantally with time ?). Perhaps little is yet really known about distant space or really about gravity.

A signal-response physics might perhaps more easily explain both Dark Matter and Dark Energy. Hence 1. at edge-of-galaxy distances gravity weakness, responses to gravity may somewhat strengthen with distance mimicking Dark Matter, and 2. at further-from-galaxies distances further gravity weakness, signal strength may weaken with more distance mimicking Dark Energy. And similar effects might possibly be predicted also for other forces besides gravity. But there has perhaps been relatively little real basic experimenting on gravity, with major physics funding since the 1940s going to atom-smashing experiments. With the existential problem of global warming looking to be beyond the abilities of current technology, we could maybe really need a big new breakthrough in basic science to deal with it. Encouraging more appropriate experiment funding might usefully help.

A 'Grand old Duke of York' experiment

There is an interesting old English nursery rhyme about Richard Duke of York and the War of the Roses 1640 Battle of Wakefield ;

The Grand old Duke of York.

The Grand old Duke of York he had ten thousand men
He marched them up to the top of the hill
And he marched them down again.
When they were up, they were up
And when they were down, they were down
And when they were only halfway up
They were neither up nor down.

Clearly this type of march-to-signals repeated action would leave the men exhausted, but could there be a physical forces equivalent ? Are bodies actually being worked on by external forces or are bodies themselves working in response to mere signals ? It should be possible to do a decisive physical 'Grand old Duke of York' experiment as using a Galileo inclined-plane with a steady gravity pulling a steel ball downhill and a steady electromagnet switching on periodically pulling the steel ball uphill again repeatedly. If all the work is being done by the gravity and electromagnetism then the ball will never tire, but if all the work is actually being done by the ball itself then the ball should show a reducing responsiveness over time in losing some energy and/or losing some mass ? Some friction might need to be accounted for and gravity and magnetism might work differently but the calculation should be interesting though generally dodged. Do you know of any physicist trying to do a decisive physical 'Grand old Duke of York' experiment ? Or the calculations for such ?

Gravity and solar system instability

Newton showed that the orbits of planets and moons in our solar system under the Sun's gravitational attraction should have substantial stability. But he did not specifically consider the issue of the Sun's stability under the varying gravitational attractions of these orbiting bodies, which involves a number of factors ;

1. The total gravitational attraction exerted on the Sun.
2. The mean directionality of gravitational attraction exerted on the Sun being equatorial due to orbit planes.
3. The point gravitational attraction exerted on the Sun due to planets being discrete.
4. The time variance of point gravitational attraction exerted on the Sun due to planet orbit velocities giving varying degrees of planet conjugations.

These gravitation factors must be the chief causes of the observed instability of the Sun as shown by solar activity and its significant variation over time. This solar instability would be reduced if some of the planets orbited the Sun in a plane at 90 degrees to their present orbit planes, though somewhat strangely that seems rarely the case in natural solar systems. For more on this see our section on Solar System Problems and its Sun Pull gravity App.

You are welcome to **link** to any page on this site, eg www.new-science-theory.com/gravity.php

OR if you like this site you could maybe make a donation ;

It will help with site development, and just possibly with some key basic physics experiments long planned but never afforded.
[PS. and you may perhaps help make history for science ?]
(The fictional time-travel and multi-universe type ideas of modern physics theory have long totally discouraged certain lines of physics experiment despite there being strong reasons to believe them to be very promising if not essential lines of experiment. Some such lines of experiment considered here identified as early as the 1960s seem still to have had no work done on them and there is maybe not much more time here for this. Science funding both government and private unfortunately now all goes to basically safe standard mainstream science, and no money at all goes to any really innovative risky science though that might pay a thousand times greater.)

otherwise, if you have any view or suggestion on the content of this site, please contact :- New Science Theory
Vincent Wilmot 166 Freeman Street Grimsby Lincolnshire DN32 7AT.

© **new-science-theory.com, 2022** - taking care with your privacy, see New Science Theory HOME.

Light as a thing or as a signal, and signal response theory physics

Homepage . William Gilbert . Rene Descartes . Isaac Newton . Albert Einstein ..... Information Physics .... General Image Theory

Seeing is believing ?

From the action-at-distance signal-response physics of William Gilbert it is possible to construct a general information physics that takes all objects in the universe as being one or more of three general classes of information objects, namely; 1. Objects that are Natural Signals emitted by some source objects and naturally carrying some information regarding their source and their journey from their source. 2. Objects that are Signal Emitters or sources of Natural Signals. 3. Objects that are Signal Responders and respond to some information from some natural signals such as to motion directed towards or away from their source.

While a natural signal carries some natural information relating to its source, it may also pick up some added information or 'natural noise' relating to its particular journey from the source to the detector and it might also carry some added intentional information or 'intentional noise' given to it by some people. Of course for detection relative to the intentional signal, both of the other two informations are 'noise'. So it may be of some interest to now consider such a general information physics concerning natural signals as communication of information, starting from some consideration of the basics of signal-response theory.

Light in a signal-response theory physics

The basics of a signal-response theory physics can be taken as being that any body can be a signal, relative to some observer body that can respond to it. And that any body can be an observer, relative to some signal body to which it can respond. Response can be to any property or properties of a signal that can be taken as carrying data or information, and hence signal responses may or may not reflect the basic nature of the signal itself but might just eg reflect anything that had previously happened to the signal. A signal-response physics like William Gilbert's is necessarily an information physics, while no kind of push-physics can really be. Data or information is a possible derivative of signals, and the continued existence of a bit of information need not require the continued existence of the signals that it derived from. The simplest signal is exist/non-exist (on/off or 0/1) and even that can give more complex data or information in its eg temporal and/or spatial arrangement.

Signal-response theory generally does not assume any specific mathematics, eg signal strength need not always decrease as the square of the distance from its source and a response strength might be fixed over some wide range of signal strengths or might vary with signal strength to some non-simple law. Of course particular pieces or forms of signal-response theory can assume some specific mathematics. In nature different bodies can respond in regular ways to different natural signals, but modern technology can link almost any response however complex to almost any signal however simple - basically as the human brain can. And in many cases direct signal detection may not be possible and can only be inferred from observed response reactions though signal and response can appear quite unlike, though not necessarily excluding, action and reaction.

Hence some or all existent objects can emit signals that indicate their colour, magnetism, mass, motion or other properties of the signal emitting object. Emitted signals indicate or reflect properties of their emitters, so that responses of other objects to received signals are responses to properties of signal emitters and are caused by the signal emission to the extent that the signal is not modified in transmission or by signal reception. In signal theory signals are basically anything to which some signal detector can produce some response, so that there are two separate related phenomena being signal and response. If light is taken as being a signal, then different physical systems might be expected to show some different responses to light. This appears to be the case with at least some light-related phenomena like reflection and refraction. It has even been shown that punching holes in thin plates can increase OR decrease the amount of light that appears to penetrate a plate, see Physics World light And while in nature there are many cases of bodies affected by light as a source of heat or energy, there are many cases in nature and in technology of bodies responding to light as a signal. Of course Newton concluded that colour was a property of light itself and was not just a property of illuminated objects modifying light, which he claimed 'to have proven definitely with a crucial experiment'. But unlike William Gilbert earlier, Newton failed to publish the exact details of his experiment and so did not help with correct replication by other scientists. And also later Newton allowed that light itself might respond to some signals from objects or their atoms as to gravitational signals.

The nature of light itself

In any physics that does not take light as being a signal, light impacting different physical systems may be taken as being different behaviours of light itself. This can lead to taking reflection, refraction, diffraction, photoelectric emission, Compton emission etcetera as being light behaviours. And some of these apparent light behaviours can be taken as evidence for light itself being an ether wave, a quantal particle, or either or both. And a quantal particle might be a simple Cartesian push particle or a Gilbert-Newton attraction physics particle that can respond to force signals like gravity or magnetism.

But in a physics that takes light as being a signal, light impacting different physical systems can be taken as evoking different detector responses. This leads to taking reflection, refraction, diffraction, photoelectric emission, Compton emission etcetera as being responses to light signals. And some of these being responses can be taken as giving no evidence for light itself being of any specific nature if the nature of responses is not fully determined by the basic nature of the signals involved.

Responses to signals

Generally in signal theory the nature of responses is not fully determined by the nature of signals, but reflects only some one property or few properties of a signal. Hence some detectors can give digital quantal responses to some continuous signals, or give analog continuous response to digital signals. See eg Digital to Analog Converter or Analog to Digital Converter - though these sources may not be the best. And the different magnetic responses (as attraction, orientation and magnetization) to the same signal can operate at very different ranges, so that apparent 'signal range' can clearly be less a property of the signal than an indicator of response sensitivities.

Modern 'signal processing' is predominantly electronic and often involves systems using designed program calculation methods in producing signal responses of any designable form irrespective of the signal involved, but other physical systems can respond in various ways to different signals using only basic physical responses. And it is perhaps that kind of non-designed signal response that is of more fundamental relevance to physics. Hence mechanical clocks can respond to an analog spring pressure with ratchet-gear digital responses. (And even computational physics can be basically simple resting on 0/1 or On/Off states, so that eg atoms for some phenomena involving one signal may have two states allowing two different responses. Some recently have even proposed a physics on that basis like the New Kind of Science of Stephen Wolfram.)

Responses of light or responses to light

There are many interesting light phenomena and Isaac Newton offered one possible signal theory or 'attraction theory' explanation of light reflection and of light refraction, though involving the response of light itself to signals. (Newton light-attraction could also explain light diffraction etc - see our Newton's Principia.) Hence, as below;

The standard 'school' explanation of light reflection is as a ball-wall contact rebound -

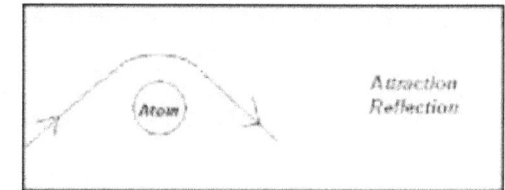

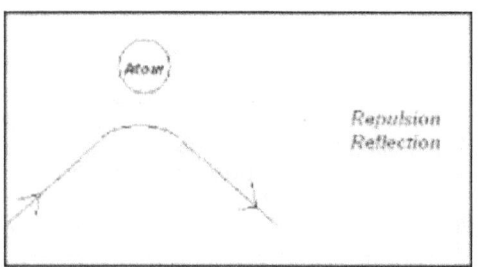

- A repulsion explanation of light reflection also looks workable.

An attraction slingshot explanation of light reflection was also considered an option by Newton -

There may be some massless bodies or energies as light that can respond to gravity but themselves produce no gravity. Gravity being as weak a force as it clearly is would make any mini-gravity effect like this practically undetectable with current technology for now at least.

So clearly Newton in his Opticks proved how action-at-distance physics allows objects to appear to collide without any actual collision. And several possible theories of light reflection, including also some absorb-photon/emit-different-photon theories and others, can give the same reflection angles - and events at the atomic level not being actually visible means that experiments cannot readily decide between a range of theories. But though classic collision and some different reflection alternatives might all have compatible macroscopic maths, they should differ in their microscopic maths where also the possibly digital nature of force signals might also need to be accommodated. While an actual collision event must occur at the time of collision, a proximity-repulsion event would start somewhat before that time and an attraction-slingshot event would start somewhat after that time and such small differences might show in a suitable large range of collider results. Some of these issues may also apply to light diffraction a seemingly simple phenomenon, taken by some as supporting maybe a wave theory or taken by some as supporting maybe a probability theory, requires the dominance of solid edges and their possible undetermined proximity forces. But photons (and elementary particles generally) do not have fully individual identification characteristics like fingerprints that might help with detailed experiment. Of course many non-light experiments involve similar interpretation problems generally with some unsubstantiated assumption being widely favoured. Newton did see most then known light behaviours as evidence more of it being a form of attraction matter than being just waves of something and requiring all space to be filled with a novel wavable no-push medium (somehow also additional to Einstein's supposed no-push spacetime-continuum also filling all space, which also perhaps fits very uneasily with the push-physics it is generally set in).

Two other interesting types of light-electron interactions that might suggest different types of signal response to light are the 'Compton Effect' and the 'Photoelectric Effect' :-
1. The Compton Effect involves light generating additional photons and deflecting electrons in proportion to overall light energy and not individual photon energy, seemingly in an analog manner or to multiple photons with no threshold-frequency.
2. The Photoelectric Effect involves atoms emitting electrons seemingly in a digital manner in response to single incoming photons of some above-threshold frequency. (however an additional lower-level of photoelectric response is also produced, apparently when two below-threshold photons are received simultaneously as Sipila et al at www.iop.org/EJ/article/1367-2630/9/10/368/njp7_10_368.html And some other new experiments are also showing that the photoelectric effect is not always a simple case of individual photons providing the amount of energy needed for individual electron escape as Einstein claimed as Reverse Electrons.)

The Compton Effect may indicate atoms or sub-atomic particles being able to add-up consecutive digital signals over time until some target figure is reached to trigger response - or that some time-spread signal property of a continuous signal triggers response. The Photoelectric Effect may indicate atoms or sub-atomic particles responding to each single case of a digital signal as per Einstein - or that some short-time signal property of a continuous signal triggers response. In both of these light phenomena a signal theory can allow of light showing either digital or analog effects without requiring the current physics duality contradiction of it itself both being a wave and not being a wave.

Some other light behaviours that are commonly taken as being responses to light, rather than properties of light, include light slowing and induced transparency - and these cases involve variable response.

Two-slit diffraction light interference is another light phenomenon for which a variety of strange explanations are currently claimed by many physicists. But in a two-slit diffraction experiment using single photons or particles, if you consider each photon or particle individually, then they seem to act 'sensibly' by each passing through one of the diffraction slits simply. However otherwise if you do not consider each photon or particle individually, then they all seem to act 'strangely' by each appearing to pass through both of the diffraction slits probabilistically or 'impossibly'. This seems to exclude any kind of simple contact-push physics mechanism controlling which passes through which slit, leaving it that the most likely actual classical physics explanation seemingly must be some version of some action-at-distance physics signal-response mechanism. But for most physicists then who wrongly dismissed classic action-at-distance physics it seemed that this required the acceptance of some illogical physics. Of course a logical and scientific classical action-at-distance physics explanation might better involve some kinds of force-signal emitted by slit-edges to which light particles somehow respond, and/or some kinds of force-signal emitted by light particles to which slit-edges somehow respond ? And what if two slits are 1mm apart or 100m apart ? Or what if slits are in glass or in a magnetic, electric or gravity field ?

Hence, much that current physics teaches as being properties of light seem clearly in fact instead response behaviours of other things to light. Clearly refraction seems the response of mediums to light, reflection seems the response of surfaces to light, and diffraction seems the response of edges to light. And maybe light itself has some responses to some other things. But unfortunately much of physics seemingly involves quite wrong preferential interpretation of phenomena and of experiments, relating to light and to gravity and to other physical phenomena.

Light as a signal for physics

A signal view of light would have some significant consequences for physics. One thing it would throw doubt on is Einstein's conclusion that photoelectric emission is evidence of light being actually quantal, but it could also cast doubt on other claimed evidence for light being also actually a wave. Waves and particles have substantially different mathematics that do not seem to be simple transforms of each other, so it seems that wave and particle theories cannot be compatible image theories of each other. But in fact a signal view of light could perhaps allow of both quantal and wave type responses without any contradictory 'duality' requirement of light itself. Digital signals can give digital or analog wave responses, and analog wave signals can give analog wave or digital responses. Contradictory appearances do not have to indicate contradictory realities.

The observed speed of sound in a given direction in moving air, like the speed of a bullet, reflects both the velocity of sound in air and the wind velocity in the given direction - also the observed speed of sound reflects any source and/or observer relative motion. For moving waves or regular beams of particles or signals from some source, there will be some 'relative velocity effect' to some observer if some measure of the distance between the source and that observer is changing. To both Doppler and Newton the applicable distance is the fixed straight-line distance, but to Einstein it is some variable spacecurve distance. But the observed speed of light is claimed not to show any such effect, and to be invariably constant. This is the basis for the main claims of Big Bang expanding-universe cosmology, which takes observed Hubble light redshifts being greater for more distant galaxies as being caused by only a Doppler relative-velocity frequency Effect. But possible alternative explanations include both large-distance slight energy loss (Tired Light), possibly due to large-distance slight gravity slowing. Hubble favoured the latter being additional to a Doppler effect.

Information in physics

Signal theory generally locates information, intentional information or unintentional natural information, in signals - but physics tends to trying to locate information either in physical bodies themselves or in ill-defined 'observers'. So a range of issues can arise such as ;
1. physical bodies either do or do not carry some natural information before an observer observes ?
2. signals either do or do not carry some natural information before they are detected ?
3. physical bodies either do or do not carry some natural information before they emit signals ?
4. signals either do or do not carry natural information reflecting the full nature of their source ?
These and other related issues have not always been properly addressed by physics theories to date. A signal physics may seem better able to handle this, though no doubt non-signal image theories of a signal theory would be compatible also.

All forces as signals

It is of course possible that forces like magnetism, electricity and gravity may also work by signal response as proposed first by William Gilbert, and so allow of a signal physics 'theory of everything'.

Responses to signals can involve issues like signal thresholds, response times, signal noise, excitation states, conditional response and signal summation. Depending on the particular signal response parameters involved, signal response systems may also be capable of looping or of hanging. And for some signal response systems a numbers of factors may vary the probability of some signal giving some response. Avoiding the use of signal theory, current physics struggles poorly to explain much, such as exactly what is light slowed and shrunk !?
(see http://physicsworld.com/cws/article/news/2009/dec/15/slowed-light-breaks-record)

Such signal theory physics goes right back to 1600 when William Gilbert explained magnetism as response to emitted signals he termed 'effluvia', and took electricity and gravity as working similarly, and this signal physics was developed further by Isaac Newton in his using 'attraction theory' as one explanation option in his blackbox theory of gravitation. Of course unintentional natural signals might include direct emissions from objects or events, as masses emitting gravitons or causing space-curvatures, or be indirect signals as from some external interaction with the object or event, with experimental interactions being intentional replicatable attempts to elicit natural signals. A signal theory physics might still have some usefulness for gravity, for light and maybe more.

The physical nature of signals

Signal theory allows that anything that can convey information can be a signal. Hence all human senses are concerned with signal detection, as in hearing, sight and touch. From this it is clear that some natural signals might take the form of waves in a medium, and others might take the form of particle objects.

Any single object's speed or velocity is simply the rate of change of its position with time in a specified direction. But waves in a medium, and sets of multiple moving objects, can have a group velocity and a phase velocity so that talk of their speed can be ambiguous. Claims that a single object is somehow a wave can involve ambiguous assumptions regarding its speed. There can also be issues with wave mathematics assuming a medium to be continuous rather than quantal or particulate.

A light beam may seem to be likely either a continuous medium-wave (if space has any medium) or a set of multiple-object photons, and maybe less likely a set of single-object photons. As zero can rarely be actually distinguished from very small, claims of light being 'massless' can maybe only be proved to be of it not having a big mass. But when discussing medium-wave or multiple-object motion, 'speed' should always be clearly specified as being either the group velocity or the phase velocity. On differing interpretations of eg 'double-slit' light experiments, see - quantum light theory.

Some things to consider when considering the question of whether a light beam is more likely a continuous-medium wave or a set of multiple-object photons ;

A. Some normal multiple-particle motion properties.
1. Particle motion is not resisted by a vacuum but is by higher-resistance mediums, and the velocity of motion of particles through a low-resistance medium like a vacuum is significantly affected by particle forces such as gravity.
2. Particles can be accelerated to some velocity as by a force like gravity, and in a low-resistance medium like a vacuum will tend to maintain their velocity. And multiple-particle a stream or beam of emitted particles being some regular stream of single particles, or may be pulsed with each pulse being some set of multiple particles.
3. For both multiple-particle source and multiple-particle detector within a low-resistance medium like a vacuum, their motion relative to each other or to the low-resistance medium changes apparent detected velocities or frequencies for the detector but does not change absolute as-emitted velocities or frequencies in the low-resistance medium.
4. In a low-resistance medium like a vacuum, no medium motion will cause multiple-particles motion in it to suffer any velocity or frequency change.

B. Some normal wave properties, as of sound waves.
1. Sound waves cannot propagate through a vacuum, and in any fixed medium the propagation of sound waves through it is not significantly affected by a particle force such as gravity. And wave motion may involve a single wave of some frequency, or may involve multiple waves of differing frequencies that will make it pulsed.
2. In any fixed medium, sound waves will propagate through it at some specific fixed velocity that is often higher for higher-resistance mediums - so the speed of sound in air is 343 m/s, in water is 1,433 m/s and in denser materials can have higher values.
3. For both sound source and sound detector within the same fixed medium, their motion relative to each other or to the medium changes apparent received sound frequency for the detector but does not change the absolute as-emitted sound frequency in the medium nor the apparent sound velocity for the detector.
4. In a medium moving at some velocity, sound will approximately propagate through it at a velocity that is the sum of its specific fixed velocity for that medium and the medium velocity - without change of absolute as-emitted sound frequency in the medium.

While some of the above wave properties and (multiple-)particle properties are logically mutually exclusive, many claim that some or all physical things (notably including light) possess some or all of both sets of properties at the same time. While duality theory generally takes the extreme position on this, some prefer the position of things changing properties between particle and wave properties in different experiment circumstances only. Hence the most interesting and useful phenomenon of Magnetic Resonance used in MRI seems best explained as response to conflicting signals, as in conflicting on/off digital signals giving alternating opposite responses. See MagneticResonance.

James Clerk Maxwell's equations for time-oscillating electric and magnetic forces are wave equations, but this really only supports time-oscillation like timed particle emission being wavelike in having a wavelike maths. It is poor support for any general wave physics theory It is certainly no real support for light being any wave, when nothing can be identified that it could be a wave of. And there seems little basis for claims that 'light is electromagnetic', when light is not affected by any steady electric or magnetic field (though electric-charged matter can produce light and show response to light). And as at least most medium-waves cannot transmit through a vacuum or through space, it is perhaps doubtful that there is any real evidence for electrical or gravitational force transmission being based on waves rather than being some as yet undetermined emission signals. (See our related signal-response Information Physics section.)

And while most modern physics theory may have no natural place for time, signal theory physics in fundamentally involving response to signals fundamentally involves time as a consequence. What basically distinguishes a response event from a signal event is simply time, with signals being causes and responses being subsequent effects. If an attraction response cannot precede an attraction signal, then the universe is not time reversible and has one-direction time inbuilt. Many other physics theories by default predict a time reversibility that is quite contrary to many very well confirmed experiments.

OR if you like this site then you could maybe make a donation ;

PayPal Donate

It will help with site development, and just possibly with some key physics experiments long planned but never afforded.
[PS. and you may perhaps help make history for science ?]
(The fictional time-travel and multi-universe type ideas of modern physics theory have long totally discouraged certain lines of physics experiment despite there being strong reasons to believe them to be very promising if not essential lines of experiment. Some such lines of experiment considered here identified as early as the 1960s seem still to have had no work done on them and there is maybe not much more time here for this. Science funding both government and private unfortunately now all goes to basically safe standard mainstream science, and no money at all goes to any really innovative risky science though that might pay a thousand times greater.)

© new-science-theory.com, 2022 - taking care with your privacy, see New Science Theory HOME.

String Theory, Quantum Mechanics, and theoretical physics today

Homepage . William Gilbert . Rene Descartes . Isaac Newton . Albert Einstein .... The Standard Model .... General Image Theory

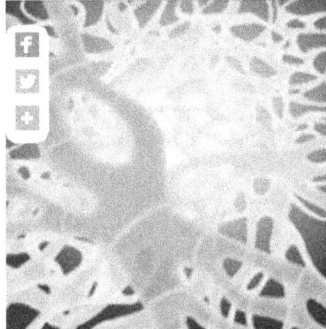

Below is a good article on the development of String Theory and on the general state of theoretical physics today. It is one physics view now, and there are of course other views around, but it is overall a reasonable view backed by numbers of modern physicists though currently a minority. Its general position that modern physics theory has become unreal is backed as in the 2016 physics book 'Fashion, Faith and Fantasy in the New Physics of the Universe', by eminent English mathematical physicist Sir Roger Penrose showing string theory as being a "fashion", quantum mechanics "faith", and cosmic inflation a "fantasy".

Following the article below are presented some other views and considerations of String Theory, M-theory, Quantum Mechanics, Uncertainty physics and Duality physics.

"We Don't Know What We Are Talking About" - Physics Nobel Laureate David Gross.

Article by Michael Strauss 2006.

Science has reached an enormous impasse. From biology to physics, astronomy to genetics, the scientific community is reaching the limits of understanding which often presage a complete rethinking of long-accepted theories. So characteristic of this new apex of modern arrogance is the inability to comprehend the obvious in physics: That we don't know what we are talking about.

Last December ('05), physicists held the 23rd Solvay Conference in Brussels, Belgium. Amongst the many topics covered in the conference was the subject matter of string theory. This theory combines the apparently irreconcilable domains of quantum physics and relativity.

David Gross a Nobel Laureate made some startling statements about the state of physics including: "We don't know what we are talking about" whilst referring to string theory as well as "The state of physics today is like it was when we were mystified by radioactivity." The Nobel Laureate is a heavyweight in this field having earned a prize for work on the strong nuclear force and he indicated that what is happening today is very similar to what happened at the 1911 Solvay meeting. Back then, radioactivity had recently been discovered and mass energy conservation was under assault because of its discovery. Quantum theory would be needed to solve these problems. Gross further commented that in 1911 "They were missing something absolutely fundamental," as well as "we are missing perhaps something as profound as they were back then."

Coming from a scientist with establishment credentials this is a damning statement about the state of current theoretical models and most notably string theory. This theoretical model is a means by which physicists replace the more commonly known particles of particle physics with one-dimensional objects which are known as strings. These bizarre objects were first detected in 1968 through the insight and work of Gabriele Veneziano who was trying to comprehend the strong nuclear force.

Whilst meditating on the strong nuclear force Veneziano detected a similarity between the Euler Beta Function, named for the famed mathematician Leonhard Euler, and the strong force. Applying the aforementioned Beta Function to the strong force he was able to validate a direct correlation between the two. Interestingly enough, no one knew why Euler's Beta worked so well in mapping the strong nuclear force data. A proposed solution to this dilemma would follow a few years later.

Almost two years later (1970), the scientists Nambu, Nielsen and Susskind provided a mathematical description which described the physical phenomena of why Euler's Beta served as a graphical outline for the strong nuclear force. By modelling the strong nuclear forces as one dimensional strings they were able to show why it all seemed to work so well. However, several troubling inconsistencies were immediately seen on the horizon. The new theory had attached to it many implications that were in direct violation of empirical analyses. In other words, routine experimentation did not back up the new theory.

Needless to say, physicists romantic fascination with string theory ended almost as fast as it had begun only to be resuscitated a few years later by another 'discovery.' The worker of the miraculous salvation of the sweet dreams of modern physicists was known as the graviton. This elementary particle allegedly communicates gravitational forces throughout the universe.

The graviton is of course a 'hypothetical' particle that appears in what are known as quantum gravity systems. Unfortunately, the graviton has never ever been detected; it is as previously indicated a 'mythical' particle that fills the mind of the theorist with dreams of golden Nobel Prizes and perhaps his or her name on the periodic table of elements.

But back to the historical record. In 1974, the scientists Schwarz, Scherk and Yoneya reexamined strings so that the textures or patterns of strings and their associated vibrational properties were connected to the aforementioned 'graviton.' As a result of these investigations was born what is now called 'bosonic string theory' which is the 'in vogue' version of this theory. Having both open and closed strings as well as many new important problems which gave rise to unforeseen instabilities.

These problematical instabilities leading to many new difficulties which render the previous thinking as confused as we were when we started this discussion. Of course this all started from undetectable gravitons which arise from other theories equally untenable and inexplicable and so on. Thus was born string theory which was hoped would provide a complete picture of the basic fundamental principles of the universe.

Scientists had believed that once the shortcomings of particle physics had been left behind by the adoption of the exotic string theory, that a grand unified theory of everything would be an easily ascertainable goal. However, what they could not anticipate is that the theory that they hoped would produce a theory of everything would leave them more confused and frustrated than they were before they departed from particle physics.

The end result of string theory is that we know less and less and are becoming more and more confused. Of course, the argument could be made that further investigations will yield more relevant data whereby we will tweak the model to an eventual perfecting of our understanding of it. Or perhaps 'We don't know what we are talking about.'

About The Author: Michael Strauss is an engineer who has an interest in this subject matter. To contact the author visit: www.relativitycollapse.com or www.relativitycollapse.net

---

AND read the general 2017 views on physics today of Edward Witten, who developed M-theory, at Duality and Information Physics.

OR below you can hear David Gross himself explain and justify some 'mainstream' modern physics theory however inadequately and despite himself accepting inadequacies as noted above - including failed predictions and the unexplained 'counter-intuitive' (or nonsense ?) claim that a strong nuclear force **increases** with distance from its source opposite to gravity and magnetism ;

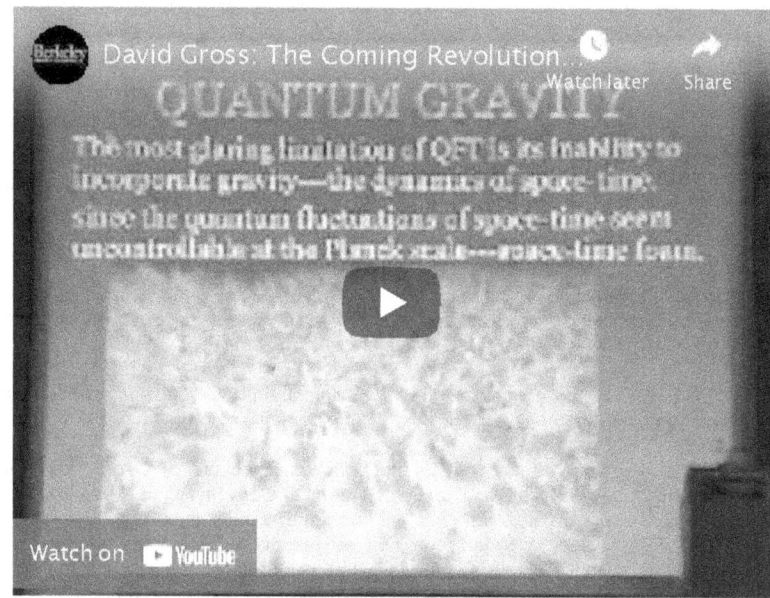

Sorry, but this dubious chunk of modern physics theory replaces what was a perhaps more interesting much-published non-scientist philosopher view that has been suppressed on claimed 'copyright issues' ridiculously by scientists objecting to themselves being quoted by a critic.

(PS. a signal physics might perhaps rather predict such an above 'counter-intuitive' strong nuclear force as involving some closer-distance stronger-signal above-threshold responses being repulsive proportional to signal strength, but some greater-distance weaker-signal below-threshold responses becoming attractive but still proportional to signal strength ? Gravity might even also work that way, below some threshold strength the attractive signal-response becoming repulsive giving universe expansion ?)

While the general sense of this 'Heisenberg-Einstein' observer approach to physics may well seem OK, it certainly looks like science with bad definition of even its basics like mass, energy and space. It also maybe looks like a physics that is a poorly defined image theory of Gilbert-Newton signal attraction physics theory where all physical objects are observers and/or signals. In comparison, 'Heisenberg-Einstein' observer physics has only anthropocentric or anthropomorphic observers in a universe in which mankind is unjustifiably totally different from the rest of the universe. In a Gilbert-Newton signal attraction physics where all physical objects are automaton observers/responders, mankind fits more naturally and has only the addition of thought to its processes. Then physical objects and mankind differ basically only to the extent that programmed computers and self-learning computers differ. Gilbert-Newton signal attraction physics can reasonably claim to better unify the physical and biological and to be the least anthropocentric physics, and certainly not the most anthropocentric and anthropomorphic as widely falsely claimed. (anthropocentrists trying to widen 'mankind' by including gods and/or alien life amounts to little real widening.) William Gilbert's experiments showing basically that rocks attract rocks is still not disproved, and it may well be that both types of theory have some valid defined place in some well defined physics.

In the 1990s, string theorists including Edward Witten, Paul Townsend and others concluded that the five versions of 10-dimension string theory current then basically describe the same thing seen from different perspectives and so were aspects of one bigger theory. Basically from considering theory-equivalences, they proposed a unifying 11-dimension string theory called 'M-theory' or 'Membrane Theory', involving multiple universes and gravity being a force that operates between each universe. Like much modern physics, the improved mathematics of M-theory seems to go with a poor physical description and no doubt its better mathematics will in the future be found to go with some one or two other better physical descriptions. The universe is unlikely to be actually constructed of strings, loops, waves, triangles or any other geometric shapes that mathematics may suggest. When these theories prove some consistencies between each other, their loose definition generally limits proved consistencies to the superficial level. And as additions to the standard 4 dimensions of space and time, the other proposed 'dimensions' of M-theory may just be physically describable as forces or energy states or signal-response states ? (also see our General Image Theory section)

Modern string theory has been for the last 30 or 40 years the most controversial big idea in physics. On the one hand, it mathematically appears to have the potential to unify some much modified Standard Model physics with General Relativity physics and give some new Theory-Of-Everything physics. But on the other hand, its predictions are varied and seem untestable and require enormous sets of assumptions that are unsupported by any actual evidence. So it is basically all theory and no experiment, so maybe no science ?

Modern physics includes theories like General Relativity theory, Standard Model theories, Quantum Mechanics theories, Loop theory, String theory, Superstring theory, M-theory and other theories which are often poorly defined and based on ridiculously weak science terminology assumptions such as 'we all know what 'mass' is'. Well no - there are actually quite a range of different physics ideas of what mass is exactly, and they will not all be consistent with a particular physics theory. Some want several of these theories to be all accepted as valid, and not needing to disprove eachother, without any substantial consistency proofs. But any science theory without exact definitions must perhaps be taken as being a weak science theory. Some of todays physicists require a spacetime continuum 'filling space' along with an electromagnetic field 'filling space' and a Higgs field also 'filling space' - each affecting particles differently yet somehow not having any effect on eachother. And some physicists today require 'filling space' to be 'continuously filling all of space', but some physicists today go with more 'nearly-filling space' as though that is equivalent though it is clearly not.

2014 saw German physicist Alexander Hartmann design a Standard Model game, called Spinglas, but it was undoubtedly not the most useless work by a modern physicist.

Quantum physics and Quantum Mechanics.

String theory is basically a quantum physics that involves the universe consisting of only one type of one-dimension 'string' body which has many different ways of vibrating within 10 'dimensions'. If the meaningfulness of 10 dimensions is doubtful, the meaningfulness of a 1-dimensional body is at least equally doubtful. String Theory seems to build a physics on an object that cannot exist.

But quantum physics started basically as the application of Heisenberg's uncertainty principle and probability to a Particle Physics, though some claim it is really only fully applicable to a Wave Mechanics with wave mathematics necessarily linking position, motion and momentum. More clearly in its early days Quantum Physics was basically a form of Descartes mechanical physics then became a form of wave energy physics with its 'wave' poorly defined, but now has mostly adopted the scary science 'Duality Principle' positing both.

Duality, claiming that everything is a wave and is not a wave, is so plainly self-contradicting that it clearly disproves itself. And that is without the additional modern requirement of waves that they are also now claimed to need no medium to wave. Support for these scary science ideas has given us an Emperors Clothes physics where none wants to risk their reputation by pointing out that these things are clearly ridiculous. The peer mob rules and maintains modern scary physics. Even Einstein bought scary duality if only for light, when it was maybe of little real use to his relativity theory which in any case had other major problems.

Like both Relativity theory and String theory, Quantum Mechanics was initially basically another form of Descartes mechanical push physics and all three of them have problems that still await satisfactory scientific solutions. They require that A forces an effect in B, unlike Gilbert-Newton attraction theory, but have no real force/push mechanism - and especially so modern quantum mechanics which allows multiple things to occupy the same space and so does not even have contact for a push or force mechanism. Some see duality as having increased the power of quantum physics, but some see it as having seriously disabled quantum physics in robbing it of real definition.

Quantum Mechanics theory has developed and is still developing in a variety of directions involving field theories and/or particle theories, as in the 'particle theory Standard Model' - though that often including 'massless particles' that are maybe better termed energy quanta and so not a particle theory in any Descartes sense. Often such theories require particles to occupy the same space and/or require forcefields or energy quanta to somehow have push abilities like mass particles though meaningful mechanisms and indeed meaningful definitions are often not offered. Claimed mechanisms include claimed exchanges of 'virtual particles', said to be unobservables and having no well defined mechanisms for their claimed probabilistic appearing or vanishing in a vacuum or in any medium. Of course a signal theory can readily allow of energy quanta signals occupying the same space and having push or pull type response effects with no problem.

Quantum mechanics also claims that evidence supports an 'entanglement' instant-communication property for some pairs of particles or photons, created as by radioactive decay, linking them no matter how far apart they are so if one particle changes spin then the other instantly changes spin oppositely. Such quantum entanglement of particles or photons, or even of atoms, looks very much like action-at-a-distance but with no explanation or mechanism at all. Einstein called it 'Spooky action-at-a-distance' though he offered no specific evidence against it and offered no alternative explanation. It being specific to only particular particles makes entanglement certainly even stranger than common at-a-distance general forces like gravity and magnetism. But a signal physics can more naturally handle multiple-signal emissions having related information without requiring any mystical 'entanglement'. The modern physics 'spooky entanglement phenomenon problem' has developed from 1 photon splitting into 2 lower-energy photons. But a general entanglement phenomenon, as "if you split something into 2 pieces, then some property of one piece may reflect some property of the other piece even if the pieces are separated by some distance", does not seem to necessarily require anything spooky or magical and looks like it might in at least some cases be explainable somehow by Newtonian physics depending on the details applying to a case. This need not imply or require any actual connection between the 2 pieces subsequent to their split, only prior to their split. Just a related creation giving related properties. Subsequent connection may well be just apparent and is not actual, and so presents no actual problem to classical Newtonian physics where appearance issues merely concern the responses of objects or observers to signals - or 'attraction theory'.

Gilbert-Newton action-at-distance by signal emission is NO real problem for a physics even if the signals are hard to detect, but instant action-at-distance with no emission involved IS obviously a killer problem for a physics requiring it as does some quantum mechanics. Of course physics is not always good at measuring actual zeros or actual infinities and so cannot always really distinguish 'instant' from 'fast'. It is easy to build a robot with anticipatory response to light that certainly appears to be faster-than-light response as near the bottom of our main section on Einstein. And in quantum mechanics physical events are claimed to be basically probabilistic. This despite the fact that the Sun always rises every morning, and a magnet always attract iron quite deterministically and not probably as is firmly established by many experiments and observations. Physical actions predominantly appear to be perfectly deterministic. Of course there certainly are some cases like radioactivity that seem to involve probabilistic action, but may simply involve an as yet unobserved determinism.

Quantum Mechanics generally incorporates Heisenberg's Uncertainty Principle at least in relation to human observers. But the Uncertainty Principle applying to ANY observer can perhaps only fully apply to a physics like William Gilbert's where all physical objects are observers in that they respond to gravity etcetera signals from other physical objects - ie. to a non-mechanical Gilbert Quantum Signal Physics ? The same should also apply for Relativity theory for ANY observer as against Einstein limiting it just for human observers ?

The unfortunately vague definition of 'observer' and 'observation' in both Relativity and Quantum Mechanics theory, with some even confusing observing with experimenting, has even allowed some physicists to conclude that 'observation' can physically affect things observed. And that has encouraged a very doubtful philosophy or religion around a claimed 'Law of Attraction' in which the human mind is supposed to be able to control the physical universe. If observation is just the reception of such signals as things emit then it cannot affect the emitter - and so experiments such as attempt to elicit such signals or responses to such signals if affecting the emitter would not be observations. And 2014 has seen Christopher Ferrie and Joshua Combes, backed by Rainer Kaltenbaek and Franco Nori, throw major doubt on Quantum Mechanics and especially its 'weak measurement' as being based on bad statistics. (see http://physicsworld.com/cws/article/news/2014/oct/09/are-weak-values-quantum-after-all)

You can read another quantum mechanics view of the issue at Many-Minds Quantum Mechanics.

Of course these physics theories have used somewhat different actual mathematics, but that does not perhaps preclude some of them being developed to use similar mathematics. Any theory that is consistent with some experiment is a theory that can give the mathematics that is consistent with that experiment. And since nobody can really prove that one object can actually touch or actually push another object, mechanical physics theories are perhaps not the only physics explanation theories possible ? Certainly modern physics has now mostly, though not entirely, abandoned the early-'victorious' Descartes matter-only physics framework for the early-'defeated' Gilbert-Newton matter-and-energy physics framework. But without acknowledging that the first big physics-war was 'won' very wrongly, and without reconsidering the basic science issues at all - wrongly taking all early physics as Cartesian physics but calling it Newtonian. Perhaps unsurprisingly the modern physics resulting is full of dispute.

In a variety of physics fields today can be found numbers of physicists who support Einstein mathematics but not the explanation given with it, or support Quantum Mechanics mathematics but not the explanation given with it, or support M Theory mathematics but not the explanation given with it - ie who are basically supporters of Black Box science in line with Newton though for post-Newtonian physics theories. With the variety of current physics 'explanation theories' being so diverse, weakly defined, and contradictory, as to perhaps offer no real explanations, maybe such a Newton-like black-box position is preferable - though maybe needing stronger agreed rules for deciding which give consistent mathematics and which does so most easily ? And while the common claim that there is now some one widely accepted 'mainstream physics theory' is far from true, modern disagreement on physics theory does maybe usefully encourage experimental physics. But the experimenting being very largely in the nuclear arena may not be the most useful experimenting possible.

2009 did see a Gilbert-Newton quantum signal attraction physics seemingly getting some modern backing from the new Hořava time-invariant quantum gravity, which was for a time at www.scientificamerican.com/article.cfm?id=splitting-time-from-space.
(On ideas relating the basics of signal theory to quantum mechanics theory see A Gersten, Annals of Physics 1998 1 and 2 at http://arxiv.org/PS_cache/physics/pdf/9911/9911018v1.pdf and http://arxiv.org/PS_cache/physics/pdf/9911/9911019v1.pdf)
A crucial part of the claimed 'proof' of Einstein's physics and various later physics theories has been their claimed 'consistency with Newton' which is largely illusionary or at least very loosely based and certainly not based on any real study of Newton or the theories that he considered his physics to be consistent with.

Of course there are other problems to trying to reconcile Einstein and post-Einstein physics with Gilbert-Newton physics. Hence while Gilbert and Newton took the mass of natural experiment and experience as showing Magnetism, Electricity and Gravity being basically similar forces, Einstein and later physics often depends on treating gravity as being entirely different and not any force. The observed behaviour of gravity is certainly very similar to that of the other forces, but does any physics fully explain both the similarities and the differences ?!

Of course both Newton and Einstein did much on gravity, though gravity is certainly basically the simplest of forces in being just attraction. They made no attempt to explain the much trickier force of magnetism as shown by William Gilbert's 1600 published experiments. And although some more recent physics theories can seem to explain parts of magnetism, chiefly its attraction and repulsion effects, none explain all of the various magnetism effects as well as does Gilbert's own action-at-distance physics theory ?

You are welcome to link to any page on this site, eg www.new-science-theory.com/string-theory.php

If you have any view or suggestion on the content of this site, please contact :- New Science Theory
Vincent Wilmot 166 Freeman Street Grimsby Lincolnshire DN32 7AT.

OR if you like this site you could maybe make a donation ;

PayPal Donate

It will help with site development, and just possibly with some key physics experiments long planned but never afforded.
[PS. and you may perhaps help make history for science ?]
(The fictional time-travel and multi-universe type ideas of modern physics theory have long totally discouraged certain lines of physics experiment despite there being strong reasons to believe them to be very promising if not essential lines of experiment. Some such lines of experiment considered here identified as early as the 1960s seem still to have had no work done on them and there is maybe not much more time here for this. Science funding both government and private unfortunately now all goes to basically safe standard mainstream science, and no money at all goes to any really innovative risky science though that might pay a thousand times greater.)

© new-science-theory.com, 2022 - taking care with your privacy, see New Science Theory HOME.

Standard Model physics theories

Homepage . William Gilbert . Rene Descartes . Isaac Newton . Albert Einstein ...... Probability Science .... General Image Theory
- Site Search at bottom v -

In current physics 'The Standard Model Theory' might maybe now be better called 'Undefined Model Theory'. And two different basic types of Standard Model theory are current - a Cartesian particle version where its 'forces' are simple particle exchanges with push-force exchanges of momentums, and a field energy version where its 'particles' are energy or field quanta with forces. But in both version theories the particles or quanta forces have poorly defined 'charges' including 'colour-charges' and other aspects that do not seem to fit any Descartes-type particle definition (only size, shape and motion), and the alternative energy or field quanta seem to be equally poorly defined energies of nothing or fields of nothing based on waves of nothing ? But some currently support a duality version of Standard Model physics, where every particle is either a fermion or a boson and they have conflicting properties. So fermions cannot occupy the same space and are matter push-force particles, while bosons can space-overlap and are non-push-force particles or force-signals. Here basically fermions are Galileo-Descartes push-particles, and bosons are Gilbert-Newton distance force signals. You then have your cake and eat it ! (But to be fair pro-signal Gilbert did seem to maybe also allow some push somehow.)

Maybe the study of the heavens is Astronomy, the study of physical matter is Chemistry and the study of physical forces is Physics. Then we should maybe talk of Standard Model Chemistry as being the extension of Periodic Table Chemistry. But the three areas do have strong real connection and are not really separate.

Standard Model physics theories.

Standard Model physics is based around matter being composed of some specified set of elementary particles (or wave-packets), taking Protons and Neutrons that were formerly considered 'elementary particles' as being compound particles or Hadrons along with some others like LHC in 2012 called Xib' and Xib*. In current Standard Model theory, elementary particles include Fermion particles involving 1 stable family pair of Quarks with 1 stable family pair of Leptons (electrons, muons and taons) plus 2 unstable family pairs of Quarks with 2 unstable family pairs of Leptons. And additional Boson particles are also commonly postulated, including 'massless' Gluon, Photon and Graviton 'particles', though some favour rather more particles and others favour somewhat less particles.

Particle .... Mass Equivalents, GeV

Fermions :
up quark 0.005000000
down quark 0.009000000
electron 0.000510000
electron neutrino .. 0 or 0.000000007 ?

charm quark 1.350000000
strange quark 0.175000000
muon 0.106000000
muon neutrino .... 0 or 0.000270000 ?

top quark 173.000000000
bottom quark 4.500000000
tau 1.780000000
tau neutrino .... 0 or 0.030000000 ?

sterile neutrino (x?) .... 0 ?
neutralino (x4) .... ?
(eg WIMP neutralino .... 7-99.000000000 ?)

Bosons :
gluon (x8) 0.000000000
photon 0.000000000
graviton .... 0.000000000 ?
graviton spin-0 ..... 0.000000001 ?
graviton spin-2 ..... 0.000000001 ?
axion .... 0.000000001 ?
W+ 60.200000000
W- 80.200000000
Z 91.200000000
higgs .... 500.000000000 ?
.... or 125.000000000 ?
X17 .... 0.017000000 ?

And with maybe many more particles suggested by different versions of Standard Model and by different versions of String Theory ?

Currently the existence of some of the above Standard Model particles is hypothetical only and not supported by experimental evidence to date, including such hypothetical particles as have been theoretically postulated most recently like the X17 Boson which is also claimed by some to produce some new force.

Gluons are claimed to have 8 'colour-charge' types being forms of red + blue + anti-red + anti-blue, or red + green + anti-red + anti-green, or blue + green + anti-blue + anti-green. And the various quarks are claimed to combine to help form neutrons, protons and other composite particles termed Hadrons.

Fermions are claimed to have half-integer 'spin' and to obey Fermi-Dirac behaviour with multiple fermions being unable to exist in the same quantum state or same space. They basically are Descartes push-particles. But some giving 'mass' to an elementary particle may really be just giving the mass-equivalence of its energy.

Bosons are claimed to have integer 'spin' and to obey Bose-Einstein behaviour in that multiple bosons can occupy the same quantum state or same space. They basically are more like energy wave packets or signals than like classical mass particles. While some bosons are claimed to have 'mass' others are claimed to not, and some bosons like photons are readily detected but others seem impossible to detect.

Bosons are generally problematic in standard model physics, as is its explanation of at-a-distance-forces as being due to 'virtual boson' exchange. Protons and Electrons are claimed to electrically attract eachother by Virtual Photon exchange in an Electrical Interaction force, and Protons and Neutrons composed of Quarks are claimed to internally bond by Virtual Gluon exchange attraction in a Strong Interaction force that increases with distance unlike other forces. Protons and Electrons are also claimed to weakly attract eachother by Virtual W and Z boson exchange in an Electroweak Interaction force. Mass particles are claimed to gravitationally attract eachother by Virtual Graviton exchange in a Gravitational Interaction force that may be mediated by the Higgs boson.

These virtual particle exchanges are said to be unobservables, and have no well defined mechanisms for their appearing or vanishing in a vacuum or in any medium. Of course normal particle exchange in a Descartes particle physics might seem a reasonable recoil explanation for a universal repulsion force if there was any such, but is trickier for the attractive forces and for the selectivity of forces actually shown by nature. Of course simple particle contact collisions could look similar to repulsions. Virtual particle exchange may seem to need some attraction mechanism as well as a signal mechanism for prompting exchanges. Forces cannot be directly shown to be due to 'force-carrying particles', since eg a photon beam does not produce electric attraction and a static-electricity charged object does not produce a photon beam. And of course photons show a wide range of variation that electric charge does not show.

Standard Model physicists Peter Higgs and Francois Englertis got a 2013 Nobel prize for their theory prediction for the Higgs Boson being that it would be around 500GeV, though the new particle being acclaimed as being the Higgs Boson is actually around 126GeV. Now 500GeV is nearly 400% of 126GeV, so modern physics theories having errors of around 400% is OK. But the same physicists claim that Newtonian physics is entirely disproved because in some cases it gives a below 1% error !

Standard Model physics uses Feynman diagrams, where only lines entering or leaving a diagram represent observable particles. Below two electrons enter a repulsion interaction, exchanging unobservable virtual photons, and then exit ;

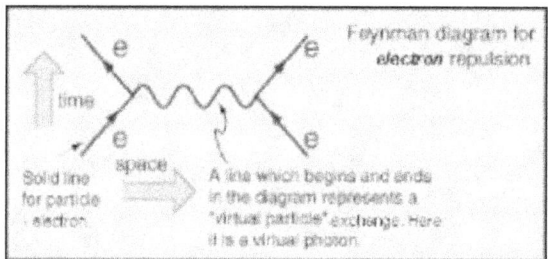

Charged fermion 'particles' are all claimed by some to have 'anti-particles' of similar mass but opposite charge that can form 'anti-matter' - eg Anti-Hydrogen composed of an Antiproton and a Positron akin to Hydrogen composed of a Proton and an Electron. But the 'charges' of matter particles and anti-matter particles are claimed to differ so as not to affect each other, and a particle and its 'oppositely charged' anti-particle are claimed to undergo spontaneous 'annihilation interactions' where both fully convert to photons. Some fermion particles are also claimed to spontaneously or magically convert into eachother. Uncharged anti-particles are generally unexplained and evidence on anti-particle behaviour is very thin, with strangely little anti-matter seeming to exist.

There are more reasonable claims that much 'dark matter' exists, probably being just uncharged free non-atomic particles like massive WIMP neutrinos or tiny Axions. Claims of 'dark energy' look weaker, as noted in our Gravity section. Multi-particle composites like atoms composed of an even number of half-spin fermions, or any number of interger-spin bosons, may have overall interger spin like bosons yet not behave as bosons. And some Standard Model particles are, like some radioactive atoms, very unstable and may be of little significance in nature.

There are four fundamental forces in Standard Model physics, the activities of which are generally defined as being ;

- Gravitational force, acting on particles termed mass particles.
- Electromagnetic force, acting on particles termed charged particles.
- Strong force, acting on particles termed coloured particles.
- Weak force, acting on particles termed left-handed particles.

This might perhaps be better redefined, explaining particle properties and better for a signal physics, as ;

- Particles that respond to Gravitational force signals are termed mass particles.
- Particles that respond to Electromagnetic force signals are termed charged particles.
- Particles that respond to Strong force signals are termed coloured particles.
- Particles that respond to Weak force signals are termed left-handed particles.

Of course signal-response systems have been built that produce several responses to one signal, or produce different responses to different signals.

So the above are not the only possible definitions of forces and/or of 'particles', and several sets of such definitions might well allow of the same force response event mathematics.

There is strong evidence that forces seem to become very digital at close distances, so sub-atomic particle bindings/ energies/ masses/ lifetimes all seem to involve very narrow and possibly specific mass/energy levels. This contrasts greatly with the apparent gradation of force effects in the universe at macroscopic levels. It is not clear if this applies to only some forces like the strong force, or to all forces including whatever collision force is. So it is not clear what the real general explanation is, or if there is one general real explanation involved or maybe more than one. It is not clear if sub-atomic force evidence favours some one general physics theory or may fit with some several general physics theories if appropriately specified. A 'counter-intuitive' or ridiculous strong nuclear force is proposed by David Gross for which a signal-response physics seems to offer a more logical explanation as in our String Theory section.

There have been some perhaps poorly defined claims that at very close distances these forces may be the same strength and effectively be just one force. But it is claimed by David Toms that the electric charge force which generally get stronger closer to its source, in fact very close to its source starts getting weaker the closer the distance - with this effect claimed to be somehow caused by gravity !? There are also claims that these forces are all due to the sending of some 'Messenger Particles' or 'Force Photons' back and forth. Of course some physicists do support Einstein's view that gravitational force differs fundamentally from the other forces.

You can listen to some interesting recent lectures by some physicists on related experiments and some interpretations of them, at http://viavca.in2p3.fr/site.html Or regarding claims for an increasing variety of unstable multi-quark hadrons such as 'charged charmoniums', see http://physicsworld.com/cws/article/news/2013/jun/18/charged-charmonium-confounds-particle-physicists

Cartesian, Field and Duality versions of 'Standard Model Physics' are supported by different physicists mostly claiming falsely that theirs is the one and only 'Standard Model Physics' and not really attempting to prove or disprove each other so really making a mockery of science. And the differing supporters of 'Einstein Physics' do exactly the same shamefully. Of course some physicists now generally support contradiction-allowed duality physics as where the 'elementary particles' both are 'wave packets' and are 'not-wave particles'. Others prefer to basically go with only one of these alternatives. And one option involving no contradiction might be taking 'elementary particles' as being multi-particle 'vibrations' composed of many standard particles allowing standard wave motion among their parts ? And anything claimed to be 'massless' can maybe only be proved to not have a big mass, since a claim that something has zero mass can be taken as requiring proof that A.) it produces zero gravity and/or proof that B.) it shows zero response to gravity. But this may be impossible to definitely prove if 'infinitely close to zero' cannot be definitely measured ? And it may be even more complicated because where gravity is stronger, some other forces may also be stronger.

Standard Model physics is mostly used by those employed in particle physics, often along with some version of Quantum Mechanics. But particle physics experiment is now often statistical experiment physics, and the real physics often boils down to statistical significance interpretation - and most physicists are poor statisticians. Modern physics 'experiment' often has the same basic statistics weakness as much modern medical 'experiment'. Some of the very different Standard Model theories maybe look like actually being image theories though no published Standard Model physicist seems to have studied that issue yet. Standard Model theories perhaps realistically represent more a promising physics awaiting a properly defined theory, and currently offer no real explanation for the strange assortments of particle masses observed to date ?

But maybe mass is just one force or signal response and is not 'solidity' and really there are no 'solid' particles, there is only a range of particles with some of a range of attractive and repulsive forces of differing strengths that can vary over distance by some range of powers ? Solidity and non-solidity may be only apparent differences, hence two different gasses can occupy the same volume without their atoms occupying the same spaces and two liquids may mix or may not mix without requiring anything about their atoms actually occupying the same space only them showing some relative location responses that remote-control boats can be made to show.

Tell a friend about this website simply,
and they will thank you for showing them the newest deepest thinking on the important basics of science ;

Type friends email address here   ...   Then click to tell your friend

OR if you like this site you could maybe make a donation ;

PayPal Donate

It will help with site development, and just possibly with some key physics experiments long planned but never afforded.
[PS. and you may perhaps help make history for science ?]
(The fictional time-travel and multi-universe type ideas of modern physics theory have long totally discouraged certain lines of physics experiment despite there being strong reasons to believe them to be very promising if not essential lines of experiment. Some such lines of experiment considered here identified as early as the 1960s seem still to have had no work done on them and there is maybe not much more time here for this. Science funding both government and private unfortunately now all goes to basically safe standard mainstream science, and no money at all goes to any really innovative risky science though that might pay a thousand times greater.)

You can do a good search of this website below ;
Search on this site www.new-science-theory.com, with .

Or do a search of the web better with DuckDuckGo -

otherwise, if you have any view or suggestion on the content of this site, please contact :- New Science Theory Vincent Wilmot 166 Freeman Street Grimsby Lincolnshire DN32 7AT.

You are welcome to **link** to any page on this site, eg www.new-science-theory.com/the-standard-model.php

© **new-science-theory.com, 2022** - taking care with your privacy, see New Science Theory HOME.

Probability Science - *in medicine and in physics*

HOME .... William Gilbert . Rene Descartes . Isaac Newton . Albert Einstein .... String Theory .... General Image Theory

Probability is increasingly used in many areas of modern science, and most notably in medicine and in physics, in scientific proof claims. But often probability is used poorly in science and really gives little or no proof of what is claimed.

In medicine, probability is now commonly used in survey data analysis, as where a 10% correlation between peoples illness and peoples behaviour statements is said to prove eg that general behaviour A always has a 10% risk of causing illness B. Often the general behaviour A has no actual effect on the illness B, but has some correlation with the use of some unidentified product C which is the actual cause correlating 100% with illness B. But the incorrect medical claim is pushed.

In physics, probability is now commonly used in experiment data analysis, as where a 95% correlation between photon emissions and some general magnetic event is said to prove eg that general magnetic event A always causes photon emission B. But the general event A may have no actual effect on the emission B, but has some strong correlation with some unidentified specific event C which is the actual cause correlating 100% with emission B. Probability is also used as the basis of Quantum Mechanics and some other physics.

Science generally attempts to discover cause-effect laws that work 100% within specified conditions, and a failing in even below 1% of cases can commonly be taken as disproving that science law. But in science today probability is widely used in different aspects of data analysis proof claims that are not reviewed by statisticians. It is used in experiment data analysis and in survey data analysis, and in both areas it is also used in error estimation. But probability is commonly used wrongly in science as noted by some major statisticians like R.A.Fisher. It is commonly used by amateur-statistician 'scientists' who are not good statisticians and consult no statistician, so much that the journal 'Nature' has now started asking statisticians to review some submitted papers.

Probability in Medicine

In medicine, probability is now used both in experiment data analysis and in survey data analysis but here we will consider chiefly the latter (below under Physics we will consider the former). The chief problem with survey data is that it always involves some limited number of selected people being asked some limited number of selected questions. It may be that an illness being studied is caused by ACME soap, but the survey had no question about ACME soap or it did but none of the people surveyed used ACME soap. But still that survey will be probability-tested for that illness, and may well give some correlations for that illness. It will be announced that some behaviours 'are a risk for the illness', while ACME soap passes unmentioned.
(PS. this is NOT a claim that ACME soap causes any illness, we use the name here only as the name of 'some hypothetical product'.)

We can now consider a hypothetical medical survey to be probability-tested regarding a hypothetical disease A ;

A hypothetical survey probability testing.

Where the unknown facts that the study seeks to discover are,
Disease A is actually caused by using too much of product C or the weaker product D.
Product C is more expensive than product D.
Product C is used more by middle-class vegetarians.
Product D is used more by working-class smokers.
Product C sells in less locations than product D, and some locations sell neither.

And where,
The survey is of pedestrians half from location X and half location Y, questions being ;
Do you own a TV ?
Do you regularly smoke cigarettes ?
Do you regularly smoke cigars ?
Do you usually drink more than 4 units of alcohol a day ?
Do you usually eat more than 2 eggs a day ?
Do you usually visit a gym more than once a week ?

This survey actually asks nothing about product C or product D, but will still give correlations for the illness caused only by these products as long as some people surveyed use either product. Hence,
TV-owners, cigar-smokers and gym-users on average may have higher incomes and to differing extents may buy more of the expensive product C than product D.
Egg-eaters on average may have the highest use of products C and D, and may have lower incomes and so buy more of the less expensive product D.
Alcohol-drinkers on average may tend to buy neither product C nor product D.
Location X on average may be more middle-class and sell more product C.

Survey question answers are used to split the survey population into sub-populations as 'TVowners' and 'non-TVowners'. Then probability testing may be done on illness rates between SOME answer sub-populations, when it should be done between ALL of the answer sub-populations - eg. between cigarette-smokers v non-smokers AND between cigarette-smokers v cigar-smokers AND between cigar-smokers v non-smokers etcetera. Of course a survey with many questions can give thousands of sub-populations, and while all should be probability tested, it is

> proper enough to publish the results only for all cases that exceed some specified significance level. (Alternatively probability testing can be done on illness rates between each answer sub-population and the total survey population, though that will dilute the probability differences and so can hide significant results)
>
> Illness rates will vary between sub-populations, such that it may be reported that **'cigar-smoking carries a 20% risk for this illness'** and **'egg-eating carries a 15% risk for this illness'**. Of course in this case we know that these behaviours do not at all cause the illness - products C and D cause the illness. So the 'scientific truths' that this study claims are not actually truthful. Whence the saying that 'There are lies, damn lies and statistics'.
>
> That holds even when probability studies are done properly, but often they are not. Hence cigarette-smokers v cigar-smokers may give a non-significant 5% while cigarette-smokers v non-smokers may give a significant 11%, but the study may have omitted to get or to publish the latter result. (as was the case for even the acclaimed Doll and Hill 1956 smoking survey study in regard to reported cigarette-lighter use -
> Doll R, Hill AB (1956) Lung cancer and other causes of death in relation to smoking. Br Med J 2: 1071).

In the early 1980s I did for a short time collaborate in a piece of nuclear-medicine related research with Dr.JSM Leung of Hong Kong that looked like disproving part of Doll and Hill's statistics. That piece of research was I believe ended by Britain's MI5 with myself jailed, as was my wife and our weeks-old first baby, and Dr Leung was also seemingly silenced somehow. I did from prison successfully do two degree applications and was accepted by the two London universities, City and LSE. I got a degree at City University with my mentor there, Andrew Mott, I believe helping somewhat to keep MI5 off my back to an extent at least though I never actually discussed that matter with him. This may seem just one minor temporary bit of science history, but that piece of research has still not been completed to date that I know by anybody. Hong Kong in the early 1980s was unusual in having a high incidence of lung cancer but with a history of a relatively low incidence of tobacco smoking, so being the place to look for any other causes of lung cancer. Of course if you did not collect evidence from those dying then, then you generally could not do so later with them dead - history can make some medical science not repeatable though that particular problem rarely if ever applies to physics. But it seems that whitewashing nuclear weapons and industries then favoured all the most powerful governments backing Doll and Hill evil-tobacco as the only cause of cancer though their statistics certainly did not prove that and later there did emerge some real evidence of some other causes also. Strong governments all acting to restrict cancer research has maybe not helped science in studying the causes and possibly cures, with the massive amounts spent on it ending somewhat wasted ?

If there is some strong evidence for any hypothesis, then additional weak evidence will now commonly be taken as confirming and strengthening that. And even if there is only weak evidence for an hypothesis, then additional weak evidence will now commonly be taken as confirming and strengthening that. But logically only strong evidence should count towards proof, and weak evidence should only ever count as an indicator of a need to look for strong evidence. Generally there are no 20% causes and so no 20% risks, mostly A actually cause B or actually does not cause B. There may commonly be dose effects, and more rarely there may be multiple causes. But much too commonly medicine is reporting, and governments spread concerns about, relatively low illness 'risks' that are not actual scientific truths like 'eating fat causes heart problems' - and scientific journal 'peer review' has tended to create and keep backing such false discipline-prejudices. They might do better having chemists review physics papers, physicists review chemistry papers and astronomers review biological papers because their discipline-peer-review just promotes prejudice science instead of real science.

Medical research in the last 50 years has often centered on the use of statistics as shown clearly in nutrition studies. Hence a couple of published nutrition studies claimed to prove that Antioxidants were very good for peoples health, but then two later studies claimed to prove they were good for younger peoples health but were bad for older peoples health. And statistics-based claims were pushed strongly for a long time that margarine was better for health than butter, so lots of people have taken to using margarine. But new statictical studies now claim that margarine is worse for health. This bad science is likely killing people, yet todays 'scientists' and governments push it regardless. Many other published statististics-based nutrition studies have made doubtful claims due to poor use of statistics and there has been little if any really good nutrition research in recent years. And about the statistical science behind the widespread recent claims that a vegetarian diet is healthier than a meat-eating diet. It seems the best information on vitamin pill takers is that they are especially non-drinker non-smoker exerciser females which happens to be a population that includes many more vegetarians. So if any vegetarians are more healthy then it is probably due to their vitamin pill taking and other healthy living rather than to their vegetarian diet ?

Probability in Experimental Physics

Probability testing in Physics and Astronomy is now commonly used in experiment data analysis or observation data analysis. This can have some of the problems seen in the use of probability testing of survey data. Hence where surveys can have omitted questions, experiment or observation can involve omissions in the factors investigated and this may have great impact in the more contentious areas like Particle Physics and Astronomy as with partial correlations between A and B being claimed as a causal proof when the true cause C was never studied.

Probability is also widely used in accuracy estimation, but often ignoring the probability fact that of several experiments or observations it is often NOT the one with the best accuracy that gives the most reliable evidence. Other significant issues are often also involved.

More recent Physics and Astronomy theories also commonly try to incorporate aspects of probability theory, correctly or incorrectly. Deductive assumptions involving infinities or limits often give false answers. So theory handling the infinitely small and infinitely large can ultimately require that the sum of an infinite set of zero probabilities add to a probability of one, which is plainly false. Physics deductions about the infinitely small or the infinitely large can generally be valid only derived correctly relative to some well proven specified finites. More recent physics theories can often involve error related to this issue.

False probability deductions can be due to a failure in specifying the data involved, or to a failure in specifying the assumed prior information involved. So there often can be no valid probability comparison between two physics theories regarding given data, if both involve assumptions about eg 'mass' but both fail to specify the prior information properties of 'mass' that their theories involve. Or phenomena that seem probabilistic may simply have some unseen or uncomputed non-probabilistic causes that may be currently unseeable or uncomputable.

Statistics based 'experiments' commonly rely on computer analyses or computer 'models' that are not fully specified and so such 'experiments' are not fully replicable to verify them or to challenge them. And replicable experiments generally though involving one set of statistical probabilities are all capable of being interpreted differently in terms of different theory paradigms. But statistics often cannot offer any valid evidence as to correctness between several alternative interpretations of an experiment. When radioactivity was discovered it soon became described as a 'causeless', 'random' or 'probabilistic' physical phenomenum, as no immediate cause could then be identified for such radioactive events. But radioactivity includes nuclear fission which occurs naturally on Earth in Uranium and Thorium ores as 'spontaneous fission' and which was found to be caused by neutrons non-randomly, so other radioactivity may well be caused by eg weak-force effects, simultaneous neutrinos or other as yet unidentified causation and not really be 'probabilistic'.

And there is some recent new evidence that at least some other apparently random probabilistic physical events may not be entirely random or probabilistic - see Quantum Events ? Any phenomenon where the cause cannot be easily seen is likely to be claimed by some at least to be causeless or random, but this is a claim that generally cannot actually be proven.

Probability in Physics Theory

For some physicists the two-slit light experiment was taken as supporting a Heisenberg probabilistic quantum mechanics, as where there is some probability that an object actually at a specified time occupies one space location and actually at the same specified time in contradiction occupies some other space location. In such a probabilistic physics universe, the universe actually behaves probabilistically whereas in a determinate physics the universe actually involves fully specifiable causes giving fully determinate effects though that may not always appear to be the case. Probabilistic physics claims to be also backed by other supporting evidence, with claims of microscopic quantum processes such as 'superposition', 'entanglement' and 'virtual particle exchange' being involved. But that some two particles having a common origin should retain some common properties is nothing surprising and continuing related probabilities does not at all prove continuing connection or 'entanglement' as claimed by some. Statictical correlation alone is not proof of causation or of simultaneous linkage and the latter is spooky nonsense anyway.

Heisenberg's Uncertainty Principle basically assumes that all possible ways of determining an objects motion and position at some instant must involve changing the objects motion or position. But the Rudolphine Tables of Kepler allow determining the position and motion of a planet at some instant by calculation alone (which has no impact on the planets position or motion), and the position and motion of a body continuously emitting light can be determined for some instant from its emitted light signals (having no impact on body position or motion but maybe limited by light having a quantal nature). It seems that there will be some cases where such determinations in principle cannot be done accurately, but also that there will be some cases where such determinations in principle can be done accurately.

Some physicists do not support probabilistic physics including Einstein who rejected probability physics "because God does not play dice" (though that is maybe no scientific disproof and Einstein still accepted duality contradiction physics). Probabilistic physics is rejected also by others like Schrodinger who reject all contradiction physics, including Einstein dualism, as in his Schrodinger's Cat probability-exposing 'thought experiment' which is perversely often quoted to help 'explain' probabilistic quantum physics. But for those who reject contradiction in science, it exposes probability physics as contradiction nonsense. Yet for those who accept contradiction in science, it helps explain probability physics !? Of course it can be said that any claimed evidence for a contradiction must be contradictory evidence, and contradictory evidence may reasonably be taken as not being valid factual evidence - eg evidence that Jane is in Paris now AND that Jane is in Tokyo now or evidence that Jane is alive now AND that Jane is dead now ?! Logically it would seem that 'evidence' for a contradiction must be data being misinterpreted. It may be more scientific to say that nature itself is NOT probabilistic, but that human consideration of nature IS probabilistic and so can make nature APPEAR to be probabilistic. But nature showing apparent statistical associations will often allow of multiple alternative causal explanations or Image Theories, and in some cases necessarily do. See http://psych-networks.com/theoretically-distinct-mechanisms-can-generate-identical-observations/?utm_content=buffercae71&utm_medium=social&utm_source=twitter.com&utm_campaign=buffer and /general-image-theory-1.php And of course A having a 10% chance of causing B, is also A having a generally or often ignored 90% chance of not causing B !

Probability methods generally are widely used in particle and quantum physics and have some use in almost all areas of physics today, even by physicists who reject actual probability physics. But where it is claimed that it has been proved that some physics is probabilistic, it is maybe best taken as meaning that it has really at most been proved that it is either probabilistic OR involves some as yet unidentified non-probabilistic causation. 2014 sees Christopher Ferrie and Joshua Combes, supported by Rainer Kaltenbaek and Franco Nori, throwing major doubt on Quantum Mechanics and especially its 'weak measurement' as being based on bad statistics. (see http://physicsworld.com/cws/article/news/2014/oct/09/are-weak-values-quantum-after-all)

While arguing for one-theory-only science, E.T.Jaynes concluded that probability theory has 'been fooled by a subtle mathematical correspondence between stochastic and dynamical phenomena'. But that rather supports multiple-theory science like Newton blackbox-theory science or perhaps preferably our General Image Theory science. See http://bayes.wustl.edu/etj/articles/prob.in.qm.pdf

Some of these physics probability issues were considered at the CERN 2007 conference 'Statistical Issues for LHC Physics', see http://physicsworld.com/cws/article/indepth/43309 Many suggest replacing the long-standing use of a probability value (p-value) of below 0.05 for 'significant' results with a stiffer p-value threshold of maybe 0.005, which should help to improve the use of probability in some areas of science though this does not affect the other issues with probability science. The probability of the Sun tomorrow not rising in the East and setting in the West is below 0.00000000001 but even that does **not** prove that the Sun orbits Earth daily, as observation strongly suggests and as used to be commonly believed though now we know that these observations are correctly explained by the fact that Earth is a sphere that revolves daily. Probabilities are probably often best used just to help identify specific issues where further real experiment are more likely to be useful. But even very good experimental science like Gregor Mendel's in genetics can have significant statistical problems as R.A.Fisher and others showed. Probabilities can be assigned to things that are not real, so probabilities can be assigned to numbers of angels sitting on a pin but that does not make that real. While there are still today some scientific physicists supporting a variety of scientific physics around different laws of nature, today increasingly there are many 'theoretical physicists' pushing various terrible lawless probability physics that are really supported by little or no experimental evidence as claimed 'science'. Of course misuse of statistics is far from the only problem with science but hard-science Physics is the leading edge of science, unfortunately long leading in bad science and only worsened by bad use of probability mathematics.

Mathematics is helpful to science chiefly insofar as it can help to increase exactitude in both experiment and reasoning proofs, but probability mathematics is basicly the mathematics of inexactitudes and so really can only help show the extent to which science proofs may be uncertain. Probabilities cannot themselves be causes of anything nor alone be proofs of any causations. So it is maybe sad to see Engand's 2020 Royal Society, David Spiegelhalter and Brian Cox pushing the false view that eminent statisticians Thomas Bayes and Sir Ronald Fisher helped build current probability science and probability physics, when Fisher chiefly concluded that probability can infer false causal effects and so can promote false science (See Brian Cox). And, without accepting Einstein's physics, the preponderance of science evidence does support laws of nature concerning nature not being probabilistic or playing dice - despite some apparent evidence for some seemingly contrary phenomena. Information is now commonly wrongly defined in relation to uncertaimties or probabilities but signal science shows many cases of information signals causing effects, and not lack of information uncertainties or probabilities, as was perhaps well demonstrated in William Gilbert's 1600 'De Magnete' or 'On The Magnet'.

You are welcome to **link** to any page on this site, eg www.new-science-theory.com/probability-science.php

Get our great Newtonian gravity Android App - 'Sun Pull' - in the Google Play app store to help you study or re-design the solar system better !
Or you can try it on here in our Solar System section, which also discusses what is probably chiefly needed for real actual contact with 'alien' people from other worlds. Hopefully more useful science Apps may follow ?!

OR if you like this site you could maybe make a donation ;

[PayPal Donate].

It will help with site development, and just possibly with some key physics experiments long planned but never afforded.
[PS. and you may perhaps help make history for science ?]
(The fictional time-travel and multi-universe type ideas of modern physics theory have long totally discouraged certain lines of physics experiment despite there being strong reasons to believe them to be very promising if not essential lines of experiment. Some such lines of experiment considered here identified as early as the 1960s seem still to have had no work done on them and there is maybe not much more time here for this. Science funding both government and private unfortunately now all goes to basically safe standard mainstream science, and no money at all goes to any really innovative risky science though that might pay a thousand times greater.)

© **new-science-theory.com, 2022** - taking care with your privacy, see Sitemap.

Solar System problems, and Sun Pull app.

Homepage . William Gilbert . Rene Descartes . Isaac Newton . Albert Einstein .... General Image Theory
- Site Search at bottom v -

Instabilities affecting the Sun and the Earth

Solar systems are commonly flat discs with planets orbiting a star in one plane, and some planets have one or more moons orbiting them. Isaac Newton did a partial study of this, only sufficient to conclude that the planetary bodies in our solar system have a degree of orbit stability that should maintain their orbits for a long time. But he did not consider other solar system stability issues, and since solar system bodies exert gravitational pulls on each other, the normal structure of a solar system can involve some instabilities, which in the case of our own solar system would chiefly seem to be ;

1. Our spherical Sun with its spherical structure and functioning would be more stable if the planetary gravitational pulls on it were basically distributed spherically. The fact that they are now distributed in one plane only, exerts destabilising pulls on the Sun. Were some planets to orbit the Sun in a plane at 90% to the present planetary orbits then this problem would be much reduced.

2. Our Earth with its spherical structure and functioning would also be more stable if gravitational pulls on it were basically distributed spherically. The chief factor going against that is our having the Moon orbiting Earth. William Gilbert before 1600 concluded that the Moon was pulling our seas and so causing tides, and there is no doubt that the Moon also pulls the land and must help encourage volcano eruptions and earthquakes and continental movement that destabilises Earth. A thin flat disc artificial moon would have little gravity and so should reduce such problems if it replaced the Moon. Earth's gravity has set its Moon's spin to equal the Moon's orbit time of 27 days (as have most other planets set their moons' spins) basically due to Moons not being homogenous spheres.

3. Both the Sun and the Earth would also be more stable if gravitational pulls on them were less from point sources, eg if the Earth's one moon was split into several smaller moons or if the Sun's few planets were split into a larger number of planets. Then the gravitational pulls on the Earth and the Sun would be less concentrated directionally.

4. Both the Sun and the Earth would be still more stable if planets did not all have separate orbits with different orbit speeds allowing intermittent alignment conjugation of their gravity pulls.

Our very unstable flat solar system   **A less unstable spherical solar system**

 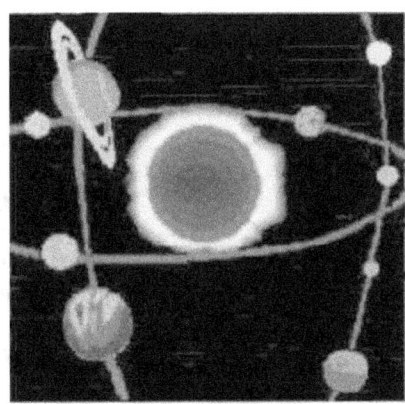

Orbits in one plane at different speeds   Orbits in 90% planes at one speed

Clearly our solar system may not be quite as stable a system as many have imagined. And in particular the Sun and the Earth do have real gravitational instability problems.

The relative gravitational pulls of the planets on the Sun at present are about - Mercury=0.37, Venus=1.57, Earth=1.00, Mars=0.05, Jupiter=11.75, Saturn=1.05, Uranus=0.04, Neptune=0.02. Jupiter, Venus, Saturn, Earth and Mercury exert the strongest pulls on the Sun. If the planets were in two orbits at 90% with orbit diameters near the present orbit diameters of Mars and Jupiter then their total pulls on the Sun would be about the same as now but with much reduced equatorial effect and much reduced conjugation effect. Of course asteroids, comets and moons have some additional effects.

Another general solar system problem of course is the large number of rogue rocks hurtling around the solar system, many coming out of the asteroid belt because of its gravitational instabilities from the type 3 and 4 affects above. And there is the general solar radiation problem made severe periodically by increased flare activity as the Sun is affected by its gravitational instabilities.

The Earth today is affected most by the Moon's gravity, though the instabilities of the Sun can and do also have significant effects on the Earth - both mostly impacting our weather system and helping to cause periodic ice ages or global warmings (and probably also helping prompt volcano eruptions and earthquakes that are more affected by moon gravity). Of course to date mankind has been able to do little or nothing about any of these solar system problems, and there are some other lesser problems also. Of course the moon's night light does have some useful effects, and even its gravity is claimed to somewhat moderate the comings and goings of ice ages by stabilising Earth's spin alignment though its pullings may well indirectly actually destabilise such to some degree. For some more on this see our section on Gravity.

In our solar system it seems that a planet is more likely to retain an atmosphere if it larger, if it is further from the sun, or if it has a stronger magnetic field. So to make an Earth-size planet habitable would seem to require it to have at least an Earth-strength magnetic field for it to be in an Earth-distance or less orbit around the sun, and with a weaker magnetic field would require it to be in a greater than Earth-distance orbit around the sun. Of course it would be good to have some other planet in Earth's orbit with no moon, or to make realistic working robot gravity models of Earth with its Moon to study as discussed in our Gravity section (akin to William Gilbert's magnetic Terrella experiments).

The Suns heat is produced by a process called Fusion where two light atoms like Hydrogen fuse to make a heavier atom like Helium, caused chiefly by the very strong gravity and/or extremely high pressure generated by it with the Suns mass being about 333,000 times the mass of Earth. The many physicists trying to cause Fusion using extremely high temperature alone are almost certainly wasting science time and money, as Fusion almost certainly needs extremely high gravity and/or pressure (with high temperature being mainly a byproduct of Fusion rather than a cause of Fusion). Of course technology that can generate gravity has still not yet been developed.

Our gravity-pull App called 'SUN PULL' can help you study or re-design the solar system to reduce solar instability, and you can try it below.

This was an Android App loading with the 2013 solar system and taking the total gravitational pull of its planets and moons on the Sun as 100.
Orbits run from the Sun, as Mercury(ME), Venus(V), Earth(E), Mars(MA), Jupiter(J), Saturn(S), Uranus(U), Neptune(N).
If a planet has multiple moons then the App uses their total mass.
Green bodies are active or present in the solar system, and Orange bodies are inactive or absent from the solar system.
Click one or more bodies to change their status, and the App gives the new gravitational pull of planets and moons on the Sun.

When the App loads showing 100, clicking the green Jupiter(J) gives a new pull value of 25.066 showing the contribution of Jupiter to the total gravitational pull of planets and moons on the Sun as being 74.934%. This can be done for any planet or their moons. Click green bodies to move them out of the solar system, or click orange bodies to add them to the solar system. This App also works at least approximately for other orbital gravitational systems that involve proportionate forces and orbits.

Moving both Mercury and Venus into Earth's orbit cuts the Sun Pull to 93.812, and then moving Mars into Earth's orbit makes it 94.203. Current solar system planet orbits are basically all in one plane, but this App allows modelling moving planets to orbit in two planes at 90 degrees by simply running it for the planets of each plane separately. Of course this App looks at the pull of planetary bodies on the Sun, not the more common looking at the pull of the Sun on planetary bodies - but obviously that is just action and reaction which are simply equal and opposite for this app.

If you do not actually have the ability to move planets and moons, this gravity-pull App may only be useful to somebody working in Science Fiction but it is used and has been liked. This interesting gravity App was available from the Google Play app store but it does have limitations and other related Apps may well follow. But below you can run solar system re-designs by clicking planetary bodies ;

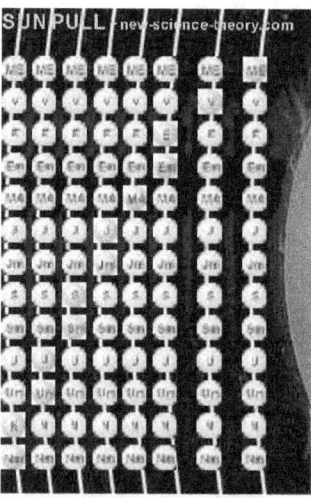

Do galaxy orbit speeds require Dark Matter ?

Imagine a solar system where instead of one planet in each orbit there are many planets in each orbit, so that each orbit approximates a mass-ring. Would the planet orbit speeds be due only to the pull of the Sun as for the planets now ? Or should not the outer planets be subject to the pull of the Sun augmented by the pull of the inner mass-rings, so that their orbit speeds would be augmented without us having to assume any additional Dark Matter ? And might this approximate to the gravity scene in at least some galaxies ?

Is there a big Ninth Planet in the outer solar system ?

Some astronomers theorise that there should be a Ninth Planet about 10 times the mass of Earth in the outer solar system about 1000 times further from the Sun than Earth. The supporting evidence being offered is based on analysing periodicities in the Suns light emission and assuming them to directly related to planet orbit periodicities varying gravitational pulls on the Sun although its light emission is not simply related to such pulls. If you use the 'Sun Pull' Android App free above, by just clicking the green E in Earth's EM orbit, to switch off Earth, then you will see that Earth's gravitational pull on the Sun is about 6.424% of the pull of all the planets which is how much that pull falls. So adding a Ninth Planet 10 times Earth's mass at 1000 times Earth's distance from the Sun means its gravitatational pull on the Sun being 1/100000th of Earth's pull [1/(1000x1000/10)]. A small asteriod falling into the Sun would have a greater impact on its light production than the pull of a Ninth Planet, and some planet conjunctions may have a periodicity similar to that of its predicted orbit. So current evidence for this Ninth Planet is probabilistic and uncertain, but that is not stopping some astronomers from continuing a search for it.

And of course as the planets pull on the Sun, so also do eg the 60+ moons of Jupiter pull on Jupiter. Currently little is known about the exact significance of these pulls, so for now at best some educated guesses only are possible on these issues. But at present we do not have sufficiently accurate or complete information on Jupiter and its many moons to make a useful Jupiter Pull app. And another interesting question, that somehow modern physics seems to have ignored, is does the Sun's gravitational pull on the Earth at all diminish during a lunar eclipse as light diminishes or do the pulls of Sun and Moon then perfectly add ? Does the Sun's pull maybe diminish by one billionth or less, or even increase slightly ?

Contact with 'alien' people from other worlds, hypothesis :

There being probably a large number of other planets similar to our Earth, it seems almost certain that some of them must have some kind of people living on them. So the possibility of contact between people of different worlds becomes an issue of some interest. Occasional trivial or insubstantial contact may be of interest to many people, but it is surely regular official trading contact that should be of most concern. With regard to that, the chief practical difference between such peoples should be their possession or non-possession of good advanced space travel technology and advanced science. This perhaps suggests the following hypothesis ;

1. Some less advanced civilizations may unreasonably see a possible danger in uncontrolled contact with more advanced civilizations - as in such contact saying 'We are mugs, come and mug us'. And most less advanced civilizations by definition may have technology capable of at most insubstantial contact or trade in any case.

2. Most more advanced civilizations may reasonably see an ethical issue in uncontrolled contact with less advanced civilizations - as in it subverting self-determination for the development of the less advanced civilizations. So more advanced civilizations may see less advanced civilizations as having a 'right to self-determination'.
This is in line with the science fiction Prime Directive of 'Star Trek' :
"As the right of each sentient species to live in accordance with its normal cultural evolution is considered sacred, no Star Fleet personnel may interfere with the normal and healthy development of alien life and culture. Such interference includes introducing superior knowledge, strength, or technology to a world whose society is incapable of handling such advantages wisely ... This directive takes precedence over any and all other considerations and carries with it the highest moral obligation."
And only more advanced civilizations by definition have technology allowing substantial regular trade contact anyway.

3. These considerations would seem to favour substantial regular contact, as involving trade relations, only between more advanced civilizations with more advanced science. And the Earth to date has clearly not yet developed an advanced science or technology that would allow it to be invited to join an advanced-species trading club. Other-world individuals would most probably be basically similar to an average human, friendly except if you greatly annoy them ? And other-world governments that have advanced science should be able to fix any big problems they might face so they should be able to govern easily and moderately and be basically friendly except if you greatly annoy them ?

The above is only hypothesis based only on reasoning and unsupported by any actual evidence to make it a scientific theory still less any substantial evidence that might make it fact. However it may have some reasonable probability.

But for this hypothesis regular trading contact with advanced science people from other worlds should chiefly need Earth having an advanced science as allowing eg fast space travel and gravity control, and well certainly it seems that Earth physics has, for now at least, really ended - but maybe there may be some small chance of a fresh physics restart happening sometime if more rebel against current useless physics ? This is probably the chief need for any real 'alien contact'. But though there seems little sign of it now, maybe our somewhat primitive science might somehow make that big breakthrough soon ? Promising experiments need funding, as below.

IF you like this site or its science ideas then you could maybe make a donation ;

It will help with site development, and just possibly with some key physics experiments long planned but never afforded.
[PS. and you may perhaps help make history for science ?]

(The fictional time-travel and multi-universe type ideas of modern physics theory have long totally discouraged certain lines of physics experiment despite there being strong reasons to believe them to be very promising if not essential lines of experiment. Some such lines of experiment considered here identified as early as the 1960s seem still to have had no work done on them and there is maybe not much more time here for this. Science funding both government and private unfortunately now all goes to basically safe standard mainstream science, and no money at all goes to any really innovative risky science though that might pay a thousand times greater.)

You can do a good search of this website, or of the web, below ;

Search on this site www.new-science-theory.com, with Google.
Search over all websites on the Web, with Google.

For enquiries, or if you have any view or suggestion on the content of this site, please contact :-
New Science Theory (e-mail:-vincent@new-science-theory.com)
Vincent Wilmot 166 Freeman Street Grimsby N.E.Lincs UK DN32 7AT.

© **new-science-theory.com, 2022** - taking care with your privacy, see New Science Theory HOME.

A General Image Theory of science theories.

1. The basics of logical thought and of thinking.

Homepage . William Gilbert . Rene Descartes . Isaac Newton . Albert Einstein .... GIT 2 . GIT 3 . GIT 4 . Sitemap

Site Search at bottom v

This General Image Theory of science theories, first published here from 2008, dates from 1964/1965 and study of the history of physics and of science philosophy. It basicly supports a logical scientific realism requiring neither irrational theory dualisms nor total theory exclusivity, and its physics is basically presented on this site.

In the spirit of William Gilbert this is not addressed to that crass multitude of career-scientists content to kick around the narrow range of ideas that science journals today consider fashionable, but to the honest studious reader or free spirit happy to labour hard and dig deep to find real science truth. Lies can be easy to produce and to swallow, the truth can be hard - and a science truth whatever its opponents is a science truth.

Any science theory is basically an attempted description of a universe, or more often of some part or aspect of a universe, and a valid science theory has generally been considered one that includes no logical inconsistencies and is consistent with observation and experiment. And as a description, a science theory will use language and may include some mathematical description and involve some logical reasoning.

In more concrete terms, a theory is some set of intentional signals intended to convey the information that constitutes that theory to some observer(s). These intentional signals may be in a printed book, in spoken sound waves or other intentional signal form - and will have such information issues as the extents of completeness, accuracy and noise.

On science theory logical reasoning, the most rigorous logical reasoning (as with Euclid) has often been in the field of mathematics - and the basis of mathematics is the equation. While the most basic equation is an identity equation, such as 2=2, mathematical equations are more commonly 'image' equations such as 2=1+1 where the two sides of an equation declare two things to be images of eachother. For some the truth described in maths by 2=1+1 may seem better described by 1+1=2, but certainly the reverse of any valid mathematical equation is also a valid mathematical equation. Hence, in mathematics there can always be at least two different valid descriptions of one thing having precisely the same real meaning. But different people can think differently and so some may not see a reversed equation as meaning exactly the same. And much in the universe is actually relative to the observer, really allowing different observers to reach different valid conclusions and the differences may seem small or large.

In physics theory Einstein's celebrated equation $E=mc^2$ if valid is as valid reversed, or as eg $m=E/c^2$. And while mathematics clearly is basically 'image manipulation', so also is language. While language is really much more complex than mathematics, it always includes "one and one is two" and "John is a big boy" basically being "John=big boy John" image equating. Even "You go !" can be interpreted as "you=future you not here" image handling. Words do not have intrinsic meanings, but only meanings that some using community assigns them or meanings that somebody wants them to have. Hence in today's common English the two phrases "That is cool" and "That is hot" are commonly taken as having basically the same meaning, but in a science setting are generally taken as having basically opposite meanings. But certainly in language, as in mathematics, there can always be more than one valid description of one thing.

The key imaging for natural language is that of a word being an image of an actual or conceivable thing, eg "bed" = bed word or "Adam" = Adam word. It is only from this base that natural language can give its universe descriptions as eg "the bed is big" being 'bed = big bed'. That the two sides of common language or mathematical equations necessarily involve an assumed identity, involves an assumed logical consistency though maybe not always an assumed actual truth.

Indeed that this image manipulation aspect of both mathematics and language must reflect the basic nature of human thought is indicated by some language disorders. People with some language disorders will commonly say "big" when they mean to say "small" - and normal people will even make that kind of slip sometimes indicating that the mind deals with balanced images as "small thing=not big thing". One of the most basic psychological tests is the Word Association test, where people most commonly associate opposite word pairs - eg "Light" with "Dark" and "Big" with "Small" or often effectively "Light" with "not Light" and "Big" with "not Big".

So it is maybe not just coincidence that Gilbert's early 'active matter' physics was soon challenged by Descartes' 'not-active matter' physics and that Newton concluded that either or both might fit with his mathematical laws and with the facts known at the time. And though science has commonly supported theory exclusivity requiring that only one theory can be true, some experimental evidence has made that hard to support. But logic also makes it hard to support theory dualisms requiring that each of two or more contadictory theories be true.

The history of science theory clearly shows that different people can think differently about the same thing. Hence observer conceptual relativity can allow description relativity and that should allow valid theory relativity. In this respect a General Image Theory that allows of some multiple alternative compatible theories looks a valid development of Newton's blackbox science theory and can avoid the modern scary-science of self-contradicting unreasoned 'Duality Physics' or 'Multi-theory Physics'.

Even one person can form different views of the same thing. Among physicists this was shown most clearly by Johannes Kepler producing three different physics. He started from a view of the universe having been created by a God choosing to create eg. a musical universe or a mathematical universe. Hence early Kepler creationist physics included a Geometry Mathematics Physics and a Music Mathematics Physics. But of course he later rejected these and produced his Descartes-style push causation physics.

But that one thing can clearly have more than one description, conflicts directly with what <u>seems</u> an equally clearly valid claim of science about science theory descriptions - that for any area of science there can be only **one** valid theory and it will disprove all other theories. But one thing or one reality can be described in multiple different ways, yet till recently science has generally followed logic philosophy in allowing only ONE valid theory description. And some now claim that multiple theories should be ALL accepted regardless of logical inconsistency between them. These most basic logical conflicts are the key issues addressed by and resolved by this General Image Theory of science theories.

While arguing for one-theory-only science, E.T.Jaynes concluded that probability theory has 'been fooled by a subtle mathematical correspondence between stochastic and dynamical phenomena'. But that rather supports limited multiple-theory science like Newton blackbox-theory science or perhaps preferably our General Image Theory science allowing some multiple self-consistent Image Theories that are different compatible descriptions of the SAME reality. Theories of different realities are clearly different theories about different things and are not image theories even if they happen to appear compatible when it would be those realities that were compatible image realities if such were conceivably possible. See http://bayes.wustl.edu/etj/articles/prob.in.qm.pdf

PS. The 2010 Stephen Hawking and Leonard Mlodinow book 'The Grand Design' notes that theories "dual" to others, where a mathematical transformation makes one theory look like another, suggests that they may just be two descriptions of the same thing. This is what developed String Theory and Supergravity (Mem)Brane Theory into the equally poorly defined 'multiple-universes' M-theory that they support, and is essentially the basic premise of General Image Theory where alone it is properly developed. (also see our String Theory section)

Tell a friend about this website simply,
and they will thank you for showing them the newest deepest thinking on the important basics of science ;
Type friends email address here   ...   Then click to tell your friend
NOTE : You can do this with confidence as we do not share and do not store this information at all.

OR if you like this site you could maybe make a donation ,

It will help with site development, and just possibly with some key basic physics experiments long planned but never afforded.
[PS. and you may perhaps help make history for science ?]
(The fictional time-travel and multi-universe type ideas of modern physics theory have long totally discouraged certain lines of physics experiment despite there being strong reasons to believe them to be very promising if not essential lines of experiment. Some such lines of experiment considered here identified as early as the 1960s seem still to have had no work done on them and there is maybe not much more time here for this. Science funding both government and private unfortunately now all goes to basically safe standard mainstream science, and no money at all goes to any really innovative risky science though that might pay a thousand times greater.)

You can do a good search of this website below ;

Search on this site www.new-science-theory.com, with Google.

Or do a search of the web better with DuckDuckGo - Type web search then Enter

For enquiries, or if you have any view or suggestion on the content of this site, please contact :- New Science Theory
(e-mail:-vincent@new-science-theory.com), or write Vincent Wilmot 166 Freeman Street Grimsby Lincs UK DN327AT

(This General Image Theory of Science Theories is by Vincent Wilmot, for a brief autobiography see Vincent Wilmot.)

You are welcome to **link** to any page on this site, eg www.new-science-theory.com/general-image-theory-1.php

---

© new-science-theory.com, 2022 - taking care with your privacy, see New Science Theory HOME.

A General Image Theory of science theories.

2. Conflicts around 'There is only one valid theory'.

Homepage . William Gilbert . Rene Descartes . Isaac Newton . Albert Einstein .... GIT 1 . GIT 3 . GIT 4 . Sitemap

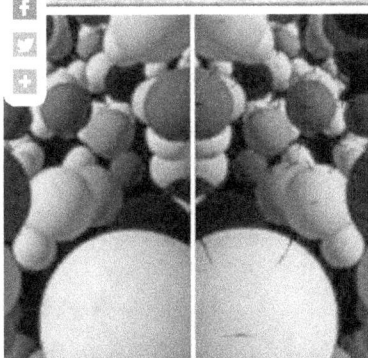

This 'General Image Theory of science theories' challenges the most basic principle of science, the claim that there can be only one valid theory and it must disprove all others. Yet this most basic challenge is undoubtedly correct, despite the only-one-theory principle being supported by almost every scientist ever to date.

In the spirit of William Gilbert this site is not addressed to the crass multitude of grant-funded scientists content to kick around the narrow range of ideas that today's science journals consider fashionable, but to the free spirit happy to labour hard and dig deep to find real truths and not to foolishly believe them to be easily found on Wikipedia or Discovery Channel.

Given that that one thing can clearly have more than one description, and that any science theory is basically an attempted description of some aspect of a universe, it seems clear that any valid science theory should allow of some other valid compatible image theory or theories.

Yet all four major scientists especially considered on this website, and indeed every scientist to date, have all basically claimed that there can only be one valid theory and it disproves all other theories. But it is to be noted that there have been some science ideas like wave-particle duality theory, and to a lesser degree blackbox theory, that in fact indicate some scientific unease with the 'only one valid theory' principle.

Isaac Newton hit what he saw as a major dilemma in finding that the two basic physics theories of action-at-distance William Gilbert and of push-physics Rene Descartes failed to disprove the other and that both seemed basically consistent with the known mathematical laws of physics of the time. Newton side-stepped that dilemma by claiming that science is really limited to blackbox mathematical laws concerning 'seens', so that the Gilbert and Descartes explanation theories based on different 'unseens' were really philosophical hypotheses including untestable unseens that could not be validated and so were outside science in philosophy where 'only one valid' need not apply. Newton was acutely concerned about this dilemma and saw his blackbox science position as essential if science itself was to hold to the 'only one valid theory' principle to which he was really fully committed. He concluded that some one form of either Gilbert physics or Descartes physics must be true - though it might never be possible to prove which.

Modern physics blindly ignores Newton's Dilemma by wrongly taking his and all previous physics theory as disproved. And another physics theory dilemma, that Newton had a small issue with, has also persisted and expanded around wave theory vs particle theory. This dilemma began with light theory, which in Newton's time had both a particle theory (Newton's 'corpuscular' theory) and a wave theory. Newton felt that only the maths mattered, and the different explanations might be only untestable philosophic hypotheses. But the wave theory of light seemed to prevail perhaps without actually disproving the particle theory. Then Einstein showed that some experimental light behaviour was particulate, or 'quantal', and claimed that light both actually was a wave and actually was not a wave but a particle. Several formulations of this wave-particle duality theory have not given anything widely agreeable, and some experiments claiming to follow light paths may involve light absorption and re-emission or combine responses to light with responses to some other signal emitted by light photons ? Variously formulated 'dualist' theories of light have been extended to all particles, now claimed to all be also waves, so that what should be two different theories are claimed to be some one 'dualist theory' accepting contradiction. Things are something, and are also not. So physics now can hold on to 'only one theory' but only by allowing basic contradictions within it which in both logic and in classical science disproves any theory.

Bohr's strange principle of complementarity, that the observation of two properties such as position and momentum requires mutually exclusive experimental arrangements, has been taken as meaning that mutually exclusive modes of language or theories (such as the language or theory of particles and the language or theory of waves are assumed to be) can be used in the description of an object, but not simultaneously. Of course some like Heisenberg have taken it as only meaning that no description or theory of an object can be certain and the only valid description or theory must be a probabilistic one.

It is certainly clear that at least modern physics theory does contain substantial logical conflicts, and that some of these can be resolved by a General Image Theory of Science Theories that allows of some sets of valid compatible image theories instead of doggedly trying to hold to the clearly false 'only one valid theory' principle.

For enquiries, or if you have any view or suggestion on the content of this site, please contact :-
New Science Theory (e-mail:-vincent@new-science-theory.com)
Vincent Wilmot 166 Freeman Street Grimsby N.E.Lincs UK DN32 7AT.

© new-science-theory.com, 2022 - taking care with your privacy, see New Science Theory HOME.

A General Image Theory of science theories.

3. What constitute sets of valid image theories in science ?

Homepage . William Gilbert . Rene Descartes . Isaac Newton . Albert Einstein .... GIT 1 . GIT 2 . GIT 4 . Sitemap

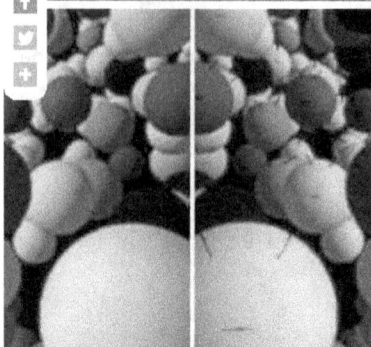

This 'General Image Theory of science theories' seeks to dispose of the most fundamental principle of science to date - the false claim that in science there can be only one valid theory. G.I.T theory seeks to replace that false assumption with a science truth that is much more useful.

In the spirit of William Gilbert this is not addressed to that crass multitude of so-called-scientists content to kick around the narrow range of ideas that science journals today consider fashionable, but to those free spirits happy to labour hard and dig deep to find real truth.

Since any science theory is basically a description of a universe or of some part or aspect of a universe, it must use language. Science developed when the language of scholars was Latin, and then most science was published in Latin until after Newton's time. That helped theory comparability though not everyone was good with Latin and translations were often problematic as with William Gilbert not being translated into English until 300 years after publication. Of course scientists including physicists even then tended to write up their theory in different ad hoc manners that make it hard to directly compare theories. And over time the use of different native languages in science theory replaced the universal use of Latin. But more recently English has become dominant in science.

Gilbert and Newton basically wrote up their physics theories in one book, with Newton's 'Principia' being the better organised and rather more complete. Later physicists published their theories in ad hoc articles, encouraged by government funders and science journals wanting newsworthy briefs. But science theory write-ups need to be comparable to show where they are compatible or incompatible to identify their proof issues. Trying to compare several physics theories now is almost impossible. All physics theories should have 'Principia' style write-ups of at least their basics to allow better theory comparison.

Language has always been a significant problem for science theory and has allowed ranges of interpretations of some theories that can be far from the intention of the theories originators. And on top of these language issues, science to date has had an as yet unrecognised theory description problem in being stubbornly stuck to the 'only one valid theory' principle, so that there have been no attempts to produce sets of valid image theories allowed by the fact that one thing can clearly have more than one valid description.

With the current 'only one valid theory' principle discarded, a General Image Theory of Science Theories would have a number of requirements that limited valid sets of image theories would have to comply with.

1. Each of a set of valid image theories would have to deal with the same universe or part or aspect of such universe.

2. Each of a set of valid image theories would have to be logically self-consistent and be consistent with current knowledge of the universe or part or aspect of the universe that they cover.

3. Each of a set of valid image theories would have to use at least some 'unique descriptions' that differ from those used by other theories of that set, with its 'unique descriptions' covering both language and mathematics terms.

4. Each of a set of valid image theories would have to not be fully logically consistent with another image theory, so if one image theory says 'A moves B' then another image theory must say something contradicting that as that 'A moves itself in response to B' - they cannot both say the same as 'A moves B' for all aspects of the theory.

5. Each of a set of valid image theories would have to be translatable into others of the set, as common languages basically are translatable, unique terms word for word or phrase for phrase, by means of a suitable translation dictionary including applicable mathematics.

The requirement that each of a set of valid image theories must not be fully logically consistent with another image theory identifies cases of differing descriptions that are the same image theory, as Gilbert's De Magnete in Latin and in a 'perfect translation' English whose meaning and mathematics are the same. A mathematics being a logically rigourous form of a description, the requirement that each of a set of valid image theories must use at least some 'unique descriptions' requires that different image theories of the same thing must allow of somewhat different but translatable mathematics though much of the mathematics might be the same.

As an example of a possible pair of valid image theories in a science, consider the following summary Descartes and Gilbert versions of Newton's laws of motion as basically specified below;

Laws of Motion a la Descartes.

1. A body will remain in its state of rest, or of constant velocity in a straight

Laws of Motion a la Gilbert.

1. A body will remain in its state of rest, or of constant velocity in a straight

line, unless a push or pull force is applied to it.	line, unless it receives repulsion or attraction signals.
2. A body accelerates in proportion to the amount of push or pull force applied to it, and in inverse proportion to its own mass, in a straight line in the direction in which the force is applied.	2. A body accelerates in proportion to the strength of signal received by it, and in inverse proportion to its own mass, in a straight line in the direction or in the opposite direction from which it receives the signal.
3. If one body applies a push or pull force to a second body, then an equal and directionally opposite push or pull force is applied to the first body.	3. If one body responds to repulsion or attraction signals from a second body, then the second body will respond equally and directionally opposite to signals from the first body.
Descartes saw action-at-distance or remote-control 'forces' like gravity and magnetism as involving currently unseen particle contact, though common contact is really also an unseen as bodies may not really contact but show close-proximity response.	Gilbertian 'repulsion or attraction signals', including electrical. gravitational and very short range proximity 'contact' signals, are currently unseen.

Isaac Newton concluded basically that the above theories were both consistent with what was known at the time about motion, including its mathematics as defined at the time. And the above summaries meet all of the five requirements for image theories given above. Of course they require that motion actually have quite different causal mechanisms, though both mechanisms involving what at the time were unseens allowed both to be compatible with what was then known of motion. These are two image theories that seem not just semantically different, so that it is possible that one or both be proved wrong by new knowledge or experiment. But it easy to produce two versions of each of the above theories that ARE just semantically different, as eg by producing 'A causes B' and 'B is caused by A' type versions. With no contradictions involved these are versions of the same image theory whose difference is entirely semantic, as with Latin and English versions, and disproving one would disprove both of course.

It may well also be possible to produce valid image theories of eg an Einstein relativistic theory or of a probabilistic theory. The only real issue for science is whether a new image theory might be likely to be of use to anybody. But if anybody makes an advance in one image theory, then it could easily translate into an advance in other image theories of that set and so help other scientists that are using those theories.

Only in such an Image Theory science are requirements regarding logical consistency set realistically. Logical **consistency** is a requirement within any valid image theory, but logical **inconsistency** is also a requirement between different valid image theories !

For enquiries, or if you have any view or suggestion on the content of this site, please contact :-
New Science Theory (e-mail:-vincent@new-science-theory.com)
Vincent Wilmot 166 Freeman Street Grimsby N.E.Lincs UK DN32 7AT.

You are welcome to link to any page on this site, eg www.new-science-theory.com/albert-einstein.php

IF you like this site then you could maybe make a donation ;

It will help with site development, and just possibly with some key physics experiments long planned but never afforded.
[PS. and you may perhaps help make history for science ?]
(Anomalies regarding modern physics theory have long totally discouraged certain lines of physics experiment despite there being strong reasons to believe them to be very promising if not essential lines of experiment. Some such lines of experiment considered here identified as early as the 1960s seem still to have had no work done on them and there is maybe not much more time here for this.)

© new-science-theory.com, 2022 - taking care with your privacy, see New Science Theory HOME.

A General Image Theory of science theories.

4. Why have sets of image theories in science ?

Homepage . William Gilbert . Rene Descartes . Isaac Newton . Albert Einstein .... Information Physics......Sitemap

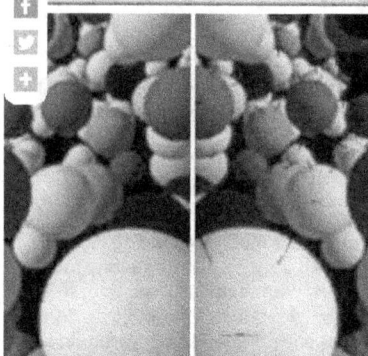

This 'General Image Theory of science theories' involved substantial studies of especially science history, philosophy, language theory and signal theory. More from those studies is being presented over time as this website progresses.

In the spirit of William Gilbert this is not addressed to the crass multitude of mere-theoriser-scientists content to kick around the narrow range of ideas that science journals today consider fashionable, but to the free spirit happy to labour hard and dig deep to find real scientific truth. Because unfortunately today Wikipedia and Discovery Channel do have some good bits of truth, but with big chunks of rubbish mixed in - though not quite as bad as the anti-science History Channel with its repeatedly claimed false 'proofs of aliens' and 'proofs of conspiracies'.

Science to date has stubbornly stuck to an 'only one valid theory' principle, so there have been no attempts to produce sets of valid image theories allowed by the fact that one thing can clearly have more than one valid description. Instead today many calling themselves scientists prefer to support 'multiple realities' when the experimental facts equally support a more logical multiple theories science.

Yet science is centrally concerned with describing causation, and both physics and philosophy have produced some basically differing theories of causation. In physics the active-matter causation of William Gilbert was opposed by the dead-matter causation of Rene Descartes. And similarly in philosophy George Berkeley's 'No matter' theory opposed Descartes' 'Never mind' theory, and 'determinism' theory opposed 'free will' theory. Could these opposed ways of thinking be, or be related to, one or more pairs of compatible valid image theories ? And how might we determine what kinds of science theories might make compatible valid image theories ?

One area deserving some study is how language and mathematics deal with causation. Hence in English we have eg ;

1.
"A causes B."
"A makes B move."

2.
"B is caused by A."
"B moves because of A."

Now some might use these two sets of description as having identical meaning and describing the same actual causal event - especially so for the causes/is-caused-by case. But somebody might use the '1' descriptions intending that B is passive or 'dead' and all action is in A. And somebody might use the '2' descriptions intending that B is active in responding to A. (the latter especially so for the makes-move/moves-because case) Eg ;

1.
"A pushes B, so making B move."

2.
"B responds to A by moving itself."

Or consider a dead bird on the ground and a child walking past sees the dead bird and is shocked and begins to cry. Many might say 'the dead bird caused the child to cry' although knowing that the dead bird did nothing. All might agree that the child responded to seeing the dead bird by beginning to cry so that 'the child seeing the dead bird caused the child to cry' or the child alone was active and was the cause. Of course it can be said that the dead bird did not do nothing in that it emitted dead-bird signals and the child receiving those signals was what caused the child to respond by crying. But that does still allow of alternative valid descriptions of the causal event.

Rene Descartes physics is clearly a '1' type physics, while William Gilbert's physics is clearly a '2' type physics. And equally clearly they are claiming proof of two quite different actual causal mechanisms. As such they did not intend to just produce different descriptions of the same causal mechanisms, and did not intend to produce 'image theories'. Descartes could claim that when you push something you clearly feel the contact push that makes it move, and that when a magnet repels another magnet that must work the same way by the contact of some particles pushing. And Gilbert working with magnets could claim that you can clearly see a magnet responding, without direct push contact, to signals received from another magnet by moving itself, and that when you push something there must actually be no contact but a response to proximity signals by the thing moving itself in the same way. Emitted signals when received establish a 'contact' without any pushing being involved. Though the two theories claim to describe very different causal mechanisms, both are basically attempted descriptions of the same universe so there is an issue of whether some modification of one or both theories might in fact make a pair of compatible image theories.

Some have produced modifications of these theories of 'dead matter' vs 'active matter', to 'matter' vs 'mind' and even to 'determinism' vs 'free will' theories - but that is perhaps going beyond science. And like Einstein and Newton, both Gilbert's and Descartes' physics theories are fully determinist.

In physics theory, the same question also arises perhaps less obviously with Wave Theory and Particle Theory, and the attempted merger of that contradictory pair in a Duality Theory, covering a smaller area of physics. Clearly the Wave and Particle theories are claiming proof of two quite different actual causal mechanisms. However these two theories are again basically attempted descriptions of the same bit of the same universe so there is an issue of whether some modification of one or both theories might in fact make a pair of compatible image theories also. Light theory looks a promising area for producing and testing a set of valid image theories, though that has not been done to date. They would need to be written up in a comparable manner, and might allow several versions :- waves-in-media, waves-without-media, simple-particles and responding-particles maybe ?

In sciences other than physics, there also seem to be possibilities of image theories as eg in animal behaviour with reflex theory vs learning theory ? Of our four major physicists only Rene Descartes ventured outside physics successfully to any extent, with his biological push theory of sensation, nerve action and animal behaviour. Hence Descartes basically claimed that light coming from food punched the animal eye, that punch travelled along nerves to the brain and then to the muscles giving a reflex behaviour. Biologists at first went with that theory, but later dropped it in favour of a William Gilbert style signal response theory. This was partly because Descartes push theory seemed not able to deal with memory and learning, and partly because nerve transmission was found to be electrical. Of course Descartes push theory did give a mechanism for electrical type behaviour, if not memory. Interestingly Gilbert's signal theory experiments did include magnetic induction which allows inanimate matter memory and became the basis of some modern computer memory and recording methods. But while advance in biological theory involved moving to signal theory, in physics signal theory got sidelined mainly by Descartes supporters falsely claiming that it assigned mind to matter to discredit it as they could not disprove it.

The important thing for science theory generally is that not merely can one thing be validly described in more than one way, but that different people tend to thinking differently or have different aptitudes so that one person might work best using one image theory while another person might work best using a different image theory. But this unlike much philosophical analysis is not just a sterile word game but is real philosophy of science and is significant actual science about actual experiment interpretation. Given different possible interpretations of any science experiment, it should be possible to define tests that will establish which of these interpretations if any are valid interpretations and which of these interpretations if any are invalid interpretations. Such testability makes this piece of 'philosophy of science' science and not just philosophy. And a science that uses several valid image theories should get more from more scientists than a science that uses only one valid theory. And a General Image Theory looks like giving the only reasonable resolution of Newton's classic Blackbox Dilemma and the more recent physics Duality Dilemma?

For enquiries, or if you have any view or suggestion on the content of this site, please contact ;
New Science Theory (e-mail:-vincent@new-science-theory.com)
Vincent Wilmot 166 Freeman Street Grimsby N.E.Lincs UK DN32 7AT.

You are welcome to link to any page on this site, eg www.new-science-theory.com/albert-einstein.php

IF you like this site then you could maybe make a donation ;

PayPal Donate

It will help with site development, and just possibly with some key physics experiments long planned but never afforded.
[PS. and you may perhaps help make history for science ?]
(Anomalies regarding modern physics theory have long totally discouraged certain lines of physics experiment despite there being strong reasons to believe them to be very promising if not essential lines of experiment. Some such lines of experiment considered here identified as early as the 1960s seem still to have had no work done on them and there is maybe not much more time here for this.)

© **new-science-theory.com, 2022** - taking care with your privacy, see New Science Theory HOME.

New Science Theory, sitemap + basic physics and basic universe facts

SITE MAP : (Updated since 6 April 2022 = ¹)

Home .... William Gilbert ..... Gilbert's De Magnete ... Gilbert's De Magnete + ... Galileo Galilei .. Johannes Kepler
.... Rene Descartes .... Descartes' Principles ..... Descartes' The World
.... Isaac Newton ...... Newton's Principia .. Newton v Descartes ..... Nikola Tesla
.... Albert Einstein ..... Einstein's continuum .. Blackbox Einstein
.... Science History .. Science Philosophy .. Information Physics .. Gravity .. Light
.... String Theory .. The Standard Model .. Probability Science .. Solar System
.... General Image Theory 1 ... GIT 2 ... GIT 3 ... GIT 4 .. About Us .. Privacy .. Sitemap(here) ..

Books......... De Magnete, New English .... De Magnete, Latin PDF .... Principia, PDF .... Opticks, PDF ..... Electromagnetism, PDF

Get this website as a Zoomable, Searchable and Printable pdf Ebook with helpful Bookmarks for just £2 - or for £9 get the nice A4 paperback version - both at New Science Theory book.

Homepage . William Gilbert . Rene Descartes . Isaac Newton . Albert Einstein .... General Image Theory

The universe is estimated to have at least about 70,000,000,000,000,000,000,000 stars, and to have a diameter of at least about 30,000,000,000 light years. It is split up into various types of galaxies and other components.

Our Milky Way galaxy is estimated to contain about 200,000,000,000 stars, and to have a diameter of about 100,000 light years with our sun about 26,000 light years from its centre. If even 1 star in a million allows the evolution of intelligent life, then humans should be far from alone.

Knowledge and physics

Knowledge of the universe has grown and continues to grow, but currently we can basically say ;

1. We have quite a lot of experience of our own planet, with quite a lot of experiments having been done regarding it.
2. We have a bit of experience of our own solar system, with some experiments having been done regarding it.
3. We have very little experience beyond our solar system, with very few experiments having been done regarding it.

Classic physics was built on 1 and 2, but modern physics ideas tend to rely more on 3.

Modern physics is fragmented and contains some real problems, as ;

A. Though the centres of galaxies are clearly very bright, it is claimed that they surround a large black hole ?
B. Though space looks clearly empty, it is claimed that it is full ?
C. Despite strong disproofs of Descartes push-physics, we have push-physics claims of physics with no actual push ?

And there are certainly other issues being strongly debated, and what are the real priority issues for physics now is far from agreed.

Space and Orbits

One of the most common forms of motion in the universe is orbital motion, mainly of smaller bodies orbiting around more massive bodies. Orbits of bodies in space can generally be taken as being determined chiefly by gravitation and so by Newton's laws of motion and gravitation.

Orbits around a massive body of some mass M require some speed that is below Escape Velocity vE but above Circular Velocity vC. For a distance r from the centre of gravity of a mass M, where r also needs to be larger than the radius of the massive body, $vE^2=2GM/r$ and $vC^2=GM/r$. At the Earth's surface vE = 11.2 km/sec (40,300 km/hr) and vC = 7.9 km/sec (28,400 km/hr).

The factor M/r means required velocities are bigger for orbits around more massive bodies, and for a particular massive body required velocities are smaller for farther orbits.

The same considerations apply to orbiting for all massive bodies, as to orbits around the Earth, the Sun or Black Holes. The greater mass of a Black Hole means that only the fastest bodies will orbit close to a Black Hole, though slower bodies will orbit farther from it.

So generally objects passing a massive body at a speed between Escape Velocity vE and Circular Velocity vC will be pulled into orbit around it. But objects passing at a speed below Circular Velocity vC will be pulled into the body, and objects passing at a speed above Escape Velocity vE and Circular Velocity vC will continue past the body. The required speeds are set by the mass of the body and the pass distance. Of course at any speed a direct collision course means collision.

And if an orbiting body has a mass insignificantly small relative to the massive body then,
1. If its (orbital) speed is exactly the circular speed vC at r, the orbit will be a Circle passing through r, around the centre of the massive body.
2. If its (orbital) speed is slower than the circular speed vC at r, the orbit will be an Ellipse smaller than the circle that passes through r, with the massive body at its far focus.
3. If its (orbital) speed is faster than the circular speed vC at r, but less than the escape speed at r (vE), then the orbit will be an Ellipse larger than the circular orbit that passes through r, with the central body at its near focus.

If orbit velocities and distances are known to some accuracy then massive body mass can be estimated to some accuracy. Of course these considerations cannot be applied to massless zero-inertia objects, even if somehow attracted by gravity.

And if particles or bodies of an atom and of a solar system both actually function or orbit in basically similar information-processing ways, that could indicate that the speed of information-processing need impose only a limited difference on information-processing effects ?

Two websites to help inform you on what physicists and astronomers are up to lately are http://physicsworld.com/ and www.universetoday.com/

(For imperfect but free online Latin translation see www.translation-guide.com/free_online_translations.htm)

otherwise, if you have any view or suggestion on the content of this site, please contact :- New Science Theory
Vincent Wilmot 166 Freeman Street Grimsby Lincolnshire DN32 7AT.
Or on Twitter.com - @vwilmot

You are welcome to link to any page on this site, eg www.new-science-theory.com/physicshistory.php

IF you like this site then you could maybe make a donation ;

It will help with site development, and just possibly with some key basic physics experiments long planned but never afforded.
[PS. and you may perhaps help make history for science ?]
(The fictional time-travel and multi-universe type ideas of modern physics theory have long totally discouraged certain lines of physics experiment despite there being strong reasons to believe them to be very promising if not essential lines of experiment. Some such lines of experiment considered here identified as early as the 1960s seem still to have had no work done on them and there is maybe not much more time here for this. Science funding both government and private unfortunately now all goes to basically safe standard mainstream science, and no money at all goes to any really innovative risky science though that might pay a thousand times greater.)

© **new-science-theory.com, 2022** - taking care with your privacy, see New Science Theory HOME.

www.ingramcontent.com/pod-product-compliance
Lightning Source LLC
Chambersburg PA
CBHW081053170526
45165CB00006B/2256